Wolfram Schommers

Institute for Scientific Computing, Germany

QUANTUM PROCESSES

NEW JERSEY · LONDON · SINGAPORE · BEIJING · SHANGHAI · HONG KONG · TAIPEI · CHENNAI

Published by

World Scientific Publishing Co. Pte. Ltd.
5 Toh Tuck Link, Singapore 596224
USA office: 27 Warren Street, Suite 401-402, Hackensack, NJ 07601
UK office: 57 Shelton Street, Covent Garden, London WC2H 9HE

British Library Cataloguing-in-Publication Data
A catalogue record for this book is available from the British Library.

Cover art by Stephanie Schommers (Karlsruhe, Germany)

QUANTUM PROCESSES

Copyright © 2011 by World Scientific Publishing Co. Pte. Ltd.

All rights reserved. This book, or parts thereof, may not be reproduced in any form or by any means, electronic or mechanical, including photocopying, recording or any information storage and retrieval system now known or to be invented, without written permission from the Publisher.

For photocopying of material in this volume, please pay a copying fee through the Copyright Clearance Center, Inc., 222 Rosewood Drive, Danvers, MA 01923, USA. In this case permission to photocopy is not required from the publisher.

ISBN-13 978-981-279-656-1
ISBN-10 981-279-656-8

Typeset by Stallion Press
Email: enquiries@stallionpress.com

Printed in Singapore by Mainland Press Pte Ltd.

Foreword

Modern physics is based on two theories: Quantum Theory and Theory of Relativity. For all physical developments, from elementary particle physics to cosmological models, these conceptions are of fundamental relevance. Although each conception has its own field of description, both theories could confirm the experimental material fantastically. This is in a certain sense surprising because quantum theory and theory of relativity have been developed independently from each other. Therefore, one cannot expect, as a matter of course, that both theories describe the experimental material equally well. But they do it with high accuracy for all areas they are responsible. However, appearances are deceptive.

When we put the laws of "general theory of relativity" and of "quantum theory" together, we obtain irreconcilable differences; both theories are mutually incompatible. Concerning this point, we find in Brian Greene's "The elegant universe" the following comment.

> *Through years of research, physicists have experimentally confirmed to almost unimaginable accuracy virtually all predictions made by each of these theories (usual quantum theory and general theory of relativity). But these same theoretical tools inexorably lead to another disturbing conclusion. As they are currently formulated, general relativity and quantum mechanics, cannot both be right. The both theories underlying the tremendous progress of physics during the last hundred years — progress that has explained the expansion of the heavens and the fundamental structure of matter — are mutually incompatible.*

That is, both theories seem to work perfectly but, on the other hand, they are mutually exclusive. This is clearly reflected in the quantum-theoretical treatment of the cosmological constant which leads to a disaster when we estimate the vacuum energy by means of quantum field theory. This disaster can obviously not be eliminated by invention of certain mechanisms within the frame of one of the two theories. In other words, the discrepancy in connection with vacuum energy and cosmological constant obviously reflects a basic fact which has probably its source in the very foundations of both theories. In fact, the two conceptions contain some fundamental open questions which have not yet been solved so far. Just in connection with quantum theory there are some unresolved problems. Because these problems are relatively old and no attempt to restore the quantum formalism was convincing, we probably have to assume that they cannot be solved by adding further details to the basic formalisms, but by changing the frameworks itself. What are the problems? Let us quote here only a few of them.

It has often been argued that one of the theories, quantum theory or general theory of relativity, must be wrong or even both. In fact, each of the both conceptions has certain shortcomings. It can therefore not be excluded that the reason for the discrepancies between quantum theory and general theory of relativity are due to the fact that both theories are based on foundations that are different from each other and which are obviously incompatible. Within quantum theory two of the problematic points can be summarized as follows:

1. Schrödinger's equation belongs to one of the basics of conventional quantum theory. However, this equation has not been deduced but was assumed, that is, we do not know the structure of physical reality on which Schrödinger's equation is based. What is the origin of the of Schrödinger's equation? This question cannot be answered within conventional quantum theory.
2. Also the superposition principle belongs to the basics of quantum theory but the so-called collapse of the wave function in connection with the measuring process cannot be explained, also not by Schrödinger's equation although it belongs to the basics as well. Schrödinger's equation is even suspended during the measuring process. This is not only confusing but rather suggests a non-physical situation. Within the Copenhagen interpretation of

quantum theory, which is the most used interpretation, the collapse of the wave function and the assignment of statistical weights are not explained. Within this picture. De Witt states that

they are consequences of an external a priori metaphysics, which is allowed to intervene at this point and suspend the Schrödinger equation.[1]

There seems to be a mismatch between the various parts (concepts) which form altogether that what we call quantum theory today. This mismatch might be the reason for the fact that even approximately 80 years after the formulation of quantum theory the interpretation status of this formalism is described by Max Jammer in "The Philosophy of Quantum Mechanics" as

by far the most controversial problem in current research in the foundations of physics and divides the community of physicists and philosophers of science into numerous opposing schools of thought.

In conclusion, conventional quantum theory is by no means a perfect, final theory and there are obviously a lot of open questions. This might be the reason why theory of relativity and quantum theory are inherently incompatible. However, we can hardly expect that modifications of only one theory (theory of relativity or quantum theory) could eliminate this unacceptable situation. Therefore, we have also to point to the shortcomings of the theory of relativity, in particular those of general theory of relativity.

General theory of relativity has a serious space-time problem, and this is due to the fact that Mach's principle is not fulfilled within this theory; special theory of relativity has the same problem. Mach's principle has often been discussed in literature. After Mach, space-time can never be the source for physically real effects, that is, space-time can never act on material objects giving them certain properties as, for example, inertia. According to Mach, a particle does not move in un-accelerated motion relative to space, but relative to the centre

---
[1] Interested readers might refer to de Witt, B., (1970) Physics Today, September issue.

of all the other masses in the universe. In fact, we can never observe such a space (space-time) because its elements (coordinates x, y, z and time τ where τ is the time measured with our clocks used in everyday life) are in principle not accessible to empirical tests. We can only say something about distances in connection with masses, and time intervals in connection with physical processes. In other words, an empty space-time as physical-theoretical conception should not be existent and would be against the basic facts; no realistic space-time theory should contain an empty space-time as a solution. Einstein thought that his field equations would fulfil this important and basic condition. However, already in 1917, de Sitter gave a solution to Einstein' field equations which corresponds to an empty universe, i.e., within the framework of this solution, space-time could exist without matter, and this is against fundamental facts. Furthermore, de Sitter also pointed out that Einstein's field equations contain a solution for a lone particle in the universe which moves along a geodesic line as if was made of inertial matter. As we have pointed out above, such a solution does not fulfil Mach's principle. These few examples demonstrate that the theory of relativity has obviously a serious space-time problem. In particular, Mach's principle could not be realized within theory of relativity, although this principle reflects a fundamental and important feature.

Again, both theories, general theory of relativity and quantum theory, are inherently incompatible; we can hardly expect that modifications of only one theory (general theory of relativity or quantum theory) could eliminate this situation. Therefore, we have to search for a principle that is valid for both phenomena, that is, for gravitational and quantum phenomena. But it is obvious that such a principle must hold for all phenomena in nature (including quantum and gravitational effects), taking place at the various levels of observation. This obviously means, on the other hand, that such a principle must be very basic, anchored at the basic level of theoretical physics. In this monograph we will work on the basis of the following conception: the real world is not embedded in space-time (container principle) but it is projected on it (projection principle). The projection principle is used to deduce a lot of new aspects, in particular Schrödinger's equation in an extended form. As we have outlined above Schrödinger's equation could not be deduced within conventional quantum theory. Furthermore, Mach's principle is fulfilled automatically.

•

Space and time are probably the most important elements in physics. Within the memory of man all essential things are represented within the frame of pictures. This is the most basic statement. All what we experience in everyday life are pictures in front of us which are represented within space and time. Such pictures contain the objects (sun, moon, etc.), the space and the time. We feel the objects, that is, we observe them with our five senses and also with measuring instruments. Thus, we believe in the concrete existence of these things and we call them "material objects". But what can we say about space and time? It is normally assumed that the space is a container filled with matter and that the time is just that what we measure with our clocks. However, there are some reasons to take another viewpoint and to consider this container-conception as unrealistic, as prejudice so to say. Already the philosopher Immanuel Kant (1724–1804) pointed on this serious problem.

In connection with space and time we have to consider the following fact. We cannot put a "piece of space" on the table and we can also not put a "piece of time" on the table. Such pieces do not exist. The elements of space, the coordinates x, y, z, and those of time τ are principally not observable. We have no senses for that and, furthermore, the development of measuring methods for the detection of x, y, z and τ are even not thinkable. We can only say something about distances in connection with masses, and time intervals in connection with physical processes. There is no exception!

Because we definitely cannot see, hear, smell, taste or measure space and time, the elements x, y, z, τ may not be the source of the physically real effects. This requirement is the contents of Mach's principle which is however not fulfilled within Newton's theory but also not within Special and General Theory of Relativity.

Nevertheless, we are all familiar with the space-time phenomenon although the elements of it (x, y, z, τ) are not observable with our five senses and measuring instruments. Thus, we have to conclude that the elements x, y, z, τ must be inside the observer and it makes no sense to assume that these elements (x, y, z, τ) are also outside the observer. The famous philosopher Immanuel Kant took this viewpoint but his fundamental ideas have never been seriously used in the formulation of physical theories. After Kant, space and time are exclusively features of our brain, and this inevitably means that the world outside is projected on space and time. Then, the material objects that occupy space and time can only be geometrical pictures. Projection theory[2] is based on this conception. If that is the case, the situation would be identical with that what we do in connection with our blackboards and notebooks: We draw pictures on them.

From the principles of evolution it follows that the structures in the picture must be different from that in outside reality. The common or naive point of view assumes the following. The inside world which we feel to be outside us, actually exists in the outside world in exactly the same form as we perceive. According to this view there is only one difference between the inside world and the outside world: inside there are only geometrical positions, whereas outside there are the real material bodies instead of the geometrical positions. In other words, it is normally assumed that the geometrical positions are merely replaced by material objects. However, due to the principles of evolution we have to assume that the information in the picture is only a selected part of the outside world (basic reality). Only that information is relevant which is useful for a human or another species. Since space and time cannot be considered as elements of basic reality its structure must be different from that in the picture. In other words, from the principles of evolution we have to conclude that basic reality contains (much) more information than an observer can depict in space and time and, furthermore, that its structure is principally different from that in the picture. This space-time information is mainly that part of the outside world which a human observer needs for survival. From this it directly follows that the picture of reality must be species-dependent. In fact, there are strong indications for that.

---

[2] For details, see also Chapter 3.

Projection theory, the main subject of this monograph, is based on these features. That is, projection theory does not work within the framework of the container principle (the material world is embedded in space and time) but it is assumed here that the world outside is projected onto space-time. In particular, projection theory also takes into consideration that the structure of the world outside (basic reality) must be quite different from that what we observe in everyday life.

One of the most interesting results is that projection theory automatically leads to a new aspect with respect to the notion "time". Here we have not only the usual time τ which is an external parameter and is just the time that we measure with our clocks, but we get a system-specific time t which is not known in conventional physics. Within all forms of conventional physics only the time τ appears.

Within projection theory "reality" is projected onto space-time [$(\mathbf{r}, t)$-space] and we obtain "pictures of reality". The description of these pictures is done on the basis of "fictitious realities" which are constructed on the space-time elements $\mathbf{r}$ and t. The variables of a fictitious reality are given by the momentum $\mathbf{p}$ and the energy E. Thus, we have two spaces in projection theory: $(\mathbf{r}, t)$-space and $(\mathbf{p}, E)$-space, and both spaces are connected by a Fourier transform.

The physical processes take place in $(\mathbf{p}, E)$-space and this information is projected onto $(\mathbf{r}, t)$-space. $(\mathbf{r}, t)$-space does not contain real material objects but only geometrical structures. In particular, $(\mathbf{r}, t)$-space cannot be an element of the outside world but is exclusively positioned in the brain of the observer and appears simultaneously with the objects (or more precisely with the geometrical structures of them). Again, this viewpoint is close to that which the philosopher Immanuel Kant proposed.

Interactions exclusively take place in $(\mathbf{p}, E)$-space in the form of $\mathbf{p}, E$-fluctuations. In the case of free systems, the momentum $\mathbf{p} = \mathbf{p}_0$ and the energy $E = E_0$ remain constant in the course of time τ, and there are no $\mathbf{p}, E$-fluctuations in the case of free systems ($\Delta \mathbf{p} = 0$, $\Delta E = 0$). Such non-interacting systems, are completely useless and their existence would be against the principles of evolution. The results of Chapter 3 actually show that this is the case. We proved mathematically that free (non-interacting) systems cannot exist within projection theory. Thus, only interacting systems are of relevance, that is, quantum processes are of fundamental importance for the material existence of the world. In other words, within the theoretical structures of projection theory everything is a matter of processes and the existence of static, non-interacting objects in space have been eliminated. Like the philosophers Whitehead and Bergson, within projection theory it is argued for the primacy of process.

The projection principle leads to fundamental new aspects in connection with the notion "interaction". It is important to underline that we have here to distinguish between distance-dependent and distance-independent correlations (calculated by distance-dependent and distance-independent interactions). Whereas the distance-independent interactions define the form (shape) of a system (that is, its geometrical structure), distance-dependent interactions describe the correlations between the various systems in $(\mathbf{r}, t)$-space. In this connection, the following is relevant. Not only the space-structure at time τ is described within projection theory but also effects with respect to the system-specific time t, that is, the interactions of the subsystems with past and future events. The formalism of usual physics only allows the formulation of the space-structure at time τ because the system-specific time t is not defined here. Within projection theory the formalism is extended and the space-time structure at time τ is concerned.

Let us finally mention that the book also deals with various other topics such as, for example, the famous particle–wave dualism. Here the situation is different from that of conventional quantum theory: We do not need an experimental arrangement that decides about the nature of a system. The system is completely described by the interaction of the system with its environment. No interaction means within projection theory that the system does not exist.

Wolfram Schommers
Karlsruhe, Germany

Contents

Foreword		v
1.	**Conventional Quantum Theory**	**1**
	1.1. Classical Description	1
	1.2. Schrödinger's Equations	2
	1.2.1. Operator Treatment of Schrödinger's Equation	4
	1.2.2. Momentum Representation	5
	1.3. Uncertainty Relations	7
	1.4. Individuals	9
	1.5. Conclusion	14
	1.6. Aspects	14
	1.6.1. The Principle of Complementarity	14
	1.6.2. Objectivity	16
	1.7. Remarks on the Superposition Principle	16
	1.8. Basic New Experiments	20
	1.8.1. General Remarks	20
	1.8.2. Conclusion	23
2.	**Projection Theory**	**25**
	2.1. Preliminary Remarks	25
	2.2. The Projection Principle	27
	2.2.1. The Elements of Space and Time	27
	2.2.2. Relationship between Matter and Space-Time	28
	2.2.3. Two Relevant Features	29
	2.2.4. Two Kinds of "Objects"	31

	2.2.5.	Perception Processes	31
	2.2.6.	Inside World and Outside World	35
	2.2.7.	The Influence of Evolution	36
	2.2.8.	Information in the Picture versus Information in Basic Reality (Outside Reality)	38
	2.2.9.	Other Biological Systems	39
	2.2.10.	Summary	42
2.3.	Projections		42
	2.3.1.	Principal Remarks	42
	2.3.2.	Mach's Principle	44
	2.3.3.	Conclusion	45
	2.3.4.	Other Spaces	45
		2.3.4.1. Fourier-space	45
		2.3.4.2. The influence of Planck's constant	46
		2.3.4.3. Reality and its picture	48
		2.3.4.4. Remark	50
	2.3.5.	Basic Properties	51
		2.3.5.1. Operators	51
		2.3.5.2. Conclusion	53
	2.3.6.	Basic Transformation Effects	55
		2.3.6.1. Particles	55
		2.3.6.2. Role of time t	56
		2.3.6.3. Non-local effects	57
		2.3.6.4. Conclusion	58
	2.3.7.	Operator Equations	59
		2.3.7.1. Determination of $\Psi(\mathbf{r}, t)$ and $\Psi(\mathbf{p}, E)$	59
		2.3.7.2. Remarks	60
		2.3.7.3. Space-specific formulation	61
		2.3.7.4. Discussion concerning equations (2.35) and (2.50)	64
		2.3.7.5. Other representations	67
		2.3.7.6. Superposition principle	70

	2.3.8.	Processes	....	74
		2.3.8.1.	General remarks	74
		2.3.8.2.	Description of properties and appearances	75
		2.3.8.3.	The meaning of the wave function	78
		2.3.8.4.	Properties of probability distributions	81
		2.3.8.5.	Does god play dice?	83
	2.3.9.	Time		83
		2.3.9.1.	Reference time and selection processes	84
		2.3.9.2.	Structure of reference time	86
		2.3.9.3.	Selections	88
		2.3.9.4.	Information inside, information outside	95
		2.3.9.5.	Reality outside	96
		2.3.9.6.	Constancy phenomena	96
		2.3.9.7.	Schrödinger's equation and its limitations	97
		2.3.9.8.	Real situation	105
		2.3.9.9.	τ-Dependent systems	107
		2.3.9.10.	Some additional remarks	108
		2.3.9.11.	Uncertainty relation for time and energy	110
		2.3.9.12.	Time within special theory of relativity	110
2.4.	Summary			115

3. Free, Non-interacting Systems (Particles) — 117

3.1. General Remarks 117
3.2. The Behaviour of the Basic Equations 119
    3.2.1. The Case $f(\mathbf{p}_0, E_0)_{free} = \infty$ 121
    3.2.2. On the Relationship Between $\mathbf{p}_0$ and E_0 . 122
    3.2.3. The Case $f(\mathbf{p}_0, E_0)_{free} \neq \infty$ 125
    3.2.4. Conclusion 128

3.3.	Classical and Quantum-Theoretical Elements . . .	128
3.4.	Behaviour of the Wave Function in $(\mathbf{r}, t)$-Space and $(\mathbf{p}, E)$-Space	129
3.5.	Probability Considerations in Connection with $\Psi(\mathbf{p}_0, E_0)$	133
3.6.	Normalization Condition	134
3.7.	Mean Values for the Momentum and the Energy .	137
3.8.	The $\mathbf{p}, E$-Pool	140
3.9.	Free, Elementary Systems do not Exist	142
3.10.	No Equation for the Determination of the Wave Function $\Psi(\mathbf{p}_0, E_0)$	144
	3.10.1. Additional Physically Relevant Conditions?	144
	3.10.2. Multi-valuedness of the Wave Function $\Psi(\mathbf{p}_0, E_0)$	145
	3.10.3. Existence and Non-Existence	146
	3.10.4. Summary	147
3.11.	Principle of Usefulness	147
3.12.	Further General Remarks	148
3.13.	Rest Mass Effect	149
3.14.	Summary	152

Appendix 3.A. 153
Free System within Usual Quantum Theory 153
    3.A.1. Superposition Ansatz 153
    3.A.2. Mean Momentum for a Free System . . . 155
    3.A.3. Usual Quantum Theory and Projection Theory 156

Appendix 3.B. 160

Appendix 3.C. 163

Appendix 3.D. 166
The Stationary Case 166
    3.D.1. Definition 166
    3.D.2. Relevant Properties 167
    3.D.3. The Behaviour of the Wave Functions . . 170
        3.D.3.1. Singularities 170

		3.D.3.2.	The probability argument . . .	172
		3.D.3.3.	More details concerning the potential $V(x,y,z)$	173
		3.D.3.4.	Mean value for the energy . . .	177
		3.D.3.5.	Normalization condition	179
	3.D.4.	Stationary Systems do not Exist	180
	3.D.5.	Final Remarks	182

Appendix 3.E. 183
Dependence of Mass on Velocity 183

4. Interactions 189

4.1. Interactions within Projection Theory 189
4.2. What does Interaction Mean within Projection Theory? 190
    4.2.1. Relationships 191
    4.2.2. Fourier-Effects 192
4.3. How Basic is the Notion "Interaction"? 194
    4.3.1. Classical Force Laws 194
    4.3.2. Equivalent Conceptions 196
    4.3.3. Further Remarks 197
    4.3.4. Remarks Concerning Quantum Field Theory and the Theory of Strings (Branes) 198
    4.3.5. Delocalised Systems in $(\mathbf{p}, E)$-Space . . . 198
    4.3.6. Summary 199
4.4. Description of Interactions within Projection Theory: Principal Remarks 200
    4.4.1. Space-Time Limiting Interactions 200
    4.4.2. Mutual (Distance-Dependent) Interactions 203
    4.4.3. Specific Treatment in Connection with the Exchange of Momentum and Energy . . . 209
    4.4.4. The $\mathbf{p}, E$-Concert 214
    4.4.5. Individual Processes 217
    4.4.6. Analogy to Conventional Physics 219
    4.4.7. Total Momentum and Total Energy . . . 220
4.5. Pair Distributions 221

- 4.5.1. Information About the Interaction 222
- 4.5.2. Collective Effects in Connection with $\Psi(\mathbf{p}, E)$ 224
- 4.5.3. Analysis in $(\mathbf{r}, t)$-Space 225
- 4.5.4. Example for $N = 4$ 230
- 4.5.5. Further Discussions 233
- 4.6. Basic Equations 234
  - 4.6.1. The Main Features 234
  - 4.6.2. Some Additional Statements 239
  - 4.6.3. Classical Formulation 243
  - 4.6.4. Introduction of Pair Potentials for Certain Configurations 246
  - 4.6.5. Interaction Effects 252
- 4.7. Energy Levels 254
  - 4.7.1. Treatment of the Problem 256
  - 4.7.2. Specific Properties 258
  - 4.7.3. Conditional Wave Functions 259
  - 4.7.4. E_α-Fluctuations 262
  - 4.7.5. Extension to N Subsystems 269
  - 4.7.6. Summary 271
- 4.8. Distance-Independent Interactions 271
  - 4.8.1. Principal Remarks 272
  - 4.8.2. Some Minor Changes 273
  - 4.8.3. Some Basic Features of Distance-Independent Interactions 274
  - 4.8.4. Absolute Space-Time Positions 276
  - 4.8.5. Arbitrary Jumps 278
  - 4.8.6. Effective Velocities 280
  - 4.8.7. Space-Effects 285
  - 4.8.8. Arbitrary Jumps and $\mathbf{p}, E$-States 287
  - 4.8.9. Resting and Moving Frames 290
  - 4.8.10. Arbitrary Jumps within Single Systems .. 294
- 4.9. The Meaning of the Potential Functions 300
  - 4.9.1. Introduction of a Potential Function in the Case of Distance-Independent Interactions (Form Interactions) 301

	4.9.2. Interaction within Conventional Physics	302
	4.9.3. Interaction Potentials are Auxiliary Elements	303
	4.9.4. Conventional Physics: What Mechanism is Behind Interaction?	304
	4.9.5. "Gravity... an Occult Quality"	305
	4.9.6. Phenomena in Usual Quantum Theory	306
	4.9.7. Summary	307
4.10.	Further Basic Features	308
	4.10.1. Can Systems be Elementary in Character?	308
	4.10.2. Self-Creating Interaction Processes	309
4.11.	Absolute Space-Time Conceptions	312
	4.11.1. Mach's Principle	312
	4.11.2. The Effect of Inertia within Newton's Theory	313
	4.11.3. Mach's Principle and Theory of Relativity	314
	4.11.4. Final Remarks	320
4.12.	Relativistic Effects	321
	4.12.1. General Remarks	321
	4.12.2. Frames of Reference within Projection Theory	322
	4.12.3. Transformation Formulas	325
	4.12.4. Arbitrary Jumps of the Entire Complex in Space-Time	327
4.13.	Hierarchy of the Parts in a Part	328
	4.13.1. Conventional Physics	329
	4.13.2. Is this Principle Realizable within Projection Theory?	330
	4.13.2.1. Pictures and $\mathbf{p}$, E-fluctuations	331
	4.13.2.2. No static building blocks	332
	4.13.2.3. No fluctuations of fluctuations!	335
	4.13.2.4. Independent $\mathbf{p}$, E-fluctuations	337
	4.13.2.5. Conclusion	339
4.14.	Granular Space-Time Structures	340
	4.14.1. Combined Interactions	341
	4.14.2. Selections	343

	4.14.3. The Unified Whole	344
	4.14.4. Simple Cosmological Considerations	347
	4.14.5. Arbitrary Motions through Space and Time	348
	4.14.6. Decomposition of the Cosmos	352
4.15.	Summary and Final Remarks	352

5. Some Basic Questions — 361

- 5.1. The Particle-Wave Question 361
  - 5.1.1. No Need for an Experimental Arrangement 362
  - 5.1.2. What do We Measure? 365
    - 5.1.2.1. Situation in conventional quantum theory 365
    - 5.1.2.2. Situation in projection theory . 371
- 5.2. The Role of the Observer 372
  - 5.2.1. Compatible with the Principles of Evolution 372
  - 5.2.2. Configurations in Space-Time 375
- 5.3. Summary 378

6. Summary — 381

Bibliography — 393

Index — 395

1

Conventional Quantum Theory

1.1. Classical Description

Within the frame work of classical mechanics the motions of objects — more or less localized in space — are considered, and it is assumed that these classical motions confirm the "principle of least action". (This principle contains Newton's mechanics as a special case.) Equations of motion are obtained which exist in various forms: Lagrange's equations, Hamilton's equations, Newton's equations. By the solution of the equations of motion we obtain generalized coordinates $q_1, q_2, \ldots$ and generalized momenta $p_1, p_2, \ldots$ as a function of time τ.[1] For example, in the case of N objects we have

$$
\begin{aligned}
& q_1(\tau), p_1(\tau) \\
& q_2(\tau), p_2(\tau) \\
& \quad \vdots \\
& q_N(\tau), p_N(\tau)
\end{aligned}
\tag{1.1}
$$

That is, $6N$ quantities specify at time τ the state of N (interacting) objects completely, and we obtain N trajectories in space and time.

It is generally accepted that the description of quantum phenomena cannot be based on this classical frame work of physics, i.e., a quantum-theoretical picture cannot be obtained by adding additional facts to the framework of classical mechanics but only by changing the framework itself.

---
[1] Instead of t we use in this section the Greek letter τ for the time. We will introduce the letter t in connection with projection theory. See Chapter 2.

Quantum theory destroyed the mechanistic world view. Quantum reality is rationally comprehensible but cannot be visualized like Newton's reality. The laws of quantum theory were not easy to recognize. The most important of them were discovered in the first decades of the last century by an international group of physicists. The essential features of the mathematical formalism of (non-relativistic) quantum theory have been constructed by Heisenberg and Schrödinger in 1925 and 1926, respectively. On the basis of this quantum-theoretical formalism, an enormous number of effects in atomic physics, chemistry, solid state physics, etc., could be predicted and explained. However, approximately 85 years after its formulation the interpretation status of this formalism is described by Jammar in his book to be

> *by far the most controversial problem in current research in the foundations of physics and divides the community of physicists and philosophers into numerous opposing schools of thought.*[2]

There is an immense diversity of opinion and a huge variety of interpretations. Most of the interpretations lead to completely different conceptions of the world. For example, the Copenhagen interpretation of quantum theory proposed by Bohr in 1927 is quite different from the many-worlds theory proposed by Everett III in 1957.

In this chapter we would like to collect some of the main results of conventional quantum theory. In particular, we want to discuss critically the following items: Schrödinger's equation, uncertainties, individuals, principle of complementarity, objectivity, and the superposition principle. All these points are of fundamental relevance for the understanding of quantum phenomena.

1.2. Schrödinger's Equations

The time-independent Schrödinger equation

$$E\,\Psi(\mathbf{r}, E) = -\frac{\hbar^2}{2m_0}\Delta\Psi(\mathbf{r}, E) + U(x, y, z)\Psi(\mathbf{r}, E), \qquad (1.2)$$

---

[2] Jammer, M., (1974) The philosophy of quantum mechanics: Wiley.

[$\mathbf{r} = (x, y, z)$] and its time-dependent version

$$i\hbar \frac{\partial}{\partial \tau} \Psi(\mathbf{r}, \tau) = -\frac{\hbar^2}{2m_0} \Delta \Psi(\mathbf{r}, \tau) + U(x, y, z, \tau) \Psi(\mathbf{r}, \tau) \quad (1.3)$$

could not be derived but were assumed on the basis of "reasonable arguments". They can be validated only in connection with experiments, and here their success is without any doubt impressive but, unfortunately, we do not really know which physical reality is reflected by these equations. Thus, it is not or almost not possible to develop these equations systematically further. This is however desirable just in connection with time τ, since τ is still a classical parameter within Schrödinger's theory. This point has often been discussed in literature and we will come back to this point below, in particular in Chapter 2.

Although equation (1.2) is also often called "wave equation" it is not simply an equation for a wave, and this is because we need in the formulation of the potential energy $U(x, y, z)$ of the particle its position $\mathbf{r}$. For example, in the case of the hydrogen atom, $\mathbf{r}$ is the distance between the fixed position $\mathbf{r}_M$ of the nucleus and the position $\mathbf{r}_m$ of the electron, and the potential energy $U(x, y, z)$ is given in this case by the Coulomb potential

$$U(x, y, z) = \frac{e^2}{|\mathbf{r}_m - \mathbf{r}_M|}. \quad (1.4)$$

Thus, equation (1.3) reflects particle behaviour because in the expression for the potential energy, equation (1.4), particle positions appear. That is, Schrödinger's equation reflects the particle–wave duality[3] and has (almost) nothing to do with the usual wave equation of classical physics

$$\frac{1}{u^2} \frac{\partial^2 \psi_W(\mathbf{r}, \tau)}{\partial \tau^2} = \Delta \psi_W(\mathbf{r}, \tau), \quad (1.5)$$

where $\psi_W(\mathbf{r}, \tau)$ is a function describing a wave disturbance propagating with phase velocity u. Equation (1.5) was the starting point in Schrödinger's original discussion. In other words, Schrödinger's equation is obviously more complex than the usual wave equation (1.5).

---

[3]This is already pointed out by Schäfer. Refer to Schäfer, C., (1951) Einführung in die theoretische Physik: Walter de Gruyter.

1.2.1. Operator Treatment of Schrödinger's Equation

In connection with Schrödinger's equation, the following question arises: What reality hide behind this equation? In order to come closer to an answer let us introduce some postulates for the introduction of Schrödinger's equation (its *time-dependent* as well as its *time-independent* form).

Schrödinger found his equation on the basis of the *classical* wave equation (1.5) and de Broglie's wave–matter postulate. Another possibility for the introduction of Schrödinger's equations is to use the classical expression for the energy of an interacting particle

$$E = \frac{1}{2m_0}(p_x^2 + p_y^2 + p_z^2) + U(x,y,z) \qquad (1.6)$$

and the following postulates.[4]

(i) The *time-independent* Schrödinger equation (1.2) can be obtained if the variables

$$E,$$
$$p_x, p_y, p_z, \qquad (1.7)$$
$$x, y, z$$

of the classical equation (1.6) are replaced by

$$E$$
$$\hat{p}_x = -i\hbar\frac{\partial}{\partial x}, \hat{p}_y = -i\hbar\frac{\partial}{\partial y}, \hat{p}_z = -i\hbar\frac{\partial}{\partial z}. \qquad (1.8)$$
$$x, y, z$$

Hence, postulate (i) is stated as "keep the energy and the coordinates as they are and replace momenta by *operators*".

(ii) On the other hand, the *time-dependent* Schrödinger equation (1.3) can be obtained if $U(x,y,z)$ in equation (1.6) is replaced by

---

[4] Interested readers might refer to Finkelnburg, W., (1962) Einführung in die Atomphysik: Springer-Verlag, and Rubinowics, A., (1968) Quantum mechanics: Elsevier.

$U(x, y, z, \tau)$ and if the variables (1.7) of (1.6) are exchanged by

$$\hat{E} = i\hbar \frac{\partial}{\partial \tau},$$
$$\hat{p}_x = -i\hbar \frac{\partial}{\partial x}, \hat{p}_y = -i\hbar \frac{\partial}{\partial y}, \hat{p}_z = -i\hbar \frac{\partial}{\partial z}, \quad (1.9)$$
$$x, y, z.$$

Hence, postulate (ii) is stated as "keep the coordinates as they are and replace the energy and momenta by *operators*".

The source of these two postulates is unknown. Within usual quantum theory, we cannot recognize what physical reality hide behind these laws. It is, for example, not at all clear why the coordinates x, y, z are not treated on an even footing with the momenta p_x, p_y, p_z. We could also regard the coordinates as being operators equal to the coordinates, i.e.,

$$\hat{x} = x, \hat{y} = y, \hat{z} = z. \quad (1.10)$$

However, we would still have to calculate with these operators in just the same way as with ordinary numbers.[5]

Role of Time

Within Schrödinger's theory, time τ is still a classical quantity, that is, time remains unchanged when we go from classical mechanics to quantum theory and behaves like an external parameter as in Newton's physics. Schrödinger tried to modify his apparatus but without success. This is an important and principal point and we would like to give some remarks concerning this subject in Chapter 2.

1.2.2. Momentum Representation

Equation (1.3) is the Schrödinger equation in terms of *coordinates*. But there also exists a momentum representation,[6] that is, a representation in terms of *momenta*. Both representations are equivalent. Let us briefly

---

[5] See footnote 4.
[6] Interested readers might refer to Landan, L.D., and Lifschitz, E.M., (1965) Quantum mechanics: Pergamon.

discuss the situation for an open system, that is, for a system with a time-dependent potential

$$U = U(x, y, z, \tau), \tag{1.11}$$

instead of $U = U(x, y, z)$.[7] Using equations (1.6), (1.11) and the rules expressed by equations (1.9) (extended by τ), Schrödinger's equation in terms of coordinates takes the well-known form

$$i\hbar \frac{\partial \psi(\mathbf{r}, \tau)}{\partial \tau} = -\frac{\hbar^2}{2m_0} \Delta \psi(\mathbf{r}, \tau) + U(x, y, z, \tau) \psi(\mathbf{r}, \tau), \tag{1.12}$$

that is, nothing is changed compared with equation (1.3). In the case of momentum representation, we have to use equations (1.6) and (1.11) and, instead of the rules given by equation (1.8), we have to introduce a further postulate, stated as "*keep the momenta and time as they are and replace the energy and coordinates by operators*":

$$\hat{E} = i\hbar \frac{\partial}{\partial \tau}$$

$$p_x, p_y, p_z,$$

$$\tau$$

$$\hat{x} = i\hbar \frac{\partial}{\partial p_x}, \hat{y} = i\hbar \frac{\partial}{\partial p_y}, \hat{z} = i\hbar \frac{\partial}{\partial p_z}. \tag{1.13}$$

Then, Schrödinger's equation in the momentum representation is given by

$$i\hbar \frac{\partial \psi(\mathbf{p}, \tau)}{\partial \tau} = \frac{\mathbf{p}^2}{2m_0} \psi(\mathbf{p}, \tau) + U\left[i\hbar \frac{\partial}{\partial p_x}, i\hbar \frac{\partial}{\partial p_y}, i\hbar \frac{\partial}{\partial p_z}, \tau\right] \psi(\mathbf{p}, \tau). \tag{1.14}$$

As is well-known, equations (1.3) and (1.14) are completely equivalent, and no other representations exist. Then, we have the following situation: Energy E, momenta p_x, p_y, p_z and coordinates x, y, z appear as *operators*, either in equations (1.12) or (1.14), but the *time* τ is

---

[7] See equation (1.6).

always a simple *number* and is never an operator:

$$E \to i\hbar \frac{\partial}{\partial \tau}, \quad \mathbf{p} \to -i\hbar \frac{\partial}{\partial x}, \quad -i\hbar \frac{\partial}{\partial y}, \quad -i\hbar \frac{\partial}{\partial z},$$

$$\mathbf{r} \to i\hbar \frac{\partial}{\partial p_x}, \quad i\hbar \frac{\partial}{\partial p_y}, \quad i\hbar \frac{\partial}{\partial p_z},$$

$$\tau \to \tau. \tag{1.15}$$

This is a fundamental point and suggests that usual quantum theory is possibly only a restricted representation of quantum phenomena because time τ remains a *classical* quantity. This was the reason why Schrödinger tried to introduce an operator for the time but without success.

Furthermore, the role of time τ within usual quantum theory is obviously also not in accordance with the basic ideas of modern space-time theories, in particular, when we consider special theory of relativity where space and time are completely symmetrical to each other. Symmetry requires that also time τ changes its character when we go from classical mechanics to quantum theory. This is not only the case for Schrödinger's equation, which has Newton's mechanics as the starting point,[8] but is also a typical feature of the relativistic Dirac equation. Let us briefly point out the consequences in connection with the uncertainty relations.

1.3. Uncertainty Relations

In atomic physics, solid state physics, nanoscience etc., it is sufficient that quantum theory conforms to the special theory of relativity because phenomena within the frame of these topics are hardly influenced by the laws of general theory of relativity, that is, by gravitation.

Since quantum theory and special theory of relativity have been developed independently from each other, the following question is of relevance: Can quantum phenomena be treated fully relativistically in accordance with the basic laws of the special theory of relativity?

---

[8]See equation (1.3).

Although the relativistic wave equations (e.g., Dirac's equation for the electron) are invariant under Lorentz-transformation, the space coordinates x, y, z and time τ are in its physical content not symmetrical to each other[9] and this is in contrast to the fundamental results of special theory of relativity. This is due to the following facts[10]:

1. Whereas the coordinates are *statistical* quantities, time does not behave statistically. We already mentioned above that time remains unchanged when we go from classical mechanics to quantum theory. This is clearly reflected by the fact that the coordinates can be *operators* but time is always a simple *parameter*.
2. The determination of the eigenfunctions and eigenvalues is restricted on space and time and is not involved in this process. In the case of stationary, systems, equation (1.2) is valid and time τ does not appear.

That is in close connection with the fact that there is no uncertainty relation for the energy and time which would agree in its physical content with the uncertainty relation for the coordinates and momenta:

$$\delta p_x \delta x \geq \frac{\hbar}{2}, \delta p_y \delta y \geq \frac{\hbar}{2}, \delta p_z \delta z \geq \frac{\hbar}{2}. \qquad (1.16)$$

An analogous relation for the time t and a quantity which has the dimensions of energy is required from the point of view of special theory of relativity. The significance of the well-known relation[11]

$$\Delta E \Delta \tau \geq \frac{\hbar}{2} \qquad (1.17)$$

is entirely different from that of relations (1.16). In the relations (1.16), the quantities $\delta p_x, \ldots$ and $\delta x, \ldots$ are the uncertainties in the values of the momenta and the coordinates at the same instant. As it is well-known, this uncertainty means that the coordinates and momenta can never have entirely definite values simultaneously. The energy E, on

---

[9] See Section 1.1.
[10] Interested readers might refer to De Broglie, L., (1943) Die Elementarteilchen: Goverts.
[11] See, for example, Schommers, W., (1989) Quantum theory and pictures of reality: Springer-Verlag.

the other hand, can be measured to any degree of accuracy at any instant. The quantity ΔE in relation (1.17) is the difference between two exactly measured values of the energy at two different instants and is not the uncertainty in the value of the energy at a given instant.[12]

In conclusion, the required symmetry between space and time (special theory of relativity) is definitely not fulfilled within the framework of usual quantum theory (both, in its relativistic and its non-relativistic versions). In particular, time τ remains a simple parameter in quantum mechanics. Thus, we may state that time is still a *classical* quantity within conventional quantum theory. We will see in Chapter 3 that within "projection theory" the time is no longer a simple classical parameter but becomes a real quantum variable.

1.4. Individuals

Not only in connection with time τ conventional quantum theory behaves classically, but, in a certain sense, classical elements obviously come also into play when we analyze the basic quantum events in *space*, that is, when we consider such events which are necessary for the understanding of the probability interpretation. Let us briefly explain why.

What does Heisenberg's momentum-position uncertainty relation (1.16) mean in connection with a trajectory of a particle? The answer is straightforward: Because the quantities $\delta p_x, \ldots$ and $\delta x, \ldots$ reflect the uncertainties in the values of the momenta and the coordinates at the same instant, relation (1.16) implies that there cannot exist a *trajectory* for a particle. Since the initial values of **r** and **p** are inherently uncertain, so is the future trajectory of the particle underdetermined; it is not defined. This has two essential consequences:

1. The existence of a particle (local existent) without a trajectory means that, within the frame of conventional quantum theory, the motion must be accidental leading to the probability interpretation proposed by Born.

---

[12] For more information, see the discussion in Landau, L.D., (1965) Quantum mechanics: Pergamon, and De Broglie, L., (1943) Die Elementarteilchen: Goverts.

2. One of the consequences of this picture is that there can be no such concept as the *velocity* of a particle in the classical sense of the word, i.e., the limit to which the difference of the two coordinates at two instants, divided by the interval $\Delta\tau$ between these instants, tends as $\Delta\tau$ tends to zero.

Born proposed to interpret the wave function $\psi = \psi(\mathbf{r},\tau)$ not as true material wave but as a *probability wave*. According to him, the probability of finding a real quantum object (assumed to be a point-like particle) in the volume element dV at a certain position $\mathbf{r}$ and time t is given by

$$\psi^*(\mathbf{r},\tau)\psi(\mathbf{r},\tau)dV. \qquad (1.18)$$

Thus, $\psi^*\psi$ plays the role of a probability density. Born's probability interpretation is one of the fundamentals of usual quantum theory. However, within relativistic quantum mechanics the wave function cannot be used for the definition of a probability density for a single particle as in the case of the (non-relativistic) Schrödinger equation.[13]

In summary, we have a particle without trajectory, and the classical concept of velocity cannot be used here. Therefore, the particle motion must be accidentally leading to the probability interpretation for the wave function $\psi(\mathbf{r},\tau)$. In other words, there is no longer a physical law that tells us *when* and *where* a particle jumps. Wolf wrote in his book that within classical mechanics motion is a

> *continuous blend of changing positions. The object moves in a flow from one point to another. Quantum mechanics failed to reinforce that picture. In fact, it indicated that motion could not take place in that way. They jumped from one place to another, seemingly without effort and without bothering to go between two places.*[14]

---

[13] Interested readers might refer to Schommers, W. (ed.), (1989) Quantum theories and pictures of reality: Springer-Verlag, to Schommers, W., (1995) Symbols, pictures and quantum reality: World Scientific.

[14] Interested readers might refer to Wolf, F.A., (1981) Taking the quantum leap: Harper and Row.

This text suggests that we should hesitate to connect such a behavior with a particle in the conventional (classical) sense of the word. It is possibly better to talk only of an "event" without imaging the details. However, it is not conceivable how the classical particle concept can be given up within usual quantum theory. On the other hand, we have to ask whether it makes sense at all to assume that in nature actually exist localized particles without trajectories.

These arguments become more accessible when consider the particle-problem in terms of typical observations in connection with the probability density $\psi^*(\mathbf{r}, \tau)\psi(\mathbf{r}, \tau)$. Let $\psi^*(\mathbf{r}, \tau)\psi(\mathbf{r}, \tau) \neq 0$ in the range Δx and let A and B be two positions within Δx. At time τ_A we register the particle with a detector at position A. At time τ_B we register the same particle at position B. Because there is no physical law that tells us *when* the real particle with mass m_0 arrives at position B after it has left position A, the time interval $\Delta \tau = \tau_B - \tau_A$ can take any value, for example, it may be infinitesimal small but different from zero. Then, the seeming velocity $v = \Delta x/\Delta \tau$ can take any value, in particular, it may be larger than the velocity of light c.

However, as we have outlined above, due to the momentum-position uncertainty relation (1.16), a particle *velocity* is not defined. On the other hand, the individual was at time τ_A at position A and the *same* individual was at time τ_B at position B. The particle cannot simply move from position A to position B in the classical sense of the word, and this is because also no *trajectory* can exist in the quantum-mechanical case, as we have remarked above.

How can we understand that? Does the individual leave space? The answer could be the following. According to Bohr, it is not allowed to make pictures of the process between the two detections, that is, for time τ with $\tau_A < \tau < \tau_B$. From this point of view, questions like "Does the individual leave the space?" are wrong questions. The only knowledge about the particle is due to its detection: We know that the particle was at time τ_A at position A and we also know that the same particle was at time τ_B at position B. But also in connection with these detections no *velocity* and no *trajectory* may be defined. This has the following consequence: A trajectory and velocity can obviously only be avoided if the particle (for example, detected at position B) is *simultaneously* existent and non-existent at position B. The problem obviously only appears when we connect the detection with an *individual* (point-like particle). An individual demands a minimum description since this notion has been chosen with respect to our intuitive demands for visualizability. The individual arrives position B at time τ_B and it leaves it at exactly the same time τ_B. Otherwise a trajectory would be defined as a horizontal line in the (x, τ)-diagram. The corresponding velocity would be zero but is defined. But this picture reflects a paradoxical situation as we have outlined above. The particle is *simultaneously* existent and non-existent! As in the case of "time",[15] also here usual quantum theory leads to an unsatisfactory situation. Let us discuss this point in somewhat more detail.

Remark

The paradoxical situation obviously only appears when we connect the detection with a local existent (here a point-like particle). As we have already mentioned above, a local existent demands a minimum description since this notion has been chosen with respect to our intuitive demands for visualizability which are based on our observations in everyday life. This conception requires that the real something (here the point-like particle) is unequivocally connected with a position, say $\mathbf{r}$, at time τ. If not, we do not talk from a point-like particle in the form of a local existent. Since the arrival and the departure of the particle at position $\mathbf{r}$ take place simultaneously (at time τ), it is problematic to state that the particle is unequivocally connected with a definite position $\mathbf{r}$, and this is because we cannot unequivocally say that the particle is at position $\mathbf{r}$. In other words, the definition of a particle as local existent becomes questionable.

However, also in the case of a classical trajectory the real particle is not unequivocally connected with a position $\mathbf{r}$. Also here the arrival and the departure in connection with $\mathbf{r}$ take place simultaneously (at time τ). When we connect "arrival" with the notion "existence" and the "departure" with the notion "non-existence" the particle is existent and simultaneously non-existent at each position $\mathbf{r}$. (like

---

[15] See Sections 1.1 and 1.2.

Zenon's arrow). But in the case of the classical motion there is a strict correlation between the effect of arrival and the effect of departure, and these correlations are responsible for the fact that at each point **r** a velocity is unequivocally defined creating the trajectory. In other words, the real existence of the point-like particle is expressed for the observer by the trajectory but not through a point **r**. That is, the existence of the trajectory justifies the definition of a particle in space (space-time). The observers imagination of a particle (point-like or not) is always connected to a concrete picture in space (space-time), and this is because we transfer our observations in everyday life to the quantum-mechanical level.

As in classical physics, also in quantum theory a particle is existent and simultaneously non-existent at each position **r**, but we have here, in contrast to classical mechanics, no correlation between the effect of arrival and the effect of departure and, therefore, a velocity at point **r** is not definable and also not a trajectory. Thus, the observers imagination of such a particle is restricted to a certain position **r** (at a certain time τ), i.e., for him the particle can only be existent and must be simultaneously non-existent. In other words, we are confronted with paradoxical situation which is unsatisfactory.

Is the concept of particle, as an entity localized in space, still a classical notion? Quite generally, an adequate description at the quantum-mechanical level possibly requires us to give up all classical notions. However, there are still quantities within usual quantum theory which are in accord with intuitive concepts valid at the macroscopic (classical) level. We have learned above that time τ is such a quantity. But also the picture of a localized particle is an intuitive concept which was transferred from the macroscopic (classical) to the microscopic level. Gribbin, in his book, states

> *But the interpretation in terms of particles is all in mind, and may be no more than a consistent delusion.*[16]

---

[16] Interested readers might refer to Gribbin, J., (1984) In search of Schrödinger's cat: Bantam books.

1.5. Conclusion

Schrödinger's equation has not been deduced but was assumed, that is, we do not know the structure of physical reality on which Schrödinger's equation is based. What is the origin of the postulates formulated by equations (1.8) and (1.9)? This question cannot be answered within the frame of usual quantum theory. A possible answer is given below within the framework of projection theory. It is therefore not surprising that usual quantum theory does not always lead to consistent pictures, in particular then when we use notions which are typical requisites of classical physics. We have discussed this point by means of time and the concept of particle as an entity localized in space. Within new developments, beyond usual quantum theory, these problems in connection with *time* and *localized particle* should be eliminated. In Chapter 3, we will discuss these points successfully on the basis of projection theory.

1.6. Aspects

1.6.1. *The Principle of Complementarity*

The more accurately we measure the position (the statistical average of many position measurements) of an object, the smaller $\delta x, \delta y, \delta z$ will be, and the more convinced we shall be that it is a particle. That is, the measured signal at a definite point in space is identified with a point-like object. However, according to Heisenberg's uncertainty relation (1.16), in the case of

$$\delta x, \delta y, \delta z \to 0. \tag{1.19}$$

We have

$$\delta p_x, \delta p_y, \delta p_z \to \infty \tag{1.20}$$

and the wavelength

$$\lambda = \frac{h}{p}, \tag{1.21}$$

the most essential characteristic of a wave, will not be determined at all. That is, as the *corpuscular* character becomes more marked, the *wave* character of the phenomenon becomes more vague.

On the other hand, if we measure the wavelength λ of matter with great precision, $\delta p_x, \delta p_y, \delta p_z$ will be small, and we shall be convinced that matter is a wave phenomenon. However, according to relation (1.16), in the case of

$$\delta p_x, \delta p_y, \delta p_z \to 0, \qquad (1.22)$$

we have

$$\delta x, \delta y, \delta z \to \infty, \qquad (1.23)$$

and the corpuscular aspect will not be evident at all. In other words, the *wave* character becomes more marked and, on the other hand, the *corpuscular* aspect of the phenomenon becomes more vague.

We are confronted with the following dilemma: Is matter (or light) a wave or a corpuscular phenomenon? The answer can be given as follows:

According to Bohr, the reason why no way out of this dilemma could be found is that the question had been posed wrong. For, it implies that only one of the two incompatible possibilities is realized, whereas, according to Bohr, it should be assumed that in nature both light and elementary particles can appear as waves or corpuscles, depending on the type of experiment performed on them. According to Bohr, wave and corpuscular aspects of physical phenomena are not incompatible, but on the contrary, are — as Bohr expressed it himself — are complementary.[17]

The principle of complementarity seems to be an artificial tool in order to rescue the apparatus of conventional (usual) quantum theory. This principle postulates that there is something in space and time that cannot be characterized without experiment. In Section 1.8, we

---

[17] Interested readers might refer to Rubinowics, A., (1968) Quantum mechanics: Elsevier.

will discuss new experiments that demonstrate the breakdown of the principle of complementarity.

Since for every pair of canonically conjugate variables a suitable uncertainty relation exists, we should regard such physical quantities as complementary. *Uncertainty* and *complementarity* belong to the foundations of the *Copenhagen Interpretation of Quantum Theory*.

1.6.2. *Objectivity*

According to the principle of complementarity, it is meaningless to talk about physical properties (e.g., wave or particle) of quantum objects without precisely specifying the experimental arrangement which determines them. Such a concept allows a description of quantum objects without logical contradiction in ways which are mutually exclusive. Note, that the experimental arrangements that determine those properties must be similarly mutually exclusive; otherwise the principle of complementarity principle breaks down. For example, different experimental arrangements are needed to measure the position and the momentum of an object.

In other words, a phenomenon (e.g., a wave or a particle) is always an *observed* phenomenon (Copenhagen interpretation). Without observation it is meaningless to talk about a phenomenon. It is therefore incorrect to regard a certain property of quantum objects as a property of the quantum object itself. It is an attribute which must be assigned to *both* the quantum object and the experimental arrangement. Since the choice of the experimental arrangement is purely a matter of human intention, the properties of quantum objects cannot be considered as objective. That is, the human intention influences the structure of physical reality.

In Section 1.8, we will discuss interesting new experiments which show that all these (interesting) constructions are possibly more than doubtful, in particular, the principle of complementarity.

1.7. Remarks on the Superposition Principle

Within usual quantum theory the properties of a system are characterized by linear operators. (For example, the Hamiltonian $\hat{H}$ is a linear operator.) Thus, the eigenfunctions $\psi_n, n = 1, 2, \ldots, m$, of an operator $\hat{F}$ can be used for the representation of an arbitrary wave function ψ, which is a linear combination of these eigenfunctions and is expressed by

$$\psi = \sum_n a_n \psi_n, \qquad (1.24)$$

where the coefficient a_n are constants. Equation (1.24) reflects the so-called superposition principle, which is one of the first principles of usual quantum theory. The use of only linear operators in usual quantum theory is a

very far-reaching restriction, just as if we were to decide to use only linear functions in a certain area of mathematics.[18]

In fact, not only this restriction is remarkable, but also the serious problems which appear in connection with the superposition principle. These problems are well-known and, therefore, we would like to summarize here only the main facts.

According to the superposition principle, the arbitrary wave function ψ in equation (1.24) has the following meaning: ψ gives the state of the quantum system in which a measurement leads with certainty to one of the m possible eigenvalues $f_n, n = 1, 2, \ldots, m$ of the operator $\hat{F}$ which belongs to the eigenfunctions $\psi_n, n = 1, 2, \ldots, m$. The squared modulus $|a_n|^2$ of each coefficient in equation (1.24) determines the probability that the measurement leads to the eigenvalue f_n. But once we have obtained a given eigenvalue, say f_n, we know that the system is necessarily in the state ψ_n. That is, the measurement process transforms the system from the arbitrary state ψ to one of the eigenstates, and we have a mixture.[19] The process is often called "reduction of the wave function".

Before a measurement the properties of a quantum object are not independent realities, and it is meaningless to talk about the physical

---

[18] Interested readers might refer to Rubinowics, A., (1968) Quantum mechanics: Elsevier.
[19] Interested readers might refer to Jammer, M., (1974) The philosophy of quantum mechanics: Wiley.

reality. They are not "either-or" alternative worlds, but there is a superposition of all possible properties (worlds). The effect of the measurement is to chop the overlapping worlds apart into disconnected worlds. In other words, the measurement process transforms the pure state (represented by the wave function ψ or by the density matrix) into a mixture (represented by the density matrix).

Within the Copenhagen interpretation of quantum theory, which is the most used interpretation, the collapse of the wave function and the assignment of statistical weights are not explained. Within this picture

they are consequences of an external a priori metaphysics, which is allowed to intervene at this point and suspend the Schrödinger equation.[20]

In connection with linear operators, which are exclusively used in conventional quantum theory, two points are remarkable:

1. Usual quantum theory is not able to describe single events (measured with a detector) which indicates that the system under investigation is in a definite state ψ_n with eigenvalue f_n. Usual quantum theory only describes ψ[21] but cannot describe the transition to ψ_n, which is relevant in connection with observations.
2. The use of only linear operators in usual quantum theory is a very far-reaching restriction in the description of nature.[22]

Again, within the standard *Copenhagen Interpretation* (proposed by Bohr in 1927) the world (or any system) consists of *options* that are equally *unreal*. By the *act of observation*, a system is forced to select one of its options and this becomes real, that is, within this interpretation reality is produced by the act of observation, so that any real system (for example, an electron) cannot be thought of as having independent existence. We know nothing about what it is doing when we are not looking at it. Within the Copenhagen Interpretation, nothing is real unless we look at it. As soon as we stop looking, it ceases to be real.

---

[20] Interested readers might refer to De Witt, B., (1970) Physics Today, September issue.
[21] See equation (1.24).
[22] Interested readers might refer to Rubinowics, A., (1968) Quantum mechanics: Elsevier.

On the other hand, within the *Many-World Theory* there is not just one world, but as many alternative real worlds as options exist within the Copenhagen Interpretation.

What happens when we make a measurement at the quantum level is that we are forced by the process of observation to select one of these alternatives, which becomes part of what we see as real world; the act of observation cuts the ties that bind alternative realities together, and allows them to go on their own separate ways through superspace, each alternative reality containing its own (real) observer who has made the same observation but got a different quantum answer...[23]

In other words, within the Many-Worlds Theory there is a real splitting into real worlds, and the theory can explain why no observer can be aware of the slitting process.

Both, the Copenhagen Interpretation and the Many-World Theory give predictions in accord with experience and, therefore, they are equivalent from the experimental point of view. However, both interpretations lead to pictures of reality which are completely different from each other, and it is legitimate at the present stage of quantum theory to choose the picture of reality which one finds satisfactory and pleasing.

New effects can be expected by the use of non-linear operators and we have to check their real existence by experiments. On the other hand, in the case of non-linear operators "the collapse of the wave function", which has to be considered as a metaphysical concept, is no longer needed. How do we get from the quantum reality of multitude possibilities to the definite events we experience in our life and in connection with experiments?

Decoherence effects cannot really solve this problem. Decoherence is always active when a system interacts with another. The effect becomes however a very fast process for macroscopic objects since the decoherence rate depends on the size of the system. Macroscopic

---

[23] Interested readers might refer to Gribbin, J., (1984) In search of Schrödinger's cat: Bantam books.

objects interact with their natural environment with a lot of microscopic objects and this is the reason why decoherence is a very fast process for such systems we observe in everyday life and, therefore, it explains why we do not observe quantum behaviour in everyday life in connection with macroscopic objects. In other words, decoherence helps us to understand how classical behaviour emerges from quantum mechanics. It is therefore important for the understanding of the famous measurement problem. However, the decoherence effect does not lead to the wave-function collapse and does not really solve the measurement problem,[24] and this is because all components of ψ[25] still exist in connection with the combined system (system under investigation and its interaction partner), that is, there is still a global superposition. However, decoherence explains why these entanglements (coherences) no longer exist for local observers. Since the decoherence rate depends on the size of the system it nicely explains why we observe quantum coherence for electrons, atoms and small molecules but never for objects of everyday life.

Within the frame of *Projection Theory*[26] the operators become nonlinear in a most natural way, and there is no superposition principle defined within projection theory.

1.8. Basic New Experiments

1.8.1. *General Remarks*

As already mentioned, the Copenhagen Interpretation is considered today as the standard interpretation of usual quantum theory, although it is not really recognized as the final step in the understanding of quantum phenomena. The "picture of reality", on which the Copenhagen Interpretation is based, seems to be artificial and not convincing in the opinion of many scientists, and this is obviously due

---

[24] For the collapse of the wave function, see also Section 1.7.
[25] See (1.24).
[26] See Chapters 2–5.

to the fact that we do not know the origin of the quantum laws.[27] The situation is unsatisfactory and can be summarized as follows:

> *The Copenhagen Interpretation held away for more than 50 years, from 1930 until well into the 1980s, almost unopposed by the vast majority of physicists. They did not care about the deep philosophical puzzles associated with the Copenhagen Interpretation — indeed, many still do not care — provided that it could be used as a practical tool for predicting the outcome of experiments. But in recent years there has been growing unease about what quantum theory "means", and increasing efforts have been made to find alternative interpretations.*[28]

However, in the opinion of the author, it will obviously hardly be possible to formulate a new interpretation without contradictions on the basis of the present formalism of quantum mechanics. There are some problems which have not been solved up to now. We discussed some of them above. The unsatisfactory role that time plays within the present form of quantum, the particle concept (particles are assumed to be local existents) seems also to reflect an unclear situation, and of course the superposition principle in connection with the collapse of the wave function. All these points indicate that the formalism of usual quantum theory cannot be considered as the final solution. Therefore, it will probably not be enough to re-interpret the Copenhagen Interpretation or one of the any other proposals made in this field. On the other hand, a changed or extended quantum formalism could lead to completely new perspectives in connection with applications, in particular, within nanoscience.

It is therefore of fundamental importance that there are new experimental results which clearly indicate that the principle of complementarity[29] cannot be considered as a general principle. These

---

[27] Schrödinger's equation has been assumed and could not be deduced. See Section 1.1.
[28] Interested readers might refer to Gribbin, J., (1995) Schrödinger's kittens and the search for reality: Little, Brown and Company.
[29] See section 1.6.1

experiments show that Bohr's viewpoint in connection with wave–particle duality is obviously not a general feature of nature, and if that is really the case, the situation is changed fundamentally. The question was:[29] "Is matter (or light) a wave or a particle phenomenon?" In answering this question, Bohr argued as follows: Only one of the two incompatible possibilities is realized, depending on the type of experiment performed on them, and we came to the notion of complementarity. Either we perform an experiment typical for waves or we choose an experimental arrangement that exclusively marks the particle aspect. Only in this way, wave and corpuscular aspects of physical phenomena are not incompatible. In other words, not only are wave and particle concepts mutually exclusive but also the corresponding experimental arrangements. However, this point is obviously not true as demonstrated in the case of light.

There exist optical experimental arrangements that can be used for the *simultaneous* detection of particle and wave aspects. The experiments have been proposed by Ghose, Home and Agarwal[30] and Ghose.[31] The first experiment in this field has been performed by Mizobuchi and Ohtake[32] which was based on the proposal given previously. Although this experiment is without any doubt important, there were experimental limitations (mainly due to low statistics) in connection with this study, outlined by Unnikrishnan and Murphy.[33] In order to overcome the limitations in connection with the Mizobuchi-Ohtake experiment Brida, Genovese, Gramegna and Predazzi performed with success a new birefringent experiment[34] leading to a clear conclusion with respect to a new understanding concerning the wave–particle duality which is in contrast to

---

[30] Interested readers might refer to Ghose, P., Home, D., and Agarwal, G.S., (1991) Phys. Lett. A **153**, 403.
[31] Interested readers might refer to Ghose, P., (1999) Testing quantum mechanics on a new ground: Cambridge University Press.
[32] Interested readers might refer to Mizobuchi, Y., and Ohtake, Y., (1992) Phys. Lett. A **168**, 1.
[33] Interested readers might refer to Unnikrishnan, C.S., and Murphy, S.A., (1996) Phys. Lett. A **221**, 1.
[34] Interested readers might refer to Brida, G., Genovese, M., Gramegna, M., and Predazzi, E., (1996) Phys. Lett. A **328**, 313.

Bohr's view, namely the *simultaneous* detection of particle and wave aspects.

1.8.2. Conclusion

Both experimental groups[34],[35] came to the same conclusion: Light showed both particle and wave aspects simultaneously, and this is in contrast to the principle of complementarity which is one of the very basics of modern quantum theory (Copenhagen Interpretation). Thus, the principle of complementarity is obviously not a general law, and this must inevitably have consequences for the whole picture we presently have of usual quantum theory.

Bohr's law "a phenomenon is always an observed phenomenon"[36] is just based on the principle of complementarity, and this has been formulated due to the problem scientists had in connection with the wave–particle phenomenon. We need the principle of complementarity within usual quantum theory in order to avoid a logical contradiction.

According to the principle of complementarity, only *one* of the two incompatible possibilities (wave, particle) can be realized and this implies that the experimental arrangements that determine those properties must be similarly mutually exclusive. Such a concept allows a description of quantum objects (particle, wave) without logical contradiction in ways which are mutually exclusive. The new experiments[34],[35] show just the opposite: Wave and particle aspects have been observed simultaneously with only one experimental arrangement, that is, the principle of complementarity obviously breaks down.

On the other hand, if Bohr's law "a phenomenon is always an observed phenomenon" should no longer be valid, the properties of a system should be also defined between observations and not only in connection with observations, for example, in the case of the double-slit experiment.

---

[35] Also, Interested readers might refer to Mizobuchi, Y., and Ohtake, Y., (1992) Phys. Lett. A **168**, 1, and to Unnikrishnan, C.S., and Murphy, S.A., (1996) Phys. Lett. A **221**, 1.
[36] See Section 1.6.1.

Our statements concerning *objectivity* must possibly revised. Within usual quantum theory, it is meaningless to talk about a phenomenon without observation. However, because the choice of the experimental arrangement is purely a matter of human intention, we may not consider the properties of quantum systems as objective since the human intention influences the structure of physical reality. This view becomes questionable without the principle of complementarity.

2

Projection Theory

2.1. Preliminary Remarks

There are obviously some specific serious problems in connection with usual quantum theory. These problems are often ignored, and this is mainly due to the fact that the empirical success of usual quantum theory is impressive. But due to the new experiments,[1] which have been performed by two independent groups, we are confronted with a new situation, and we have to analyze critically the basics of conventional quantum theory (its mathematical apparatus as well as its interpretation). In Chapter 2, we already have quoted some critical points which appear in connection with usual quantum theory. The main facts can be summarized as follows:

1. Schrödinger's equation could not be deduced in usual quantum theory but was assumed on the basis of "reasonable arguments". In other words, we do not really know the physical reality on which Schrödinger's equation is based.
2. The *collapse of the wave function* cannot be explained in usual quantum theory. One would expect that Schrödinger's equation (the fundamental tool within non-relativistic quantum theory) was able to describe such a process but this is unfortunately not the case. The collapse of the wave function and the assignment of statistical weights are not explained. Within usual quantum theory, de Witt wrote

---

[1] Interested readers might refer to Mizobuchi, Y., and Othake, Y., (1992) Phys. Lett. A **168**, 1, and to Brida, G., Genovese, M., Gramegna, M., and Predazzi, E., (1996) Phys. Lett. A **328**, 313.

> *they are consequences of an external a priori metaphysics, which is allowed us to intervene at this point and suspend the Schrödinger equation.*[2]

In other words, the situation in connection with the collapse of the wave function is unsatisfactory and has to be solved.

3. Time τ is still a classical parameter in usual quantum theory. While the coordinates are *statistical* quantities, time does not behave statistically. Time τ remains unchanged when we go from classical mechanics to quantum theory. This is clearly reflected in the fact that the coordinates can be *operators*, while time is always a simple *parameter*. This is the reason why there is no uncertainty relation for the time τ and the energy E which would agree in its physical content with the position–momentum uncertainty relation.[3]

4. The use of a particle defined as a local existent (which is assumed within the frame of usual quantum theory) seem also to be problematic and this is probably due to the fact that we transfer a picture used in everyday life to the microscopic realm. Can a particle (point-like individual) exist without trajectory (usual quantum theory)? We argued that such a concept is problematic because an individual demands a minimum description since this notion has been chosen with respect to our intuitive demands for *visualizability* (due to our experiences in everyday life). The concept of a point-like individual without trajectory obviously leads to a paradoxical situation.

In the next sections, we will discuss these points in connection with the so-called "projection theory". It will turn out that the point of view of conventional quantum theory cannot simply be extended, but we have to include further fundamental facts, so far not directly considered in the basic description of physical systems. We will in particular see that we can learn something new about the relationship between the "observer" and that what we call "reality" when we consider certain facts from biological evolution and behavior research.

---

[2] Refer to de Witt, B., (1970) Physics Today, September.
[3] See Section 2.2.

First, we discuss the so-called projection principle. The projection principle is based on the insight that the material objects (sun, moon, etc.) are not embedded in space (space-time), but the physical reality is projected onto space (space-time) within the frame of the projection principle. After that we use this principle for the mathematical formulation of the relevant quantum laws. In this way new features come into play. In particular, we get a completely new formulation for the notion "time" within projection theory. Time becomes a real quantum variable, and it is no longer an external parameter as in classical physics and conventional quantum theory.

2.2. The Projection Principle

2.2.1. *The Elements of Space and Time*

Within the memory of man all essential things are represented within the frame of pictures. This is in our opinion the most basic statement. Such pictures contain the objects (sun, moon, etc.), the space and the time. We feel the objects, that is, we observe them with our five senses and also with measuring instruments. Thus, we believe in the concrete existence of these things and we call them "material objects". But what about space and what about time? It is not easy to answer this question but only because these notions, space and time, are burdened with a serious prejudice: We are firmly convinced that all the material objects around us are embedded in space and space-time, respectively. It is normally assumed that the space is a container filled with matter. However, there are some reasons to take another viewpoint and to consider this container-conception as unrealistic, as prejudice so to say. Already the philosopher Immanuel Kant pointed on this serious problem.[4]

All that what we can grasp with our five senses and/or with measuring instruments can be considered as really existing things in the outside reality. However, can we seize space and time with our five senses or with specific measuring instruments? No, we cannot.

---

[4]Interested readers might refer to Schommers, W., (2008) Advanced Science Letters **1**, 59, and to Schommers, W., Cosmic Secrets: World Scientific, in preparation.

Therefore, we may conclude that space and time do not exist in the reality outside. Although this conclusion is straightforward it must be more justified and underpinned with further arguments and more details.

The situation with respect to space and time is very clear. We cannot put a "piece of space" on the table and we can also not put a "piece of time" on the table. Such pieces do not exist. We are not able to observe the elements of space, that is, its coordinates x, y, z and we are also not able to observe the elements of time which we have denoted in Chapter 1 by the letter τ (normally the letter t is used). We have no senses for that and, furthermore, the development of measuring methods for the detection of x, y, z and τ are even not thinkable.

Within Newton's mechanics space and time are independent of each other and also independent of matter. This concept has been developed further by Einstein. Within Theory of Relativity, space and time are no longer independent of each other but form a space-time. Then, the following question arises: Can we put a "piece of space-time" on the table? No, we definitely cannot. As in the cases of space and time, also a piece of space-time does not exist in the sense of an observable quantity.

Nevertheless, we are all familiar with the space-time phenomenon. Thus, we have to conclude that the elements x, y, z, τ must be inside the observer and, as we will recognize in the course of our debate, it makes no sense to assume that these elements (x, y, z, τ) are also outside the observer. As we have already mentioned above, the philosopher Immanuel Kant took this viewpoint but his fundamental ideas have never been seriously used in the formulation of physical theories.

2.2.2. Relationship between Matter and Space-Time

In what form do we experience the space-time phenomenon? The answer is quite clear: We never observe isolated space-time points which we characterize by x, y, z, τ but without exception only distances of real objects and time intervals in connection with physically real processes. In other words, space and time (or distances of them) never appear without objects and processes. In all cases "matter" and "space and time" are closely linked. Neither of them is able to exist without the other. Nobody is able to observe isolated space-time points x, y, z, τ without objects and processes, respectively.

Is this important experimental finding fulfilled within modern physical theories? No, it is not. Within Newton's mechanics, space and time are independent of each other and both elements may exist independent of matter. But also within the Theory of Relativity, space-time may appear without material objects.[5]

2.2.3. Two Relevant Features

In summary, the relationship between "matter" and "space-time" is given by the following facts and features, respectively:

Feature 1

We definitely cannot see, hear, smell, or taste space and time, that is, space and time (absolute or non-absolute) are not accessible to our senses. Also measuring instruments for the experimental determination of the space-time points x, y, z, and τ are not known and even not imaginable.

Therefore, space and space-time, respectively, may not be the source of physically real effects. This requirement is the contents of Mach's principle which is however not fulfilled within Newton's theory but also not within the Special and General Theory of Relativity. It is obviously not possible to fulfil Mach's principle when we assume that the material bodies are embedded in space (space-time). We will discuss Mach's principle in more detail in Chapter 4.

Feature 2

We can only say something about *distances in connection with masses*, and *time intervals in connection with physical processes*. There is no exception!

---

[5] De Sitter already showed in 1917 that an empty space-time is possible within General Theory of Relativity.

Remarks concerning feature 1

Space and time should never be the source for physically real effects as, for example, inertia. This is not fulfilled within Newton's theory but also not within Einstein's theory. It is still possible within Theory of Relativity to talk about the rotation of the entire mass of the world relative to absolute space. Within Gödel's universe,[6] an absolute space-time is defined, and in this case the space is the source of inertia.

Remarks concerning feature 2

1. From feature 2, it directly follows that "matter" and "space-time" are closely linked. Neither should be able to exist without the other. In other words, an *empty* space-time as physical-theoretical conception should not exist.
2. When we define a certain distance in space by two material bodies, then there must be a relationship between the two masses and space. The material bodies can have, for example, a *constant* space distance which already expresses a certain kind of relationship. But what kind of relationship? This cannot be due to an interaction between the material bodies and space (space coordinates). The reason is given by feature 1. There can be no interaction between space-time and the sense organs or measuring instruments which are made of material bodies, that is, no interaction between material bodies (masses) and space-time.
3. But what kind of relationship exists between the two material bodies that determine a certain space distance? In the following we will answer this question.

In conclusion, space and time are not accessible by the human senses and no measuring instrument is able to make space-time-specific "clicks" but, on the other hand, we experience space-time in connection with material bodies and processes as a concrete phenomenon.

---

[6]Refer to, in particular, Schommers, W. (2008) Advanced Science Letters **1**, 59, and to Schommers, W., Cosmic Secrets: World Scientific, in preparation, and to Gödel, K., (1947) Rev. Mod. Phys. **21**, 447.

2.2.4. Two Kinds of "Objects"

Due to feature 1, space-time has to be considered as a "nothing" in the physical sense but, on the other hand, we experience space and time as a real phenomenon. Clearly, this "nothing" can hardly be occupied by physically real objects. It is problematic to embed the real world into such a metaphysical space-time. Within all modern physical theories (Newton's mechanics, Theory of Relativity, conventional quantum theory, etc.) real material bodies are the contents of an "object" (space-time) that has to be considered as a nothing or as a metaphysical substratum different from matter.

This seems to be a contradiction. Two kinds of objects are mixed which are obviously mutually incompatible. We can overcome this bad situation if we take the position of the philosopher Immanuel Kant: "Space and time are exclusively features of our brain and the world outside is projected on it". Then, the material objects that occupy space and time can only be geometrical pictures.[7] Projection theory[8] is based on this conception. If that is the case, the situation would be identical with that what we do in connection with our blackboards and notebooks: we draw pictures on them.

2.2.5. Perception Processes

Everything that we see is primarily in our head; it is not outside us. Persons, cars, aeroplanes, the sun, moon and stars are "pictures of reality" in our brain. We have only the impression that all these things are located outside us.[9] This conclusion is supported by Bruckner:

> *We have devices in the cerebral cortex which—comparable with a television screen—produce "pictures" in our awareness from the nerve-excitations coming from the retina. It is characteristic for the sight-process that our awareness does not register the picture of a candle on the retina inside the eye, but we have the impression that we are standing opposite the candle-light which is located in*

---

[7] See Figure 2.1.
[8] See also Chapter 4 and Chapter 5.
[9] See also Figure 2.2.

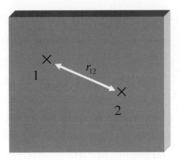

Fig. 2.1. Within the memory of man all essential things are represented within the frame of pictures. This is obviously the most basic statement. That what we always have in front of us in our observations in everyday life are geometrical objects which are embedded in space and time; not the material objects themselves are embedded in space and time but "only" their geometrical positions. In the figure, the geometrical positions $(x_1, y_1, z_1, x_2, y_2, z_2)$ of two objects 1 and 2 are represented by crosses which have the distance $r_{12} = [(x_2 - x_1)^2 + (y_2 - y_1)^2 + (z_2 - z_1)^2]^{1/2}$ at a certain time τ.

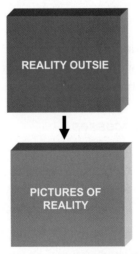

Fig. 2.2. Information about the "reality outside" flow through our sense organs in the brain and forms a "picture of reality". We never observe reality outside directly but "only" pictures of it.

the space outside, not standing on its head but upright. We see "real objects" in front of us and around us. Within this act of perception, the eye, the optic nerves and the brain work together. To see without the brain is as impossible as to see without eyes.[10]

---

[10] Refer to Brückner, R., (1977) Das schielende Kind: Schwabe Verlag.

We normally assume that these sensations produced by the brain are identical with reality itself,[11] but this should not be the case as we have argued above since space-time cannot be outside the brain because space-time has to be considered as an auxiliary element for the representation of physically real processes. In other words, the outside world, the material bodies, cannot be embedded in space-time. That in particular means that not only the things in front us (cars, houses, trees, etc.) are in our head but also space-time, where all these things are positioned. We have only the impression that all these "hard objects", together with space-time, are located outside us. Space and time are obviously elements of the brain; they come into existence due to specific brain functions. This is the reason why we cannot put space and time on the table as we can in the case of matter.

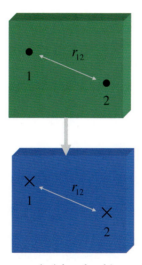

Fig. 2.3. Most people assume *a priori* that the things which they observe and which appear spontaneously in front of them is the material reality itself, and those which are conscious of the fact that it is only a picture, normally assume as a matter of course that the structures in the picture are identical with the material reality outside, that is, it is assumed that in reality outside the geometrical positions in the picture (crosses in the figure) are replaced by material objects (full points in the figure). This is however not the case, it is obviously a fallacy, and is supported by a large number of indications.

---

[11] See Figure 2.3.

This is because we have in general to assume that there is no similarity between the structures and characteristics in the picture[12] and those in the actual reality outside. How reality outside is constructed can principally not be said because a picture-independent view for the observer does not exist.[13] Thus, instead of Figure 2.3 we obtain Figure 2.4.

What about feature 2?[14] This feature is expressed by the following fact: *We can only say something about distances in connection with masses, and time intervals in connection with physical processes.* The solution for this finding is relatively simple and can be understood within projection

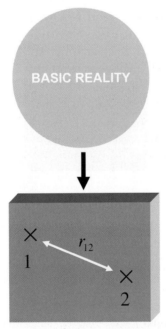

Fig. 2.4. The full circle symbolizes basic (objective) reality. It cannot be observed directly; no statements can be made about basic reality but certain characteristics can only be recognized in the form of pictures which appear in front of us and which are represented in space and time. Space and time do not appear within basic reality.

---

[12] See Figure 2.1.
[13] This view is outlined in Schommers, W., (1994) Space and Time, Matter and Mind: World Scientific.
[14] See Section 2.2.3.

theory by the conception that space and time only appear (in the brain) when there is actually something (objects of basic reality) to picture. Then, matter is closely linked to space and time. Why has evolution chosen this way? Also here the answer is straightforward:

Each species developed in connection with the laws of evolution[15] and these laws are dictated by the "principle of usefulness". An observation of an empty space (space-time) would not be useful because it does not contain any information for the observer needs. The principle of usefulness is obviously only fulfilled if the selected objects are closely linked to space and time.

2.2.6. *Inside World and Outside World*

We experience the world by our sense organs, that is, the observer interacts with reality outside: information about reality outside flow via our senses into the body, and the brain forms a picture of it. In this way we obtain a "picture of reality". This is definitely a projection. In other words, we have a *reality outside* and an *inner picture* (picture of reality), and we are firmly convinced that the inner picture is identical with reality outside. For example, the well-known psychologist C.G. Jung wrote:

> *When one thinks about what consciousness really is, one is deeply impressed of the wonderful fact that an event that takes place in the cosmos outside, produces an inner picture, that the event also takes place inside.*[16]

This statement by C.G. Jung suggests the following. Although there is a projection of reality onto space-time, we may also state that the real world outside is embedded in space-time. But, due to feature 1, this is strictly forbidden as we have analyzed above. Strictly speaking, there can be no space-time in connection with reality outside because space-time can only be an auxiliary element for the representation of physically real processes that take place outside. It

---

[15] See Section 2.2.7.
[16] Refer to Jung, C.G., (1990) Synchronizität, akausalität und okkultismus: Deutscher Taschenbuch Verlag GmbH & Co.

gives therefore no sense to assume that space-time is also an element of reality outside. The *unobserved* reality outside is free of space and time (Figure 2.4 is correct and not Figure 2.3).

Thus, the structure of reality outside must be quite different from the structure of the inner picture that is always based on space and time. However, the inner picture is the most direct impression we have of reality outside. Arthur Eddington wrote in his paper "*Science and the unseen world*" during the Swarthmore Lecture in 1929:

> *In comparing the certainty of things, spiritual and things temporal, let us not forget this — Mind is the first and most direct thing in our experience; all else is remote inference.*

Again, the inner picture must be quite different from the structures in the outside world and this is because there can be no space-time in the outside world.

2.2.7. The Influence of Evolution

All the statements we have made in connection with reality outside and its picture is confirmed when we consider the principles of evolution. Let us briefly discuss this point.

The perception of true reality in the sense of a precise reproduction implies that we need as much information from the actual reality outside as possible. In the case of a precise reproduction we even need the complete information about the outside world. Then, evolution would have developed sense organs with the property to transmit as much information from reality as possible. However, the opposite is correct: The strategy of nature is to take up as little information from the outside world as possible. The reason is relatively simple.

The solution of specific problems does not demand a complete knowledge of the world. From the point of view of evolution, the impressions before us are not precise reproductions of reality but merely appropriate pictures of it, formed by the individual from certain pieces of information from the outside world. According to the principles of evolution the central factor is "favourable towards survival" versus "hostile towards survival". The formation of a "true" picture of the world in an absolute sense is irrelevant. An individual registers situations in the environment in certain patterns which are tailor-made for the particular needs of the species (principle of usefulness) and which are completely free of any compulsion towards precise "objectivity". In particular, we have outlined[17] that we have to assume that an event occurring in the cosmos is portrayed inside a biological system "only" in symbolical form.

Let us briefly discuss a simple example.[17] In order to find a certain place in a cinema, it is not necessary that a visitor gets at the pay desk a small but true model of the cinema, i.e., a precise reproduction of the cinema, which is reduced in size; a simple cinema ticket with the essential information is more appropriate. In this respect, the cinema ticket is the picture of the cinema.

From the principles of evolution follow that the structures in the picture must be different from that in outside reality. The common or naive point of view assumes the following. The inside world which we feel to be outside us, actually exists in the outside world in exactly the same form as we perceive. This was also the view of the psychologist C.G. Jung as we have already remarked above. According to this view there is only one difference between the inside world and the outside world: inside there are only geometrical positions, whereas outside there are the real material bodies instead of the geometrical positions. In other words, it is normally assumed that the geometrical positions are merely replaced by material objects.

But why should events in nature occur, so to speak, twice, once outside of us, and again in the form of a picture? This would be against the principles of evolution. Let us briefly outline the reason.

It would make not much sense if events in a world, which is tailored to fulfil the principle of usefulness, took place twice. We ascertained,[18] that in nature knowledge for its own sake does not play the major role but rather the recognition of the factors "hostile towards survival" and

---

[17] Refer to Schommers, W., (1994) Space and Time, Matter and Mind: World Scientific, and to Schommers, W., (1998) The visible and the invisible: World Scientific.
[18] Refer to Schommers, W., (1994) Space and Time, Matter and Mind: World Scientific, and to Schommers, W., (1998) The visible and the invisible: World Scientific.

"favourable towards survival". For this purpose a consistent picture of an event occurring in the cosmos must be produced, but not a true reproduction in a one-to-one sense. In particular, such a picture does not need to be complete, because that would unnecessarily burden the biological organism in mastering specific situations in life.

2.2.8. *Information in the Picture versus Information in Basic Reality (Outside Reality)*

From the principles of evolution we have to conclude that basic reality contains much more information than an observer can depict in space and time.[19] This space-time information is mainly that part of the outside world which a human observer needs for survival.

The contents of the space-time information, selected from basic reality, is dictated by the principles of evolution. The unselected information, on which the selected picture of reality is based, remains principally unknown. For example, we know the observer's shape in space and time but not his structure in basic reality (reality outside). However, the projection principle also says that the observer must be more than that what is pictured in space and time which exclusively shows the material part of the observer. But this material part is only the selected information about the observer, and the unselected information must necessarily be "richer" than the material part which is pictured in space and time,[20] that is, we may not conclude that the space-time pictures contain the complete information about the reality outside. In fact, we know that a human observer has mind and feelings. His intellectual abilities are accompanied by logical laws that have to be considered as real extensions of the material world. However, these logical laws etc., are not depictable in space and time.

Why can certain things be represented in space and time and others not? The answer is simple and is given by the principles of evolution. Certain facts are relevant and other facts are less relevant. Those things and processes, respectively, which are depictable in space and time are of

---

[19] See Figure 2.5.
[20] See also Figures 2.6 and 2.7.

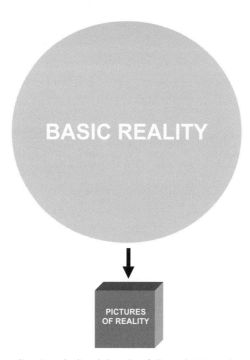

Fig. 2.5. Basic reality (symbolized by the full circle) in relation to the part of information (box with "picture of reality") an observer can experience of basic reality. The circle is much larger than that in Figure 2.4 indicting that the basic reality contains much more information than an observer can depict in space and time. It is mainly that part of information of the outside world which a human observer needs for survival.

particular importance for the survival of a human observer. Intellectual statements, that is, the products of mind, are also relevant but not in connection with processes which touch specific questions of survival. As said, here the principles of evolution are relevant here, and we introduced "levels of reality" for the unique description of all these phenomena.[21]

2.2.9. *Other Biological Systems*

The picture of reality must be species-dependent. In other words, we have to conclude that the actions of other biological systems are in

---

[21] Refer to Schommers, W., (1994) Space and Time, Matter and Mind: World Scientific, and to Schommers, W., (1998) The visible and the invisible: World Scientific.

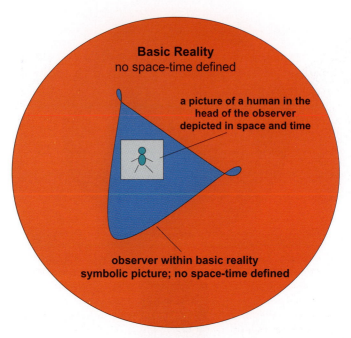

Fig. 2.6. A human observer is able to make statements about himself only in connection with space and time, but he has not the ability to conclude from the space-time picture to that which is existent of him (and of course all the other things) in basic reality. Therefore, the form of the observer in basic reality has been chosen arbitrarily. This arbitrarily chosen shape should only express that the space-time picture of the observer in his head must be different from that which is existent of him within basic reality.

general based on a picture of reality that is different from that of the human observer. How can we verify that? Wolfgang Schleidt performed some interesting experiments using a turkey, its chick and a weasel which is the turkey's deadly enemy, and he studied the behaviour of the turkey in order to learn something about the perception apparatus of the turkey.

Schleidt worked with more or less everyday methods. However, his experiments could be of such importance as certain key experiments in physics which have fundamentally changed the scientific world view. Schleidt demonstrated convincingly that the perception apparatus of a turkey must be quite different from that of human beings. These experiments led to dramatic and unexpected results and demonstrate that the turkey must experience the world optically quite differently

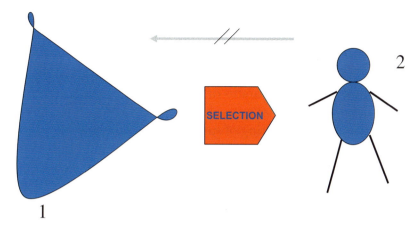

Fig. 2.7. The observer in basic reality is characterized by structure 1. His shape has been chosen arbitrarily because we can nothing say about the observer in basic reality. A picture-independent point of view is not possible within projection theory. The observer in space and time is characterized by the familiar structure 2 as he appears after a selection and projection process. The observer in basic reality (structure 1) contains more information than the part of him which is depictable in space and time (structure 2). The observer is principally not able to conclude from structure 2, familiar to all observers, to structure 2 (observer in basic reality).

from the way we do, even though the eyes of the turkey are quite similar to ours. There is obviously no similarity between what the turkey experiences and what a human being sees in the same situation.

Both systems, man and turkey, react correctly in the normal case because both species are able to exist in the world, which can only be possible from the point of view of modern principles of evolution if their particular views of the world are correct. Therefore, although the conceptions of the world of man and turkey are on the one hand different from each other, they are on the other hand correct in each case. This means that neither of these two conceptions of the world can be true in the sense that they are a faithful reproduction of nature: objective reality (basic reality) must be different from the images that are constructed by biological systems. We already came to a similar conclusion on the basis of general space-time arguments.[22,23]

---

[22] See Sections 2.2.1–2.2.5.
[23] A detailed description of these experiments together with conclusions and interpretations is given in Schommers, W., (1994) Space and time, matter and mind: World Scientific.

In summary, these experiments by Schleidt deliver essential contributions about our understanding of that what we call reality and they can help us to learn something about the relationship between reality and any kind of observer (man, turkey, etc.). However, we can only say that the perception apparatus of the turkey is different from that of humans. The details are not accessible in this way but we do need more information for the answering of such principal questions. We even do not know if such biological systems (turkeys) experience "their world" within the framework of space and time.

These results can be generalized because there is no reason to believe that turkeys have to be considered as an exception. Without doubt Schleidt's results support strongly our view concerning space and time developed above.

2.2.10. *Summary*

The material objects which we observe in everyday life are not embedded in space (space-time). Space and time have to be considered as auxiliary elements for the representation of the selected information about the outside world (basic reality). The chick experiment demonstrated that these pictures must be species-dependent, and this is confirmed by the principles of evolution. In other words, the physical reality is projected onto space and time, and the space-time elements x, y, z, τ cannot be seized with our five senses or with specific measuring instruments. These features of space and time must be considered in the formulation of physical theories. In the following, we will introduce a certain projection principle for the description of quantum phenomena.

2.3. Projections

2.3.1. *Principal Remarks*

We have pointed out in Section 2.2 that statements about the world are always species-specific, that is, any world view must always be superimposed by certain properties of the biological system (man, turkey, etc.) that makes the statements, consciously or unconsciously.

Therefore, it is principally not possible to give statements about the absolute reality since absolute reality is by definition independent of the observers biological structure, and there is obviously no way to eliminate these species-specific factors.[24]

This is in complete agreement with Wheeler's statement about this topic:

No theory of physics that deals only with physics will ever explain physics. I believe that as we go on trying to understand the universe, we are at the same time trying to understand man. [...] The physical world is in some deep sense tied to the human being.[25]

From the point of view of projection theory, species-relevant information about reality outside must be *projected* on the observer's cognition system leading to species-specific world views. The observer constructs (consciously or unconsciously) "his" picture of reality on the basis of a frame, whose elements are in the case of a human observer space and time, but this frame with its specific elements can be different for different biological systems. In other words, relevant information is selected from the outside world and is projected on the observer's frame of representation (for example, space and time). Within the frame of *projection theory*, space and time appear as auxiliary elements for the representation of certain aspects of the reality outside.[26]

This principle, strongly supported by the principles of evolution and the basic experimental results of behavior research,[27] is compatible to *Mach's Principle* which played an important role in connection with the development of theory of relativity. This is an important point, and we would like to outline the basic facts about Mach's Principle.

---

[24] Refer to Schommers, W., (1998) The Visible and the Invisible: World Scientific.
[25] Refer to von Baeyer, H.C., (2004) Information: The Language of Science: Harvard University Press.
[26] More details are given in Section 2.2 and the references therein.
[27] Interested readers might refer to Schommers, W., (1995) Symbols, pictures and quantum reality: World Scientific.

2.3.2. *Mach's Principle*

Space and time are absolute quantities in Newton's mechanics. Concerning the term "absolute" note the following:

1. Absolute space was invented by Newton for the explanation of inertia. However, we do not know of any other phenomenon for which the absolute space would be responsible. So, the hypothesis of absolute space can only proved by the phenomenon (inertia) for which it has been introduced. This is unsatisfactory and artificial.
2. The term "absolute" not only means that space is physically real,[28] but also "independent in its physical properties, having a physical effect, but not itself influenced by physical conditions". This must also be considered as unsatisfactory.

This is why Ernst Mach eliminated space as an active cause in the system of mechanics (Mach's Principle). According to him, a particle does not move in accelerated motion relative to space, but relative to the centre of all the other masses in the universe. In this way, the series of causes of mechanical phenomena was closed, in contrast to Newton's mechanics.[29]

In fact, absolute space and, of course, absolute time must be considered as metaphysical elements because they are, in principle, not accessible to empirical tests. There is no possibility of determining the space coordinates x, y, z and time τ. We can only say something about distances in connection with objects, and time intervals in connection with physical processes.[30]

Mach's Principle requires the elimination of space-time as an active cause; space-time cannot give rise to any physically real effect and cannot be influenced by any physical condition. This means

---

[28] Interested readers might refer to Schommers, W., (Ed.), (1989) Quantum Theory and Pictures of Reality: Springer-Verlag, and to Schommers, W., (1995) Symbols, Pictures and Quantum Reality: World Scientific.
[29] It should be emphasized that Mach's Principle is also not fulfilled in the theory of relativity. In order to fulfil Mach's Principle within General Theory of Relativity, one tried to extend it but without success.
[30] See, in particular, Section 2.2.

that there can be no interaction between space-time and reality, in accordance with the fact that the elements x, y, z and time τ are not accessible to empirical tests. Any change in distances (of objects) is not due to the interaction between coordinates or between coordinates and masses but is entirely caused by the interaction between the masses. Thus, space-time must be considered as an *auxiliary element* for the geometrical description of physically real processes. In other words, physically real processes are *projected* on space-time.[31] Mach's principle will be discussed in more detail in Chapter 4, also in connection with projection theory.

2.3.3. Conclusion

Within the frame of *projection theory*, space and time appear as *auxiliary elements* for the representation of certain aspects of the reality outside. Quite independent from the arguments used within *projection theory*, Mach's Principle exactly leads to same conclusion: Physically real processes are *projected* on space-time; space-time must be considered as an *auxiliary element*.

2.3.4. Other Spaces

2.3.4.1. Fourier-space

Space-time is formed by the variables $\mathbf{r} = (x, y, z)$ and τ; therefore, let us denote space-time "$(\mathbf{r}, \tau)$-space". Because space and also time play the role of an *auxiliary elements*,[32] it should be possible to "invent" other spaces on which *reality* can be projected. Such a space is, for example, the so-called Fourier-space (reciprocal space) having the variables $\mathbf{k}, \omega$ with the characteristics of a wave:

$$|\mathbf{k}| = \frac{2\pi}{\lambda}, \quad \omega = 2\pi\nu, \tag{2.1}$$

---

[31] Such a space-time can also be curved which depends on the form of description. Such a curved space-time can be associated with a mass — as within General Theory of Relativity but, on the other hand, Mach's Principle requires that such a space-time curvature cannot be due to an interaction process between the mass and the space-time elements x, y, z and time τ.
[32] See Section 2.2.

where λ is the wavelength and ν the frequency of the wave. Let $\rho(\mathbf{r}, \tau)$ be the "information" of a classical system in $(\mathbf{r}, \tau)$-space, i.e., the projection of a physically real process on $(\mathbf{r}, \tau)$-space is given by $\rho(\mathbf{r}, \tau)$. Furthermore, let $F(\mathbf{k}, \omega)$ be the projection of the *same* physically real process on $(\mathbf{k}, \omega)$-space. Then, $\rho(\mathbf{r}, \tau)$ can be obtained from $F(\mathbf{k}, \omega)$ by the Fourier transform

$$\rho(\mathbf{r}, \tau) = \int_{-\infty}^{\infty} F(\mathbf{k}, \omega) \exp[i(\mathbf{k} \cdot \mathbf{r} - \omega \tau)] \frac{d\mathbf{k}}{(2\pi)^3} d\omega. \qquad (2.2)$$

The inverse transformation enables the determination of $F(\mathbf{k}, \omega)$ from $\rho(\mathbf{r}, \tau)$. For example, in the case of classical many-particle systems,[33] $\rho(\mathbf{r}, \tau)$ is a correlation function, and the function $F(\mathbf{k}, \omega)$ can directly be measured by scattering experiments. The information concerning the many-particle system is completely transformed from the $(\mathbf{k}, \omega)$-space into the $(\mathbf{r}, \tau)$-space, and vice versa. Either the $(\mathbf{r}, \tau)$-space or the $(\mathbf{k}, \omega)$-space can be used to describe the system; both representations are equivalent.

In conclusion, besides $\mathbf{r}, \tau$ other variables (for example, $\mathbf{k}$ and ω) can be used in the description of reality; there is a coexistence of several spaces. The measurements are performed in $(\mathbf{k}, \omega)$-space[34] and the intuition of the observer is adapted to the $(\mathbf{r}, \tau)$-space.

2.3.4.2. The influence of Planck's constant

In the last section we discussed that, besides x, y, z, τ other variables can be used in the description of reality. Because of the coexistence of several spaces, the following important question appears: To what space does a given variable belong? What role, for example, is played by the momentum $\mathbf{p}$ (having the components p_x, p_y, p_z) and the energy E?

Let us introduce Planck's constant $\hbar$ by the well-known relations

$$\mathbf{p} = \hbar \mathbf{k}, \quad E = \hbar \omega. \qquad (2.3)$$

---

[33] For example, refer to van Hove, L., (1954) Phys. Rev. **95**, 249, and to Schommers, W., (1977) Phys. Rev. A **16**, 327.
[34] For example, refer to van Hove, L., (1954) Phys. Rev. **95**, 249, and to Schommers, W., (1977) Phys. Rev. A **16**, 327.

Because **p** is proportional to **k** and E to ω, we have to consider the momentum **p** and the energy E as the elements of a new space (In the following, it is denoted by $(\mathbf{p}, E)$-space.) with the characteristics of the $(\mathbf{k}, \omega)$-space, discussed above. Consequently, the $(\mathbf{r}, \tau)$-space and the $(\mathbf{p}, E)$-space are equivalent, and are connected to each other by a Fourier transform. However, we have to be careful. In connection with $\mathbf{p} = \hbar\mathbf{k}, E = \hbar\omega$[35] the time τ can be no longer identified with the external time τ of classical mechanics (and usual quantum theory) which is measured with our clocks. It will turn out that τ becomes a real quantum variable in connection with $\mathbf{p} = \hbar\mathbf{k}, E = \hbar\omega$.[35] Therefore, we would like to use the letter t instead of τ:

$$\tau \to t.$$

Then, instead of "$(\mathbf{r}, \tau)$-space" we will use the marking $(\mathbf{r}, t)$-space:

$$(\mathbf{r}, \tau)\text{-space} \to (\mathbf{r}, t)\text{-space}.$$

Nevertheless, time τ will also be used in the novel quantum-theoretical description given here as reference time (external time measured by our clocks). In other words, when we talk in the following about time τ, it is always the time which is used in classical physics as well as in usual quantum theory. The quantum-theoretical time t is completely new and is not known in conventional quantum theory.

Then, in equation (2.2) we have to replace the variables **k** and ω by **p** and E using equation (2.3) and the variable τ has to be replaced by t:

$$\varphi(\mathbf{r}, t) = \frac{1}{\hbar^4} \int_{-\infty}^{\infty} \varphi(\mathbf{p}, E) \exp\left\{i\left[\frac{\mathbf{p}}{\hbar} \cdot \mathbf{r} - \frac{E}{\hbar}t\right]\right\} \frac{d\mathbf{p}}{(2\pi)^3} dE,$$

$$\varphi(\mathbf{p}, E) = \int_{-\infty}^{\infty} \varphi(\mathbf{r}, t) \exp\left\{-i\left[\frac{\mathbf{p}}{\hbar} \cdot \mathbf{r} - \frac{E}{\hbar}t\right]\right\} \frac{dt}{2\pi} d\mathbf{r}.$$

Hence, $(\mathbf{r}, t)$-space and $(\mathbf{p}, E)$-space are equivalent, and connected to each other by a Fourier transform. In other words, due to $\hbar$ the variables

---

[35] See equation (2.3).

$\mathbf{r}$, t, $\mathbf{p}$ and E are arranged in a quite different ways than in classical mechanics, where $\hbar$ is not defined.

$\varphi(\mathbf{r}, t)$ is again a certain "information" in $(\mathbf{r}, t)$-space, and $\varphi(\mathbf{p}, E)$ is the equivalent information in $(\mathbf{p}, E)$-space. As argued,[36] the determination of $\varphi(\mathbf{r}, t)$ and $\varphi(\mathbf{p}, E)$, respectively, cannot be done on the basis of classical mechanics as in the case of $\rho(\mathbf{r}, \tau)$.

2.3.4.3. Reality and its picture

Instead of the functions $\varphi(\mathbf{r}, t)$ and $\varphi(\mathbf{p}, E)$, let use the symmetrical quantities $\Psi(\mathbf{r}, t)$ and $\Psi(\mathbf{p}, E)$ which we would like to define by

$$\Psi(\mathbf{r}, t) = \frac{1}{(2\pi\hbar)^2} \int_{-\infty}^{\infty} \Psi(\mathbf{p}, E) \exp\left\{i\left[\frac{\mathbf{p}}{\hbar} \cdot \mathbf{r} - \frac{E}{\hbar} t\right]\right\} d\mathbf{p}\, dE. \quad (2.4)$$

The inverse transformation is given by

$$\Psi(\mathbf{p}, E) = \frac{1}{(2\pi\hbar)^2} \int_{-\infty}^{\infty} \Psi(\mathbf{r}, t) \exp\left\{-i\left[\frac{\mathbf{p}}{\hbar} \cdot \mathbf{r} - \frac{E}{\hbar} t\right]\right\} dt\, d\mathbf{r}. \quad (2.5)$$

$(\mathbf{r}, t)$-space and $(\mathbf{p}, E)$-space are equivalent as far as their information is concerned. $\Psi(\mathbf{r}, t)$ and $\Psi(\mathbf{p}, E)$ will be characterized below, and we will formulate equations for their mathematical determination.

Again, the information content of $(\mathbf{r}, t)$-space and $(\mathbf{p}, E)$-space are exactly the same. But why in nature is the *same* information repeated? According to the *projection theory* and *Mach's principle*,[37] physically real processes are projected on space-time, we have the following situation.[38] The $(\mathbf{p}, E)$-space reflects the physical reality, and this reality (its information content) is completely projected on $(\mathbf{r}, t)$-space. This is consistent with the fact that we do not measure the elements $\mathbf{r}$ and t, but we measure $\mathbf{p}$ and E: A measurement provides an interaction process, and interaction means exchange of energy and momentum. On the basis of this measured information, *pictures* are formed in

---

[36] For more detail, refer to Schommers, W., (Ed.), (1989) Quantum Theory and Pictures of Reality: Springer-Verlag, and to Schommers, W., (1995) Symbols, Pictures and Quantum Reality: World Scientific.
[37] See Section 2.3.2.
[38] See also Figure 2.8.

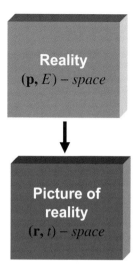

Fig. 2.8. We always measure **p** and E, but never points with **r** and t of space-time, i.e., the elements $\mathbf{r} = (x, y, z)$ and t are not accessible to empirical tests. We can only say something about distances in connection with masses and time intervals in connection with real physical processes (Section 2.2). The variables **p** and E have to be assigned to reality, whereas the variables x, y, z and t belong to the picture. In other words, reality is not embedded in space-time, but is projected onto it. In particular, we recognize that space and time are exclusively elements of the picture, that is, they do not occur in reality with the variables **p** and E. On the other hand, reality with the variables **p** and E cannot be identified with basic reality, since we cannot make statements about basic reality:[39] a picture-independent point of view is not conceivable. Therefore, basic reality principally remains hidden; this is true even in connection with specific variables for the mathematical description of basic reality. Thus, we have to consider reality with the variables **p** and E as "fictitious" reality that is constructed from the observers point of view which is based on space and time.

$(\mathbf{r}, t)$-space. That is, we obtain the following general scheme:

Reality
↓
Pictures of Reality

---

[39] Refer, also, to Schommers, W., (Ed.) (1989) Quantum Theory and Pictures of Reality: Springer-Verlag, and to Schommers, W., (1995) Symbols, Pictures and Quantum Reality: World Scientific.

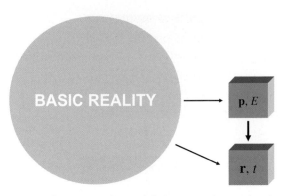

Fig. 2.9. Basic reality, fictitious reality and their connection to the pictures of reality.

This projection scheme is identical with that in Figure 2.2. However, reality outside is not identical with the reality described by the variables **p** and E. This point has to be explained in more detail.

2.3.4.4. Remark

We have stated in Section 2.2 that we cannot make any statement about basic (absolute) reality. This is true even in the case of the variables of it. On the other hand, the variables **p** and E represent reality. That is not a contradiction. The variables **p** and E have to be considered as abstract ideas, a product of the human mind, so to speak. Thus, reality that is embedded in (**p**, E)-space also has to be considered as an abstract idea, i.e., it cannot be identified with basic reality. (**p**, E)-space represents a "fictitious reality". The energy is introduced on the basis of (**r**, t)-space to which our intuition is adapted. In fact, in the unit of energy (g×length²/time²) the units of space and time explicitly occur. In other words, **p** and E do not occur in nature. This point has often been discussed in literature.[40] A good statement in connection with energy E is given by Christian von Baeyer:

> *The gradual crystallization of the concept of information during the last hundred years contrasts sharply with the birth of the equally*

---

[40] Refer, for example, to Schommers, W., (1995) Symbols, Pictures and Quantum Reality: World Scientific.

abstract quantity called energy in the middle of the nineteenth century. Then, in the brief span of twenty years, energy was invented, defined and established as a cornerstone, first in physics, then of all science. We don't know what energy is, any more than we know what information is, but as a now robust scientific concept we can describe it in precise mathematical terms, and as a commodity we can measure, market, regulate and tax it.[41]

Concerning energy, Edgar Lüscher remarked the following:

Energy is not a quantity that actually occurs in nature, but it is an abstract idea, a product of the human mind who tries to understand nature within his capabilities.[42]

Reality is characterized by the variables **p** and E and, on the other hand, the pictures of reality are expressed in terms of the variables **r** and t. Both spaces are products of mind.

2.3.5. *Basic Properties*

2.3.5.1. *Operators*

The "picture of reality" must be connected to "reality", otherwise we would not be able to fulfil the very basic fact that we experience the world in $(\mathbf{r}, t)$-space, in particular, we exclusively perform our measurements in $(\mathbf{r}, t)$-space. In other words, due to the connection between the "picture of reality" and "reality" we are able to observe the properties of the world in $(\mathbf{r}, t)$-space. This mutual dependence of the variables $\mathbf{r}, t, \mathbf{p}$ and E can be expressed by the relation $h(\mathbf{r}, t, \mathbf{p}, E) = 0$. However, due to the equivalence of $(\mathbf{r}, t)$-space and $(\mathbf{p}, E)$-space, it only makes sense to express the properties of the world in one of the two spaces, either we work in $(\mathbf{r}, t)$-space or in $(\mathbf{p}, E)$-space. Then, due to $h(\mathbf{r}, t, \mathbf{p}, E) = 0$, we have to know how the variables $\mathbf{p}$ and E are expressed in $(\mathbf{r}, t)$-space and, on the other hand, if we would like to

---

[41] Refer to von Baeyer, H.C., (2004) Information: The Language of Science: Harvard University Press.

[42] Refer to Schommers, W., (1995) Symbols, Pictures and Quantum Reality: World Scientific, and the references therein.

work in $(\mathbf{p}, E)$-space we have to know how the variables $\mathbf{r}$ and t are expressed in $(\mathbf{p}, E)$-space.

Due to the structure of the Fourier transform, it is not possible to give definite values for the coordinates x, y, z and the time t if $\mathbf{p}$ and E take definite values. On the other hand, it is not possible to give definite values for $\mathbf{p}$ and E if $\mathbf{r} = (x, y, z)$ and t take definite values. For example, for the determination of $\Psi(\mathbf{r}, t)$ for definite values $\mathbf{r}$ and t, say $\mathbf{r}_1$ and t_1, we need all possible values for $\mathbf{p}$ and E, in principle from $-\infty$ to ∞. This is a property of the Fourier transformation (2.4).

Thus, in the analysis of quantum phenomena given here the variables $\mathbf{p}$ and E, expressed in $(\mathbf{r}, t)$-space, cannot be simple numbers and, on the other hand, the variables $\mathbf{r}$ and t, expressed in $(\mathbf{p}, E)$-space, can also not be simple numbers.

But how can we express $\mathbf{p}$ and E in $(\mathbf{r}, t)$-space and, on the other hand, $\mathbf{r}$ and t in $(\mathbf{p}, E)$-space? In order to answer this question, let us consider the following identity, which can be obtained from equation (2.4):

$$-i\hbar \frac{\partial}{\partial \mathbf{r}} \Psi(\mathbf{r}, t) = \frac{1}{(2\pi\hbar)^2} \int_{-\infty}^{\infty} \mathbf{p}\Psi(\mathbf{p}, E) \exp\left\{i\left[\frac{\mathbf{p}}{\hbar} \cdot \mathbf{r} - \frac{E}{\hbar}t\right]\right\} d\mathbf{p}\, dE, \quad (2.6)$$

where

$$\frac{\partial}{\partial \mathbf{r}} = \mathbf{i}\frac{\partial}{\partial x} + \mathbf{j}\frac{\partial}{\partial y} + \mathbf{k}\frac{\partial}{\partial z}. \quad (2.7)$$

Interpretation of equation (2.6): Any information given in $(\mathbf{r}, t)$-space can be completely transformed into $(\mathbf{p}, E)$-space, and vice versa. Both information must be physically equivalent. We have two representations of the same thing. $\Psi(\mathbf{p}, E)$ in equation (2.6) is equivalent to $\Psi(\mathbf{r}, t)$, and vice versa.[43] Also, $-i\hbar\partial/\partial\mathbf{r}\Psi(\mathbf{r}, t)$ and $\mathbf{p}\Psi(\mathbf{p}, E)$ in equation (2.6) must be equivalent. Thus, the operator

$$\hat{p} = -i\hbar\frac{\partial}{\partial \mathbf{r}}, \quad (2.8)$$

---

[43] See equation (2.5).

with the components

$$\hat{p}_x = -i\hbar\frac{\partial}{\partial x}, \hat{p}_y = -i\hbar\frac{\partial}{\partial y}, \hat{p}_z = -i\hbar\frac{\partial}{\partial z}, \quad (2.9)$$

must be equivalent to the momentum $\mathbf{p}$, i.e., the momentum takes the form of an operator in $(\mathbf{r}, t)$-space.

In the same way, we can find on the basis of equations (2.4) and (2.5) operators for E, $\mathbf{r}$ and t. For example, with equation (2.5), we have

$$-i\hbar\frac{\partial}{\partial E}\Psi(\mathbf{p}, E) = \frac{1}{(2\pi\hbar)^2}\int_{-\infty}^{\infty} t\Psi(\mathbf{r}, t)\exp\left\{-i\left[\frac{\mathbf{p}}{\hbar}\cdot\mathbf{r} - \frac{E}{\hbar}t\right]\right\} d\mathbf{r}\, dt. \quad (2.10)$$

Thus, the operator

$$\hat{t} = -i\hbar\frac{\partial}{\partial E} \quad (2.11)$$

must be equivalent to the time t, i.e., the time t takes the form of an operator in $(\mathbf{p}, E)$-space.

Furthermore, it is easy to show that the operator of $(\mathbf{p}, E)$-space

$$\hat{\mathbf{r}} = i\hbar\frac{\partial}{\partial \mathbf{p}}, \quad \frac{\partial}{\partial \mathbf{p}} = \mathbf{i}\frac{\partial}{\partial p_x} + \mathbf{j}\frac{\partial}{\partial p_y} + \mathbf{k}\frac{\partial}{\partial p_z}, \quad (2.12)$$

with the components

$$\hat{x} = i\hbar\frac{\partial}{\partial p_x}, \quad \hat{y} = i\hbar\frac{\partial}{\partial p_y}, \quad \hat{z} = i\hbar\frac{\partial}{\partial p_z} \quad (2.13)$$

must be equivalent to $\mathbf{r} = (x, y, z)$, and the operator of $(\mathbf{r}, t)$-space

$$\hat{E} = i\hbar\frac{\partial}{\partial t} \quad (2.14)$$

must be equivalent to the energy E.

2.3.5.2. Conclusion

Within projection theory, we have the following situation. Both spaces, $(\mathbf{r}, t)$-space and $(\mathbf{p}, E)$-space, are equivalent concerning their information about the physical system under investigation. In particular, we have

($\mathbf{r}, t$)-space:
coordinates and time are *numbers* (x, y, z, t) momentum and energy are *operators* ($-i\hbar\partial/\partial\mathbf{r}, i\hbar\partial/\partial E$).

($\mathbf{p}, E$)-space:
coordinates and time are *operators* ($i\hbar\partial/\partial\mathbf{p}, -i\hbar\partial/\partial E$), momentum and energy are *numbers* (p_x, p_y, p_z, E).

There are two interesting points:

1. We have shown that either the coordinates or the momenta are numbers. If the coordinates are numbers, the momenta must be operators for ($\mathbf{r}, t$)-space. If the momenta are numbers, the coordinates must be operators for ($\mathbf{p}, E$)-space. These rules can only be obtained in usual quantum theory in connection with Schrödinger's equation, and this equation is an assumption, i.e., it has not been derived.[44] Therefore, also the rules concerning numbers and operators have to be considered as assumed within usual quantum theory. We already pointed out in Chapter 1 that the source of these rules is unknown within usual quantum theory and therefore become postulates here. In contrast to projection theory (physical reality is projected onto space-time), within conventional quantum theory (physical reality is embedded in space-time) we cannot recognize what physical reality hides behind these laws.
2. The time-coordinate is not always a simple number within projection theory, but can also be an operator ($-i\hbar\partial/\partial E$). This is of fundamental importance, and we will discuss the consequences below. We have pointed out in Chapter 1 that within usual quantum theory, time is always a simple parameter.

These rules determine the kind of information we can have about a quantum-mechanical system. We will apply these rules below when we formulate specific equations for the determination of certain properties of the system under investigation.

---

[44] See Chapter 1.

2.3.6. *Basic Transformation Effects*

Due to the coexistence of spaces and the connection of their variables $\mathbf{r}$, t and $\mathbf{p}$, E by a Fourier transform, remarkable properties come into play, and this is not a matter of interpretation. In other words, these specific properties can principally not be eliminated by the use of another interpretation but have to be considered as basic facts. In this section, let us briefly discuss non-local effects, the role of time and we want also to investigate whether point-like particles are definable within the frame given here.

2.3.6.1. *Particles*

Within usual quantum theory a material particle, having the energy E_0 and the momentum $\mathbf{p}_0$, is *embedded* in $(\mathbf{r}, t)$-space. As we have outlined above, the situation is different in projection theory. The material particle (if it exists) is embedded in $(\mathbf{p}, E)$-space and not in $(\mathbf{r}, t)$-space, and only its picture is represented in $(\mathbf{r}, t)$-space. The picture contains the maximum information about the particle because both spaces are equivalent concerning their information content.

Let us consider a simple system (particle) in $(\mathbf{p}, E)$-space and let the $(\mathbf{p}, E)$-distribution, say $A(\mathbf{p}, E)_\tau$, be characterized at time τ by delta-functions:

$$A(\mathbf{p}, E)_\tau = \delta(E - E_0)\delta(p - p_0), \tag{2.15}$$

where E_0 and p_0 are constant parameters for the energy and momentum at time τ (τ is the time of our clocks which we use in everyday life). In other words, the system under investigation is defined in reality by the distribution $A(\mathbf{p}, E)_\tau$. The picture of reality, represented in $(\mathbf{r}, t)$-space, contains, at time τ, exactly the same information, say $A(\mathbf{r}, t)_\tau$, and $A(\mathbf{r}, t)_\tau$ is given by the projection of $A(\mathbf{p}, E)_\tau$ onto $(\mathbf{r}, t)$-space. This projection can be performed by using the Fourier transform (2.4) and we obtain immediately

$$A(\mathbf{r}, t)_\tau = B \exp\left\{\frac{i}{\hbar}(\mathbf{p}_0 \cdot \mathbf{r} - E_0 t)\right\}, \quad B = \frac{1}{(2\pi\hbar)^2}. \tag{2.16}$$

That is, the system, having at time τ the energy E_0 and the momentum $\mathbf{p}_0$, is completely *delocalized* in $(\mathbf{r}, t)$-space, that is, it takes at time τ all possible values for $\mathbf{r} = (x, y, z)$ and t:

$$-\infty \leq \mathbf{r} \leq \infty, \qquad (2.17)$$
$$-\infty \leq t \leq \infty.$$

In other words, within the theoretical picture given here, a system with a definite energy E_0 and a definite momentum $\mathbf{p}_0$ can never be *localized* in space, that is, its picture can never represent a point-like particle in $(\mathbf{r}, t)$-space, but just the opposite is correct, i.e., it is completely delocalized, and this is due to the mathematical fact that the Fourier transform of a delta-function does not lead to a delta-function.

As already pointed out several times, within usual quantum theory material particles are embedded in $(\mathbf{r}, t)$-space and, furthermore, they are assumed to be strict localized, that is, they are point-like in character.[45]

In conclusion, the system (particle) is localized in $(\mathbf{p}, E)$-space,[46] but this space has nothing to do with the space of our experience and this is exclusively $(\mathbf{r}, t)$-space. A *localized* object with definite energy E_0 and definite momentum $\mathbf{p}_0$ is a requisite of classical mechanics and this is based on our experiences in everyday life; cars, trees etc., are localized in space in a good approximation.

2.3.6.2. Role of time t

What is the meaning of time t within the frame of the theoretical picture given here? The t-time scale cannot be identified with the τ-time scale. While time t characterizes the system under investigation in $(\mathbf{r}, t)$-space, time τ is just the time defined by our clocks.

A simple analysis shows that in the case of $t = \tau$, with $-\infty \leq t, \tau \leq \infty$, the system must have for *all* possible times τ the definite energy E_0 and the definite momentum $\mathbf{p}_0$ and not only at a certain instant (say τ_1) as assumed above.[47] However, at another instant $\tau_2 \neq \tau_2$, the same

---

[45] See also Chapter 1.
[46] See equation (2.15).
[47] See Section 2.3.4. in connection with equation (2.4).

system could have other values for the energy and the momentum, different from E_0 and $\mathbf{p}_0$.

Furthermore, identical time scales (t-time scale = τ-time scale) would imply that at time τ also all the other values for τ must also be realized, otherwise equation (2.17) would not be fulfilled. This would however be in contradiction to our assumption made in connection with equation (2.15): E_0 and p_0 are constant parameters for the energy and momentum at time τ. Thus, the correct relationship between τ and t is expressed in the case of equation (2.15) by equation (2.16): At time τ, *all* times t must be realized leading to equation (2.17). A more detailed discussion concerning time t is given below.

In conclusion, time t must be defined by the system under investigation. In our example, the system is characterized by equation (2.15). In other words, time t is a *system-specific time*. Its characteristics are different for different systems. This is a new feature. But how are these characteristics fixed? What is the relation between the various time scales and the τ-time scale? Our simple example cannot give the answer, but we will answer these questions below within a more detailed analysis using equations that are similar to Schrödinger's equation, but are generalized with respect to the system-specific time t.

It should be emphasized once more[48] that in usual quantum theory there is only one time scale and this is exclusively given by the time τ of our clocks used in everyday life. However, τ merely plays the role of an external parameter and has nothing to do with the system under investigation. The situation within projection theory is entirely new, and we will see below that the existence of a system-specific time has serious consequences.

2.3.6.3. Non-local effects

Let us now consider the same system but now at time $\tau + \Delta\tau$, where $\Delta\tau$ is an arbitrarily chosen interval. In principle, $\Delta\tau$ can be infinitesimal but should be different from zero. We would like to assume that, due to an interaction process, the system has no longer the energy E_0 and

---
[48]See also Chapter 1.

the momentum $\mathbf{p}_0$ but E_0' and $\mathbf{p}_0'$ at $\tau + \Delta\tau$. Then, we get instead of $A(\mathbf{p}, E)_\tau$[49] the following expression for $A(\mathbf{p}, E)_{\tau+\Delta\tau}$:

$$A(\mathbf{p}, E)_{\tau+\Delta\tau} = \delta(E - E_0')\delta(\mathbf{p} - \mathbf{p}_0'). \tag{2.18}$$

Using this distribution and equation (2.4), we obtain

$$A(\mathbf{r}, t)_{\tau+\Delta\tau} = B\exp\left\{\frac{i}{\hbar}(\mathbf{p}_0' \cdot \mathbf{r} - E_0' t)\right\}. \tag{2.19}$$

By comparison of equation (2.16) with equation (2.18), we immediately recognize that the properties at all points $\mathbf{r}, t$ ($-\infty \leq \mathbf{r} \leq \infty, -\infty \leq t \leq \infty$) at $\tau + \Delta\tau$ are different from those at time τ. These global changes take place instantaneously since $\Delta\tau$ can in principle be infinitesimal. In other words, we are confronted with non-local effects.

These *non-local effects* are entirely due to the projection process from reality $(\mathbf{p}, E)$-space onto $(\mathbf{r}, t)$-space, like a flashlight on a screen. There is no preferred point in $(\mathbf{r}, t)$-space, say $\mathbf{r}_1$ and t_1, which would influence instantaneously all the other space-time points $\mathbf{r}, t$ ($-\infty \leq \mathbf{r} \leq \infty, -\infty \leq t \leq \infty$) by actions through space-time. Since there are only projections, no information or material objects move through space-time. Therefore, the instantaneous transition from one global structure[50] to the other[51] is not in contradiction to Theory of Relativity.

2.3.6.4. Conclusion

In this section we have pointed out that point-like particles with definite energy and definite momentum are not definable within the frame of the investigation given here. This "point-model", used in usual quantum theory, is obviously a holdover from classical mechanics and is mainly based on our observations in everyday life.

Furthermore, there are non-local effects, that is, we observe in $(\mathbf{r}, t)$-space global instantaneous changes in the structure, and it is a feature of the projection that all space-time points $\mathbf{r}, t$ ($-\infty \leq$

---

[49] See equation (2.15).
[50] This is described by equation (2.16).
[51] This is described by equation (2.18).

$\mathbf{r} \leq \infty, -\infty \leq t \leq \infty$) are instantaneously involved. This effect is comparable with a flashlight on a screen.

Note, that this non-locality is not restricted on space but time t is involved too. We have no longer one time τ measured with our clocks, but also a *system-specific* time t. Its characteristics are different for different systems. More details are given below.

There still remain of course a lot of questions. For example, we do our measurements in $(\mathbf{r}, t)$-space but how is the contact to reality $((\mathbf{p}, E)$-space) organized? There is no particle or any other material object embedded in $(\mathbf{r}, t)$-space but everything real is located in $(\mathbf{p}, E)$-space and we have to clarify how we can fix the properties of a system in $(\mathbf{r}, t)$-space, which are expressed within projection theory as $(\mathbf{p}, E)$-distribution. These points will also be discussed below.

Remark

It should already be mentioned here that the possible values for the variables $\mathbf{r}$ and t ($\mathbf{p}$ and E) are not simultaneously existent, but play the role of *random variables*, that is, at time τ there is a certain probability to observe one of all possible $(\mathbf{r}, t)$-points $((\mathbf{p}, E)$-points). Within the frame of the example discussed in this subsection, these points are defined at time τ by the range of $A(\mathbf{r}, t)_\tau$ ($A(\mathbf{p}, E)_\tau$).

2.3.7. Operator Equations

2.3.7.1. Determination of $\psi(\mathbf{r}, t)$ and $\psi(\mathbf{p}, E)$

Let us start with the well-known classical equation for a system (particle) in an external field $U(x, y, z)$:

$$E = \frac{\mathbf{p}^2}{2m_0} + U(x, y, z), \qquad (2.20)$$

where m_0 is the mass of the particle, $\mathbf{p}$ its momentum and E its energy. Here the particle is still point-like and it is embedded in space. The effect of $\hbar$[52] is that we have no longer one space, but two co-existing spaces $((\mathbf{p}, E)$-space and $(\mathbf{r}, t)$-space) representing reality

---
[52] See Section 2.3.4.2.

and, on the other hand, its picture. Since both spaces are equivalent we may describe the physical system in $(\mathbf{p}, E)$-space as well as in $(\mathbf{r}, t)$-space. Using equation (2.20) and the rules derived in Section 2.3.5.2, we can formulate the corresponding quantum-mechanical equations for $\Psi(\mathbf{p}, E)$ and $\Psi(\mathbf{r}, t)$. First, let us give some general remarks.

2.3.7.2. Remarks

In the derivation of these equations, we have to notice the following. The function f of a variable, say x, becomes in those cases, where the variable becomes an operator $i\hbar \partial/\partial p_x$,[53] a function of this operator, i.e., we have the following equivalence:

$$f(x)\Psi(\mathbf{r}, t) \leftrightarrow f\left(i\hbar \frac{\partial}{\partial p_x}\right) \Psi(\mathbf{p}, E), \qquad (2.21)$$

where rule (2.13) has been used: $x \leftrightarrow i\hbar \partial/\partial p_x$.

Equation (2.21) can easily be verified. Expanding the function $f(x)$ in a Taylor series, we have

$$f(x) = \sum_{n=0}^{\infty} a_n x^n. \qquad (2.22)$$

With equation (2.5), we immediately obtain

$$i^n \hbar^n a_n \frac{\partial^n}{\partial p_x^n} \Psi(\mathbf{p}, E) = \frac{1}{(2\pi\hbar)^2} \int_{-\infty}^{\infty} a_n x^n \Psi(\mathbf{r}, t)$$
$$\times \exp\left\{-\frac{i}{\hbar}(\mathbf{p} \cdot \mathbf{r} - Et)\right\} d\mathbf{r}\, dt, \qquad (2.23)$$

and we recognize that the nth term of equation (2.22) takes in $(\mathbf{p}, E)$-space the form

$$i^n \hbar^n a_n \frac{\partial^n}{\partial p_x^n}, \qquad (2.24)$$

---

[53] See Section 2.3.5.

and it is equivalent to

$$a_n x^n. \qquad (2.25)$$

This argumentation is valid for all terms in equation (2.22), and we may write

$$f\left(i\hbar\frac{\partial}{\partial p_x}\right)\Psi(\mathbf{p}, E) = \frac{1}{(2\pi\hbar)^2}\int_{-\infty}^{\infty} f(x)\Psi(\mathbf{r}, t)$$
$$\times \exp\left\{\frac{i}{\hbar}(\mathbf{p}\cdot\mathbf{r} - Et)\right\} d\mathbf{r}\, dt. \qquad (2.26)$$

That is, we get the equivalence given by equation (2.21).

2.3.7.3. Space-specific formulation

In order to formulate the equations $\Psi(\mathbf{p}, E)$ and $\Psi(\mathbf{r}, t)$, we have to consider the situations in the two spaces.

$(\mathbf{p}, E)$-space:

Only the function $U(x, y, z)$ in equation (2.20) has to be transformed since the space variable $\mathbf{r} = (x, y, z)$ does not belong to $(\mathbf{p}, E)$-space. In analogy to equation (2.26), we have the equivalence

$$U(x, y, z) \leftrightarrow U\left(i\hbar\frac{\partial}{\partial p_x}, i\hbar\frac{\partial}{\partial p_y}, i\hbar\frac{\partial}{\partial p_z}\right). \qquad (2.27)$$

That is, $U(x, y, z)$ is given in $(\mathbf{p}, E)$-space by $U\left(i\hbar\frac{\partial}{\partial p_x}, i\hbar\frac{\partial}{\partial p_y}, i\hbar\frac{\partial}{\partial p_z}\right)$.

$(\mathbf{r}, t)$-space:

The terms E and $\mathbf{p}^2/2m_0$ in equation (2.20) belong to $(\mathbf{p}, E)$-space, and we have to use the corresponding relations in $(\mathbf{r}, t)$-space:[54]

$$E\Psi(\mathbf{p}, E) \leftrightarrow i\hbar\frac{\partial}{\partial t}\Psi(\mathbf{r}, t), \qquad (2.28)$$

$$\frac{\mathbf{p}^2}{2m_0}\Psi(\mathbf{p}, E) \leftrightarrow -\frac{\hbar^2}{2m_0}\Delta\Psi(\mathbf{r}, t). \qquad (2.29)$$

---

[54] See also equations (2.8) and (2.14).

Then, we can formulate the following quantum-mechanical equations for the determination of $\Psi(\mathbf{p}, E)$ and $\Psi(\mathbf{r}, t)$. For the classical system, characterized by equation (2.20), we can formulate the following quantum-mechanical equivalency by means of equation (2.4),

$$\Psi(\mathbf{r}, t) = \frac{1}{(2\pi\hbar)^2} \int_{-\infty}^{\infty} \Psi(\mathbf{p}, E) \exp\left\{i\left[\frac{\mathbf{p}}{\hbar} \cdot \mathbf{r} - \frac{E}{\hbar}t\right]\right\} d\mathbf{p}\, dE,$$

as follows:

$$\left\{i\hbar \frac{\partial}{\partial t} + \frac{\hbar^2}{2m_0}\Delta - U(x, y, z)\right\} \Psi(\mathbf{r}, t)$$

$$= \frac{1}{(2\pi\hbar)^2} \int_{-\infty}^{\infty} \left\{E - \frac{\mathbf{p}^2}{2m_0} - U\left(i\hbar\frac{\partial}{\partial p_x}, i\hbar\frac{\partial}{\partial p_y}, i\hbar\frac{\partial}{\partial p_z}\right)\right\}$$

$$\times \Psi(\mathbf{p}, E) \exp\left[\frac{i}{\hbar}(\mathbf{p}\cdot\mathbf{r} - Et)\right] d\mathbf{p}\, dE. \tag{2.30}$$

Equation (2.30) leads to

$$\left\{i\hbar\frac{\partial}{\partial t} + \frac{\hbar^2}{2m_0}\Delta - U(x, y, z)\right\} \Psi(\mathbf{r}, t) = f(\mathbf{r}, t), \tag{2.31}$$

and

$$\left\{E - \frac{\mathbf{p}^2}{2m_0} - U\left(i\hbar\frac{\partial}{\partial p_x}, i\hbar\frac{\partial}{\partial p_y}, i\hbar\frac{\partial}{\partial p_z}\right)\right\} \Psi(\mathbf{p}, E) = f(\mathbf{p}, E), \tag{2.32}$$

with

$$f(\mathbf{r}, t) = \frac{1}{(2\pi\hbar)^2}$$

$$\times \int_{-\infty}^{\infty} \left\{E - \frac{\mathbf{p}^2}{2m_0} - U\left(i\hbar\frac{\partial}{\partial p_x}, i\hbar\frac{\partial}{\partial p_y}, i\hbar\frac{\partial}{\partial p_z}\right)\right\}$$

$$\times \Psi(\mathbf{p}, E) \exp\left[\frac{i}{\hbar}(\mathbf{p}\cdot\mathbf{r} - Et)\right] d\mathbf{p}\, dE, \tag{2.33}$$

and

$$f(\mathbf{r}, t) = \frac{1}{(2\pi\hbar)^2} \int_{-\infty}^{\infty} f(\mathbf{p}, E) \exp\left[\frac{i}{\hbar}(\mathbf{p}\cdot\mathbf{r} - Et)\right] d\mathbf{p}\, dE. \tag{2.34}$$

Projection Theory

Not only the potential $U(x, y, z)$ is unknown but also the function $f(\mathbf{r}, t)$ and, therefore, we would like to combine the functions $U(x, y, z)$ and $f(\mathbf{r}, t)$ to one function which we want to call $V(x, y, z, t)$. This re-formulation leads to the final quantum-mechanical equation of the classical case (2.20),

$$i\hbar \frac{\partial}{\partial t} \Psi(\mathbf{r}, t) = -\frac{\hbar^2}{2m_0} \Delta \Psi(\mathbf{r}, t) + V(x, y, z, t) \Psi(\mathbf{r}, t), \quad (2.35)$$

formulated in $(\mathbf{r}, t)$-space, where

$$V(x, y, z, t) = V(\mathbf{r}, t) = U(\mathbf{r}) + g(\mathbf{r}, t), \quad (2.36)$$

with

$$g(\mathbf{r}, t) = \frac{f(\mathbf{r}, t)}{\Psi(\mathbf{r}, t)}. \quad (2.37)$$

$V(x, y, z, t)$ should be considered as a generalized interaction potential. In the quantum-mechanical case, we have no longer the classical static potential $U(x, y, z)$ but the time-dependent potential $V(x, y, z, t)$.

In other words, we obtain from the classical stationary case (2.20) a non-stationary quantum-mechanical equation.[55] The stationary case $U(x, y, z, t) = U(x, y, z)$,[56]

$$i\hbar \frac{\partial}{\partial t} \Psi(\mathbf{r}, t) = -\frac{\hbar^2}{2m_0} \Delta \Psi(\mathbf{r}, t) + U(x, y, z) \Psi(\mathbf{r}, t)$$

of usual quantum theory (with $t = \tau$) is eliminated here.

Using the procedure above, $V(x, y, z, t)$ takes in $(\mathbf{p}, E)$-space the form

$$V\left(i\hbar \frac{\partial}{\partial p_x}, i\hbar \frac{\partial}{\partial p_y}, i\hbar \frac{\partial}{\partial p_z}, -i\hbar \frac{\partial}{\partial E}\right), \quad (2.38)$$

---

[55] See equation (2.35).
[56] See equation (1.3) in Chapter 1.

resulting from the equivalence (using equation (2.5))

$$V\left(i\hbar\frac{\partial}{\partial p_x}, i\hbar\frac{\partial}{\partial p_y}, i\hbar\frac{\partial}{\partial p_z}, -i\hbar\frac{\partial}{\partial E}\right)\Psi(\mathbf{p}, E)$$

$$= \frac{1}{(2\pi\hbar)^2}\int_{-\infty}^{\infty} V(x, y, z, t)\Psi(\mathbf{r}, t)$$

$$\times \exp\left\{-i\left[\frac{\mathbf{p}}{\hbar}\cdot\mathbf{r} - \frac{E}{\hbar}t\right]\right\} dt\, d\mathbf{r}. \qquad (2.39)$$

Note that the time t takes in $(\mathbf{p}, E)$-space the form $-i\hbar\partial/\partial E$.[57] Using the equivalencies (2.29) and (2.30), the quantum-mechanical case of equation (2.20) is given in $(\mathbf{p}, E)$-space by

$$E\Psi(\mathbf{p}, E) = \frac{\mathbf{p}^2}{2m_0}\Psi(\mathbf{p}, E)$$

$$+ V\left(i\hbar\frac{\partial}{\partial p_x}, i\hbar\frac{\partial}{\partial p_y}, i\hbar\frac{\partial}{\partial p_z}, -i\hbar\frac{\partial}{\partial E}\right)\Psi(\mathbf{p}, E), \qquad (2.40)$$

with[58]

$$V\left(i\hbar\frac{\partial}{\partial p_x}, i\hbar\frac{\partial}{\partial p_y}, i\hbar\frac{\partial}{\partial p_z}, -i\hbar\frac{\partial}{\partial E}\right)\Psi(\mathbf{p}, E)$$

$$= U\left(i\hbar\frac{\partial}{\partial p_x}, i\hbar\frac{\partial}{\partial p_y}, i\hbar\frac{\partial}{\partial p_z}\right)\Psi(\mathbf{p}, E) + f(\mathbf{p}, E). \qquad (2.41)$$

Equation (2.40) is completely equivalent to equation (2.35): $\Psi(\mathbf{r}, t)$ is determined by equation (2.35) and $\Psi(\mathbf{p}, E)$ by equation (2.40), and both solutions ($\Psi(\mathbf{r}, t)$ and $\Psi(\mathbf{p}, E)$) are connected by equations (2.4) and (2.5), respectively.

2.3.7.4. Discussion concerning equations (2.35) and (2.50)

Comparison with Schrödinger's equation

---

[57] See equation (2.11).
[58] See equation (2.32).

We pointed out that the stationary case is eliminated within projection theory. Although we obtain Schrödinger's equation for the non-stationary case, there is, however, a big difference between the two formulations (1.3) and (2.35). Equation (2.40) does not exist in usual quantum theory and this because the operator $-i\hbar\partial/\partial E$ for the time-coordinate does not exist in usual quantum theory. Since equations (2.40) and (2.35) are completely equivalent, also equation (2.35) cannot exist in usual quantum theory. If we know, for example, $\Psi(\mathbf{p}, E)$ we can determine $\Psi(\mathbf{r}, t)$ without explicitly using equation (2.35). $\Psi(\mathbf{p}, E)$ and $\Psi(\mathbf{r}, t)$ are connected to each other by a Fourier transform.[59]

The meaning of equation (2.35) is quite different from that of usual quantum theory.[60] As we have already discussed in Section 2.3.6.2, the role of time within the investigation given here is quite different from that in usual quantum theory. We pointed out that t is a *system-specific* quantity. The time within usual quantum theory is exclusively given by the time τ and is simply defined by our clocks which we use in everyday life, that is, the time used in usual quantum theory must be identified with τ ($t = \tau$), and when we use $V(x, y, z, t) = U(x, y, z)$ we obtain in the case of equation (2.20) Schrödinger's equation of usual quantum theory (instead of Eq. (2.35)):

$$i\hbar\frac{\partial}{\partial \tau}\Psi(\mathbf{r}, \tau) = -\frac{\hbar^2}{2m_0}\Delta\Psi(\mathbf{r}, \tau) + U(x, y, z)\Psi(\mathbf{r}, \tau). \quad (2.42)$$

Some aspects of the system-specific time t have already been discussed in Section 2.3.6.2, but we will continue this discussion below in connection with equation (2.35).

($\mathbf{r}$, E)-space representation

It can be convenient to work in the intermediate ($\mathbf{r}$, E)-space. In this case, we need the Fourier transform that connects ($\mathbf{r}$, t)-space with

---
[59] See equations (2.4) and (2.5).
[60] See equation (1.3).

$(\mathbf{r}, E)$-space:

$$\Psi(\mathbf{r}, E) = \frac{1}{(2\pi\hbar)^{1/2}} \int_{-\infty}^{\infty} \Psi(\mathbf{r}, t) \exp\left\{i\frac{E}{\hbar}t\right\} dt. \tag{2.43}$$

In analogy to equation (2.30) we get

$$\left\{E + \frac{\hbar^2}{2m_0}\Delta - U(x, y, z)\right\} \Psi(\mathbf{r}, E)$$
$$= \frac{1}{(2\pi\hbar)^{1/2}} \int_{-\infty}^{\infty} \left\{i\hbar\frac{\partial \psi(\mathbf{r}, t)}{\partial t} + \frac{\hbar^2}{2m_0}\Delta - U(x, y, z)\right\}$$
$$\times \Psi(\mathbf{r}, t) \exp\left[\frac{i}{\hbar}Et\right] dt. \tag{2.44}$$

Equation (2.44) leads to equation (2.31) and to

$$\left\{E + \frac{\hbar^2}{2m_0}\Delta - U(x, y, z)\right\} \Psi(\mathbf{r}, E) = f(\mathbf{r}, E), \tag{2.45}$$

with

$$f(\mathbf{r}, E) = \frac{1}{(2\pi\hbar)^{1/2}} \int_{-\infty}^{\infty} \left\{i\hbar\frac{\partial \psi(\mathbf{r}, t)}{\partial t} + \frac{\hbar^2}{2m_0}\Delta - U(x, y, z)\right\}$$
$$\times \Psi(\mathbf{r}, t) \exp\left[\frac{i}{\hbar}Et\right] dt, \tag{2.46}$$

and

$$f(\mathbf{r}, E) = \frac{1}{(2\pi\hbar)^{1/2}} \int_{-\infty}^{\infty} f(\mathbf{r}, t) \exp\left\{i\frac{E}{\hbar}t\right\} dt. \tag{2.47}$$

Using the procedure above, $V(x, y, z, t)$[61] takes in $(\mathbf{r}, E)$-space the form

$$V\left(x, y, z, -i\hbar\frac{\partial}{\partial E}\right), \tag{2.48}$$

---

[61] See equation (2.38).

resulting from the equivalence (using equation (2.43)),

$$V\left(x, y, z, -i\hbar \frac{\partial}{\partial E}\right) \Psi(\mathbf{r}, E) = \frac{1}{(2\pi\hbar)^{1/2}} \int_{-\infty}^{\infty} V(x, y, z, t)$$
$$\times \Psi(\mathbf{r}, t) \exp\left\{i\frac{E}{\hbar}t\right\} dt. \quad (2.49)$$

Using the equivalencies (2.29) and (2.44), the quantum-mechanical case of equation (2.20) is given in $(\mathbf{r}, E)$-space by

$$E\Psi(\mathbf{r}, E) = -\frac{\hbar^2}{2m_0} \Delta\Psi(\mathbf{r}, E) + V\left(x, y, z, -i\hbar \frac{\partial}{\partial E}\right) \Psi(\mathbf{r}, E), \quad (2.50)$$

with[62]

$$V\left(x, y, z, -i\hbar \frac{\partial}{\partial E}\right) \Psi(\mathbf{r}, E) = U(x, y, z) \Psi(\mathbf{r}, E) + f(\mathbf{r}, E). \quad (2.51)$$

Also equation (2.50) is completely equivalent to equation (2.35): $\Psi(\mathbf{r}, t)$ is determined by equation (2.35) and $\Psi(\mathbf{r}, E)$ by equation (2.50), and both solutions $\Psi(\mathbf{r}, t)$ and $\Psi(\mathbf{r}, E)$ are connected by equation (2.43).

2.3.7.5. Other representations

The variables $\mathbf{p}$ and E belong to "reality" (fictitious reality) and, on the other hand, the variables $\mathbf{r}$ and t belong to the "picture of reality".[63] Since reality and its picture must necessarily be coupled, the variables $\mathbf{p}, E, \mathbf{r}$ and t must also necessarily coupled, and this is described by Fourier transforms[64]

$$\Psi(\mathbf{r}, t) \leftrightarrow \Psi(\mathbf{p}, E). \quad (2.52)$$

In the classical stationary case (2.20), we have

$$E = E(\mathbf{p}, \mathbf{r}) \quad (2.53)$$

---

[62] See equation (2.45).
[63] See also the discussion in Section 2.3.4.
[64] See equations (2.4) and (2.5).

leading to a quantum-mechanical non-stationary situation.[65] The classical equation (2.20) cannot be extended to

$$E = \frac{\mathbf{p}^2}{2m_0} + U(x, y, z, t), \tag{2.54}$$

and this is because the system-specific time t is not defined in classical mechanics. In classical mechanics, the non-stationary case is expressed by $E = \mathbf{p}^2/2m_0 + U(x, y, z, \tau)$. The situation in connection with reference time τ is discussed in Section 2.3.9. τ is an external parameter measured with our clocks and has nothing to do with the system under investigation. The system-specific, quantum-mechanical time t cannot lead to τ in the transition to the classical case.

We can also formulate the problem, outlined in the Sections 2.3.7.3–2.3.7.6, by the inverse functions of $E = E(\mathbf{p}, \mathbf{r})$:

$$\mathbf{r} = \mathbf{r}(\mathbf{p}, E), \tag{2.55}$$

$$\mathbf{p} = \mathbf{p}(\mathbf{r}, E). \tag{2.56}$$

As in the case of equation (2.53), in the quantum-theoretical description, the right-hand side of the expressions (2.55) and (2.56) must always be operators. Using the rules derived in Section 2.3.5.1 we get, for example, the following relations:

$(\mathbf{r}, E)$-space:

$$\mathbf{r} = \mathbf{r}(\mathbf{p}, E) \rightarrow \hat{R}\left(-i\hbar\frac{\partial}{\partial \mathbf{r}}, E\right). \tag{2.57}$$

$(\mathbf{p}, E)$-space:

$$\mathbf{p} = \mathbf{p}(\mathbf{r}, E) \rightarrow \hat{P}\left(i\hbar\frac{\partial}{\partial \mathbf{p}}, E\right). \tag{2.58}$$

In the case of equation (2.53) we got[66]

$$E = E(\mathbf{p}, \mathbf{r}) \rightarrow \hat{H}\left(-i\hbar\frac{\partial}{\partial \mathbf{r}}, \mathbf{r}, t\right), \tag{2.59}$$

---

[65] See also Section 2.3.7.3.
[66] See also Section 2.3.7.3.

when we work in $(\mathbf{r}, t)$-space. The Hamilton operator $\hat{H}$ is given by

$$\hat{H} = -\frac{\hbar^2}{2m_0}\Delta + V(x, y, z, t) \quad (2.60)$$

leading to equation (2.35).

We can obtain alternative equations when we use instead of the Hamiltonian $\hat{H}$ the position operator $\hat{R}$[67] or the momentum operator $\hat{P}$.[68] For the determination of the eigenvalues and wave functions for the system under investigation, all representations are equivalent concerning their physical information. For example, in the case of $\mathbf{p} = \mathbf{p}(\mathbf{r}, E)$,[69] we obtain in analogy of equations (2.31) and (2.32) the following relations:

$(\mathbf{r}, t)$-space:

$$-i\hbar\frac{\partial}{\partial \mathbf{r}}\Psi(\mathbf{r}, t) - \hat{P}\left(\mathbf{r}, i\hbar\frac{\partial}{\partial t}\right)\Psi(\mathbf{r}, t) = h(\mathbf{r}, t). \quad (2.61)$$

$(\mathbf{p}, E)$-space:

$$\mathbf{p}\Psi(\mathbf{p}, E) - \hat{P}\left(i\hbar\frac{\partial}{\partial \mathbf{p}}, E\right)\Psi(\mathbf{p}, E) = h(\mathbf{p}, E), \quad (2.62)$$

with[70]

$$h(\mathbf{r}, t) = \frac{1}{(2\pi\hbar)^2}\int_{-\infty}^{\infty} h(\mathbf{p}, E)\exp\left\{i\left[\frac{\mathbf{p}}{\hbar}\cdot\mathbf{r} - \frac{E}{\hbar}t\right]\right\}d\mathbf{p}\,dE. \quad (2.63)$$

The following is important: the wave functions $\Psi(\mathbf{r}, t) = \Psi_H(\mathbf{r}, t)$ and $\Psi(\mathbf{p}, E) = \Psi_H(\mathbf{p}, E)$, respectively, resulting from equations (2.31) and (2.32) must be the same as those resulting from

---

[67] See equation (2.57).
[68] See equation (2.58).
[69] See equation (2.56).
[70] See equation (2.4).

equations (2.61) and (2.62) ($\Psi(\mathbf{r}, t) = \Psi_P(\mathbf{r}, t), \Psi(\mathbf{p}, E) = \Psi_P(\mathbf{p}, E)$):

$$\Psi_H(\mathbf{r}, t) = \Psi_P(\mathbf{r}, t), \tag{2.64}$$

$$\Psi_H(\mathbf{p}, E) = \Psi_P(\mathbf{p}, E). \tag{2.65}$$

For example, if we have determined $\Psi(\mathbf{r}, t) = \Psi_H(\mathbf{r}, t)$ by equation (2.31), the function $h(\mathbf{r}, t)$ of equation (2.61) has to be fixed so that equation (2.64) is fulfilled.

2.3.7.6. *Superposition principle*

Within usual quantum theory both, the Copenhagen Interpretation of and the Many-world Theory[71] are of particular interest, and the *superposition principle* plays a key role. This principle is one of the basics of usual quantum theory. In Chapter 1, we have already discussed this principle, in particular in connection with the collapse of the wave function which cannot be described in usual quantum theory and this makes the superposition principle a doubtful concept. Furthermore, we have remarked in Chapter 1 that the use of only linear operators has a "very far-reaching restriction, just as if we were to decide to use only linear functions in a certain area of mathematics".[72]

The general case

Projection theory is more general than usual quantum theory because the operator $-i\hbar\partial/\partial E$ for the time-coordinate appears, which is not defined within usual quantum theory. Thus, we have to investigate whether the superposition principle is also valid within projection theory. Let us discuss this point by means of equation (2.50) which can be re-written as follows:

$$\hat{H}\left(\mathbf{r}, -i\hbar\frac{\partial}{\partial \mathbf{r}}, -i\hbar\frac{\partial}{\partial E}\right)\Psi(\mathbf{r}, E) = E\Psi(\mathbf{r}, E), \tag{2.66}$$

---

[71] See also Chapter 1.
[72] Interested readers might refer to Rubinowics, A., (1968) Quantum Mechanics: Elsevier.

with

$$\hat{H} = -\frac{\hbar^2}{2m_0}\Delta + V\left(x, y, z, -i\hbar\frac{\partial}{\partial E}\right). \qquad (2.67)$$

The Hamiltonian $\hat{H}$ in equation (2.66) is not linear. Why? Let us assume, for example, that $\Psi(\mathbf{r}, E)$ is a continuous function with respect to the variable E. In this case, a linearly composed function

$$\varphi(\mathbf{r}) = \int_{-\infty}^{\infty} a_E \Psi(\mathbf{r}, E) dE \qquad (2.68)$$

does not fulfil the relation

$$\hat{H} \int_{-\infty}^{\infty} a_E \Psi(\mathbf{r}, E) dE = \int_{-\infty}^{\infty} a_E \hat{H} \Psi(\mathbf{r}, E) dE. \qquad (2.69)$$

That is, the terms

$$\hat{H}\left(\mathbf{r}, -i\hbar\frac{\partial}{\partial \mathbf{r}}, -i\hbar\frac{\partial}{\partial E}\right) a_E \Psi(\mathbf{r}, E) \qquad (2.70)$$

are not equal to

$$a_E \hat{H}\left(\mathbf{r}, -i\hbar\frac{\partial}{\partial \mathbf{r}}, -i\hbar\frac{\partial}{\partial E}\right) \Psi(\mathbf{r}, E), \qquad (2.71)$$

and this means that one of the conditions of linearity is not satisfied.[73] The fact that the superposition principle is not applicable to the basic structures of projection theory is of course due to the appearance of the operator $-i\hbar\partial/\partial E$. In this case the coefficient a_E cannot be treated as a constant with respect to $\hat{H}$.[74]

---

[73] For example, refer to Landau, L.D., and Lifschitz, E.M., (1965) Quantum Mechanics: Pergamon.
[74] See equation (2.67).

If there is a superposition (2.68), there must also be a superposition with respect to the time variable. With[75]

$$\Psi(\mathbf{r}, E) = \frac{1}{(2\pi\hbar)^{1/2}} \int_{-\infty}^{\infty} \Psi(\mathbf{r}, t) \exp\left\{i\frac{E}{\hbar}t\right\} dt,$$

and equation (2.68), we immediately obtain

$$\varphi(\mathbf{r}) = \int_{-\infty}^{\infty} b_t \Psi(\mathbf{r}, t) dt, \qquad (2.72)$$

with

$$b_t = \frac{1}{(2\pi\hbar)^{1/2}} \int_{-\infty}^{\infty} a_E \exp\left\{i\frac{E}{\hbar}t\right\} dt. \qquad (2.73)$$

As in the case of equation (2.68) also the superposition (2.72) cannot be valid. The reason is simple. We have only to apply equation (2.72) to

$$i\hbar \frac{\partial}{\partial t} \Psi(\mathbf{r}, t) = -\frac{\hbar^2}{2m_0} \Delta \Psi(\mathbf{r}, t) + V(x, y, z, t) \Psi(\mathbf{r}, t),$$

which is the basic equation for the determination of $\Psi(\mathbf{r}, t)$ in $(\mathbf{r}, t)$-space[76] and we recognize that the superposition (2.72) does not lead to the condition

$$i\hbar \frac{\partial}{\partial t} \int_{-\infty}^{\infty} b_t \Psi(\mathbf{r}, t) dt = \int_{-\infty}^{\infty} b_t i\hbar \frac{\partial}{\partial t} \Psi(\mathbf{r}, t) dt. \qquad (2.74)$$

In other words, as in the case of superposition (2.68) also the superposition (2.72) in $(\mathbf{r}, t)$-space cannot be valid.

---

[75] See equation (2.43).
[76] See equation (2.35) and also Section 2.3.7.3.

Stationary systems

In the case of a discrete energy spectrum (stationary case), we have

$$\varphi(\mathbf{r}) = \sum_E a_E \Psi(\mathbf{r}, E) \qquad (2.75)$$

instead of equation (2.68), and we obtain again equation (2.72) with

$$b_t = \frac{1}{(2\pi\hbar)^{1/2}} \sum_E a_E \exp\left\{i\frac{E}{\hbar}t\right\}. \qquad (2.76)$$

On the other hand, in the stationary case the potential $V(x, y, z, t)$ is not dependent on time t and we have $V = V(x, y, z)$ leading to, instead of equations (2.50) and (2.35),

$$E\Psi(\mathbf{r}, E) = -\frac{\hbar^2}{2m_0}\Delta\Psi(\mathbf{r}, E) + V(x, y, z)\Psi(\mathbf{r}, E), \qquad (2.77)$$

and

$$i\hbar\frac{\partial}{\partial t}\Psi(\mathbf{r}, t) = -\frac{\hbar^2}{2m_0}\Delta\Psi(\mathbf{r}, t) + V(x, y, z)\Psi(\mathbf{r}, t). \qquad (2.78)$$

The superpositions (2.75) and (2.72) must be valid simultaneously. Since equation (2.72) cannot be applied to equation (2.78) (also here relation (2.74) is not fulfilled), relation (2.75) can also not be valid although the Hamiltonian

$$\hat{H} = -\frac{\hbar^2}{2m_0}\Delta\Psi(\mathbf{r}, t) + V(x, y, z) \qquad (2.79)$$

defined by equation (2.78) fulfils relation (2.69). We will recognize in Chapter 3 (Appendix 3.A) that the stationary case $V(x, y, z, t) = V(x, y, z)$ reflects an unphysical situation within the framework of projection theory.

Conclusion

In summary, within the investigation given here the superposition principle is not valid and this means that a "collapse of the wave function" is no longer needed. This is an important point.

2.3.8. *Processes*

2.3.8.1. *General remarks*

We will see in Chapter 3 that $\Psi(\mathbf{r}, t)$ is zero for a *stationary* system $V(x, y, z, t) = V(x, y, z)$ as well as for a *free* system $V(x, y, z, t) = 0$. Thus, the following important question arise: What kind of information is reflected by the general equations (2.35) and (2.50), respectively? These equations exclusively reflect *processes*. Thus, the functions $\Psi(\mathbf{r}, t)$ and $\Psi(\mathbf{p}, E)$ must also reflect processes, that is, interaction processes between the system under investigation and its environment. We will investigate this point in Chapter 4.

In connection with the interpretation of the wave functions $\Psi(\mathbf{r}, t)$ and $\Psi(\mathbf{p}, E)$, just the behaviour of free (non-interacting) systems is relevant. In Chapter 3, we will investigate free systems in detail. But let us anticipate the result already here: The momentum $\mathbf{p} = \mathbf{p}_0$ and the energy $E = E_0$ of a non-interacting system must remain constant in the course of time τ, and the $\mathbf{p}, E$-fluctuations $\Delta \mathbf{p}$ and ΔE are zero: $\Delta \mathbf{p} = 0, \Delta E = 0$. We will show in Chapter 3 that the corresponding wave functions $\Psi(\mathbf{r}, t)$ and $\Psi(\mathbf{p}_0, E_0)$ must be zero:

$$\Psi(\mathbf{r}, t) = 0, \quad \Psi(\mathbf{p}_0, E_0) = 0.$$

This result means that within projection theory free (static) systems are not able to exist. In fact, such static systems are not relevant for a human in connection with his actions in everyday life, which are dictated by the "principle of usefulness". Static systems are dead systems, so to say, and are superfluous from the point of view of evolution. The "principle of usefulness" does not allow the existence of superfluous objects in space and time.[77] Thus, our qualitative discussion in Section 2.2 is confirmed by the mathematical treatment in connection with the projection principle. Whether or not free systems exist in basic reality can principally not be said. Also this point has already been discussed in Section 2.2.

In conclusion, the functions $\Psi(\mathbf{r}, t)$ and $\Psi(\mathbf{p}, E)$ must exclusively reflect quantum processes and, therefore, the general equations (2.35) and (2.50) must exclusively also reflect such processes, that is,

---

[77] See, in particular, Section 2.2.

interaction processes between the system under investigation and its environment. Static, non-interacting systems are not described by $\Psi(\mathbf{r},t)$ and $\Psi(\mathbf{p},E)$ and, therefore, also not by the general equations (2.35) and (2.50).

These interactions exclusively take place in reality outside, that is, within $(\mathbf{p},E)$-space, and are characterized at any instant τ by *elementary processes* ($\Delta\mathbf{p}$ and ΔE) in connection with $\mathbf{p}$ and E. There are no other elements characterizing $(\mathbf{p},E)$-space. There is a continues exchange of momentum $\Delta\mathbf{p}$ and energy ΔE between the system under investigation and its environment.[78] Within projection theory energy and momentum are conserved. There is no reason to assume that this property should not be fulfilled.

Since no material objects are embedded in $(\mathbf{r},t)$-space, there can be no interactions in that space, but there are *correlations* between the various $\mathbf{r},t$-points. These correlations are (almost) zero in the case of an (almost) free system.

2.3.8.2. Description of properties and appearances

The function $\Psi(\mathbf{r},t)$ describes the system under investigation, which can be an atom, electron or even a macroscopic object as, for example, a stone. We know that the properties of a system (for example, a stone) are not changed when we move it from one position $\mathbf{r}$ to another $\mathbf{r}+\mathbf{r}_c$. The same holds for the time t. The properties are not dependent on the time origin, that is, the properties are not changed when we go from t to $t+t_c$. Therefore, let us require the following. *A function $D_{\mathbf{p},E}$ describes reality in $(\mathbf{p},E)$-space if it remains unchanged when we go from $\Psi(\mathbf{r},t)$ to $\Psi(\mathbf{r}+\mathbf{r}_c,t+t_c)$.* The function $D_{\mathbf{p},E}$ characterizes the appearance of the system under investigation in $(\mathbf{p},E)$-space.

With equation (2.5), we obtain

$$\Psi(\mathbf{r}+\mathbf{r}_c,t+t_c) = \frac{1}{(2\pi\hbar)^2}\int_{-\infty}^{\infty}\phi(\mathbf{p},E)\exp\left\{\frac{i}{\hbar}(\mathbf{p}\cdot\mathbf{r}-Et)\right\}d\mathbf{p}\,dE, \quad (2.80)$$

---

[78] Interested readers might refer to Schommers, W., (Ed.), (1989) Quantum Theory and Pictures of Reality: Springer-Verlag.

where

$$\phi(\mathbf{p}, E) = \Psi(\mathbf{p}, E) \exp\left\{\frac{i}{\hbar}(\mathbf{p} \cdot \mathbf{r}_c - E t_c)\right\}. \tag{2.81}$$

Then, the function $D_{\mathbf{p},E}$ with the above required feature is expressed by

$$D_{\mathbf{p},E} = \phi^*(\mathbf{p}, E)\phi(\mathbf{p}, E) = \Psi^*(\mathbf{p}, E)\Psi(\mathbf{p}, E), \tag{2.82}$$

that is, not the functions $\Psi(\mathbf{r}, t)$ and $\Psi(\mathbf{p}, E)$ are of relevance for the physical process but the quantities $\Psi^*(\mathbf{r}, t)\Psi(\mathbf{r}, t)$ and $\Psi^*(\mathbf{p}, E)\Psi(\mathbf{p}, E)$, and we have the following equivalencies:

$$\Psi^*(\mathbf{r}, t)\Psi(\mathbf{r}, t) \leftrightarrow \Psi^*(\mathbf{p}, E)\Psi(\mathbf{p}, E), \tag{2.83}$$

$$\Psi^*(\mathbf{r} + \mathbf{r}_c, t + t_c)\Psi(\mathbf{r} + \mathbf{r}_c, t + t_c) \leftrightarrow \Psi^*(\mathbf{p}, E)\Psi(\mathbf{p}, E). \tag{2.84}$$

On the other hand, since $\Psi(\mathbf{r}, t) = 0$ for any constant values of E_c and $\mathbf{p}_c$,[79] we may require the following: *A function $D_{\mathbf{r},t}$ describes reality (its picture) in $(\mathbf{r}, t)$-space if it remains unchanged when we go from $\Psi(\mathbf{p}, E)$ to $\Psi(\mathbf{p} + \mathbf{p}_c, E + E_c)$.* The function $D_{\mathbf{r},t}$ characterizes the appearance of the system under investigation in $(\mathbf{r}, t)$-space.

This property is due to the fact that only fluctuations of $\mathbf{p}$ and E are relevant[80] and not their absolute values ($\Psi(\mathbf{r}, t)$ is zero for constant values of $\mathbf{p}$ and E).

With equation (2.5), we get

$$\Psi(\mathbf{p} + \mathbf{p}_c, E + E_c) = \frac{1}{(2\pi\hbar)^2} \int_{-\infty}^{\infty} \phi(\mathbf{r}, t)$$

$$\times \exp\left\{-\frac{i}{\hbar}(\mathbf{p} \cdot \mathbf{r} - E t)\right\} d\mathbf{r} \, dt, \tag{2.85}$$

with

$$\phi(\mathbf{r}, t) = \Psi(\mathbf{r}, t) \exp\left\{-\frac{i}{\hbar}(\mathbf{p}_c \cdot \mathbf{r} - E_c t)\right\}. \tag{2.86}$$

Then, the function $D_{\mathbf{r},t}$ with the feature required above is expressed by

$$D_{\mathbf{r},t} = \phi^*(\mathbf{r}, t)\phi(\mathbf{r}, t) = \Psi^*(\mathbf{r}, t)\Psi(\mathbf{r}, t). \tag{2.87}$$

---

[79] See Chapter 3.
[80] See Chapter 3.

That is, also here the functions $\Psi(\mathbf{r}, t)$ and $\Psi(\mathbf{p}, E)$ are not of relevance for the physical process but the quantities $\Psi^*(\mathbf{r}, t)\Psi(\mathbf{r}, t)$ and $\Psi^*(\mathbf{p}, E)\Psi(\mathbf{p}, E)$. In analogy to equations (2.83) and (2.84), we obtain the following equivalencies:

$$\Psi^*(\mathbf{r}, t)\Psi(\mathbf{r}, t) \leftrightarrow \Psi^*(\mathbf{p}, E)\Psi(\mathbf{p}, E), \tag{2.88}$$

$$\Psi^*(\mathbf{r}, t)\Psi(\mathbf{r}, t) \leftrightarrow \Psi^*(\mathbf{p} + \mathbf{p}_c, E + E_c)\Psi(\mathbf{p} + \mathbf{p}_c, E + E_c). \tag{2.89}$$

Finally, with equation (2.80) and the inverse transformation formula we get the following relationships:

$$\phi(\mathbf{p}, E) = \frac{1}{(2\pi\hbar)^2} \int_{-\infty}^{\infty} \Psi(\mathbf{r} + \mathbf{r}_c, t + t_c)$$
$$\times \exp\left\{-\frac{i}{\hbar}(\mathbf{p} \cdot \mathbf{r} - Et)\right\} d\mathbf{r} \, dt, \tag{2.90}$$

$$\phi(\mathbf{p} + \mathbf{p}_c, E + E_c) = \frac{1}{(2\pi\hbar)^2} \int_{-\infty}^{\infty} \phi_c(\mathbf{r}, t)$$
$$\times \exp\left\{-\frac{i}{\hbar}(\mathbf{p} \cdot \mathbf{r} - Et)\right\} d\mathbf{r} \, dt, \tag{2.91}$$

with

$$\phi_c(\mathbf{r}, t) = \Psi(\mathbf{r} + \mathbf{r}_c, t + t_c) \exp\left\{-\frac{i}{\hbar}(\mathbf{p}_c \cdot \mathbf{r} - E_c t)\right\}. \tag{2.92}$$

On the other hand, equation (2.80) directly leads to

$$\phi(\mathbf{p} + \mathbf{p}_c, E + E_c) = \Psi(\mathbf{p} + \mathbf{p}_c, E + E_c) \exp\left\{\frac{i}{\hbar}(\mathbf{p} \cdot \mathbf{r}_c - Et_c)\right\}$$
$$\times \exp\left\{\frac{i}{\hbar}(\mathbf{p}_c \cdot \mathbf{r}_c - E_c t_c)\right\}. \tag{2.93}$$

Then, we get the following equivalency:

$$\phi^*(\mathbf{p} + \mathbf{p}_c, E + E_c)\phi(\mathbf{p} + \mathbf{p}_c, E + E_c) \leftrightarrow \phi_c^*(\mathbf{r}, t)\phi_c(\mathbf{r}, t), \tag{2.94}$$

leading to

$$\Psi^*(\mathbf{p} + \mathbf{p}_c, E + E_c)\Psi(\mathbf{p} + \mathbf{p}_c, E + E_c)$$
$$\leftrightarrow \Psi^*(\mathbf{r} + \mathbf{r}_c, t + t_c)\Psi(\mathbf{r} + \mathbf{r}_c, t + t_c), \tag{2.95}$$

where the constant parameters $\mathbf{r}_c, t_c, \mathbf{p}_c, E_c$ are independent from each other and can be chosen arbitrarily without changing the processes in $(\mathbf{p}, E)$-space, described by $\Psi^*(\mathbf{p}, E)\Psi(\mathbf{p}, E)$. Therefore, also the structures in $(\mathbf{r}, t)$-space, described by $\Psi^*(\mathbf{r}, t)\Psi(\mathbf{r}, t)$, remain unchanged. In other words, not the wave functions $\Psi(\mathbf{p}, E)$ and $\Psi(\mathbf{r}, t)$ but the quantities $\Psi^*(\mathbf{p}, E)\Psi(\mathbf{p}, E)$ and $\Psi^*(\mathbf{r}, t)\Psi(\mathbf{r}, t)$ are responsible for the appearances in both spaces.

2.3.8.3. *The meaning of the wave function*

The basic information about a system are the wave functions $\Psi(\mathbf{r}, t)$ and $\Psi(\mathbf{p}, E)$, which are completely equivalent concerning their physical content. Both functions are not independent from each other and are connected by equation (2.4). However, the quantities $\Psi^*(\mathbf{r}, t)\Psi(\mathbf{r}, t)$ and $\Psi^*(\mathbf{p}, E)\Psi(\mathbf{p}, E)$ are more suitable to interpret the role of the variables $\mathbf{r}, t, \mathbf{p}$ and E because these quantities fulfil certain natural conditions[81] which are obviously also valid at the macroscopic level.

Deterministic laws for the variables $\mathbf{r} = \mathbf{r}(\tau), t = t(\tau), \mathbf{p} = \mathbf{p}(\tau), E = E(\tau)$ are not defined and it is relatively easy to verify that such laws cannot exist if equations (2.4) and (2.5) are valid simultaneously. Time τ is again defined by our clocks which goes strictly from the past to the future. In other words, there is no longer a physical law that tells us *when* and *where* a "particle" jumps. Then, the variables $\mathbf{r}, t, \mathbf{p}$ and E must behave statistically. There is no alternative. In other words, $\mathbf{r}, t, \mathbf{p}$ and E are statistical variables and we have to find a description for the corresponding probabilities. What quantities are relevant here? We have outlined in Section 2.3.8.2 that the quantities $\Psi^*(\mathbf{p}, E)\Psi(\mathbf{p}, E)$ and $\Psi^*(\mathbf{r}, t)\Psi(\mathbf{r}, t)$ are responsible for the appearances in both spaces $((\mathbf{p}, E)$-space and $(\mathbf{r}, t)$-space). On the other hand, the appearance of the variables $\mathbf{r}, t, \mathbf{p}$ and E is the statistical behaviour. Thus, for the appearance of $\mathbf{r}, t, \mathbf{p}$ and E the functions $\Psi^*(\mathbf{p}, E)\Psi(\mathbf{p}, E)$ and $\Psi^*(\mathbf{r}, t)\Psi(\mathbf{r}, t)$ should be relevant. Let us investigate this point in more detail.

At time τ we have one value for $\mathbf{r}$ and one value for t:

$$\tau : \mathbf{r}, t. \tag{2.96}$$

---

[81] See Section 2.3.8.2.

At time τ we also have one value for $\mathbf{p}$ and one value for E:

$$\tau : \mathbf{p}, E. \tag{2.97}$$

But all the variables behave statistically, i.e., we can principally nothing say about the values $\mathbf{r}', t', \mathbf{p}', E'$ at time τ'. However, due to equations (2.4) and (2.5), the variables $\mathbf{r}, t$ are completely equivalent to $\mathbf{p}, E$. Therefore, if there is a probability distribution for $\mathbf{r}, t$ there must also be a probability distribution for $\mathbf{p}, E$. In other words, at time τ we measure one of the possible sets of $\mathbf{r}, t$ and, simultaneously, one of the possible sets of $\mathbf{p}, E$. If the functions $\Psi^*(\mathbf{r}, t)\Psi(\mathbf{r}, t)$ and $\Psi^*(\mathbf{p}, E)\Psi(\mathbf{p}, E)$ are the probability densities for these variables, i.e., for $\mathbf{r}, t$ and $\mathbf{p}, E$ the normalization integrals

$$\int_{-\infty}^{\infty} \Psi^*(\mathbf{r}, t)\Psi(\mathbf{r}, t) d\mathbf{r}\, dt,$$

and

$$\int_{-\infty}^{\infty} \Psi^*(\mathbf{p}, E)\Psi(\mathbf{p}, E) d\mathbf{p}\, dE,$$

for both functions $\Psi^*(\mathbf{r}, t)\Psi(\mathbf{r}, t)$ and $\Psi^*(\mathbf{p}, E)\Psi(\mathbf{p}, E)$ must be identical. In fact, we get with equations (2.4) and (2.5) the following relationship:

$$\int_{-\infty}^{\infty} \Psi^*(\mathbf{r}, t)\Psi(\mathbf{r}, t) d\mathbf{r}\, dt = \int_{-\infty}^{\infty} \Psi^*(\mathbf{p}, E)\Psi(\mathbf{p}, E) d\mathbf{p}\, dE. \tag{2.98}$$

Furthermore, if we set

$$\int_{-\infty}^{\infty} \Psi^*(\mathbf{r}, t)\Psi(\mathbf{r}, t) d\mathbf{r}\, dt = 1, \tag{2.99}$$

we immediately get

$$\int_{-\infty}^{\infty} \Psi^*(\mathbf{p}, E)\Psi(\mathbf{p}, E) d\mathbf{p}\, dE = 1. \tag{2.100}$$

Both probability distributions are connected, as it should be. In conclusion, due to the properties just quoted, the functions $\Psi^*(\mathbf{r}, t)\Psi(\mathbf{r}, t)$ and $\Psi^*(\mathbf{p}, E)\Psi(\mathbf{p}, E)$ have in fact to be identified with probability densities for the variables $\mathbf{r}, t$ of $(\mathbf{r}, t)$-space and for the variables $\mathbf{p}, E$

of $(\mathbf{p}, E)$-space. There is nothing else that could be applied for the definition of probability densities.

Within the theoretical picture given here, (almost) free particles[82] are never located in $(\mathbf{r}, t)$-space. Thus, it is not possible to interpret the quantity $\Psi^*(\mathbf{r}, t)\Psi(\mathbf{r}, t)$ as a probability distribution of finding a *particle* in the intervals $\mathbf{r}, \mathbf{r} + \Delta\mathbf{r}$ and $t, t + \Delta t$, and this contradicts Born's statistical interpretation of the wave function.[83] Clearly, in those cases where $\Psi^*(\mathbf{r}, t)\Psi(\mathbf{r}, t)$ has sharp maxima we can define particles in $(\mathbf{r}, t)$-space although the material object is not embedded here but exclusively in $(\mathbf{p}, E)$-space. However, such a particle definition is based on $\Psi(\mathbf{r}, t)$ and is not used for the interpretation of $\Psi(\mathbf{r}, t)$ as in Born's statistical interpretation.

When there are no local existents in $(\mathbf{r}, t)$-space, the following question arises: Of what is $\Psi^*(\mathbf{r}, t)\Psi(\mathbf{r}, t)$ the probability density? Of course, it is the probability distribution for the variables $\mathbf{r}$ and t. As we have already stated in Section 2.3.4.3[84] the measurement of the variables $\mathbf{r}$ and t is only possible in connection with real objects and real physical processes. However, within projection theory, objects and physical processes are embedded in $(\mathbf{p}, E)$-space, that is, real objects and physically real processes are described by the variables $\mathbf{p}$ and E.

Thus, in order to answer the above question (Of what is $\Psi^*(\mathbf{r}, t)\Psi(\mathbf{r}, t)$ the probability density?) we have to use equation (2.4). Accordingly, $\Psi(\mathbf{r}, t)$ is determined at location $\mathbf{r}, t$ in space-time by all possible values $\mathbf{p}$ and E ($-\infty < \mathbf{p}, E < \infty$) which are given with the probability density $\Psi^*(\mathbf{p}, E)\Psi(\mathbf{p}, E)$. Therefore, $\Psi^*(\mathbf{r}, t)\Psi(\mathbf{r}, t)$ can only be interpreted in connection with the variables $\mathbf{p}$ and E. There is no other way:

One of the possible values for $\mathbf{p}$ and for E is present in the intervals $\mathbf{r}, \mathbf{r} + \Delta\mathbf{r}$ and $t, t + \Delta t$ with the probability density of $\Psi^*(\mathbf{r}, t)\Psi(\mathbf{r}, t)$.

Since only the variables $\mathbf{p}$ and E are accessible to measurements we can also state:

---

[82] See Section 2.3.6.1, and Chapters 3 and 4 in connection with distance-independent interactions.
[83] See Chapter 1.
[84] See Figure 2.8.

The measurement of one of the possible values for **p** *and for E is done in the space-time intervals* **r**, **r** + Δ**r** *and* $t, t + \Delta t$ *with the probability density of* $\Psi^*(\mathbf{r}, t)\Psi(\mathbf{r}, t)$.

This is actually the situation in practical experiments. Signals are recorded with detectors in space and time, no more, no less. This has primarily nothing to do with a particle localized in space.

2.3.8.4. Properties of probability distributions

At time τ we have probability distributions $\{\mathbf{r}\}, \{t\}, \{\mathbf{p}\}, \{E\}$ for the variables **r**, t, **p** and E:

$$\tau : \{\mathbf{r}\}, \{t\}, \{\mathbf{p}\}, \{E\} \qquad (2.101)$$

But only *one* value of each distribution can be realized at time τ, and there is a certain probability for the existence of these values which are expressed by $\Psi^*(\mathbf{r}, t)\Psi(\mathbf{r}, t)$ and $\Psi^*(\mathbf{p}, E)\Psi(\mathbf{p}, E)$, respectively. But how many values of each variable **r**, t, **p** and E could in principle be registered within a small time interval $\Delta\tau$ different from zero?

Let $\Delta_\mathbf{r}, \Delta_t, \Delta_\mathbf{p}, \Delta_E$ be the ranges of **r**, t, **p** and E for which the distributions $\Psi^*(\mathbf{r}, t)\Psi(\mathbf{r}, t)$ and $\Psi^*(\mathbf{p}, E)\Psi(\mathbf{p}, E)$ are not zero, and let us assume that the ranges $\Delta_\mathbf{r}, \Delta_t, \Delta_\mathbf{p}, \Delta_E$ are different from infinity. How many values **r**, t, **p** and E can come into existence in the time interval $\Delta\tau = \varepsilon$, where ε is infinitesimal but different from zero? This number of values for each variable is identical with the number of τ-values in the interval ε, and this number is given by the number of all real numbers within ε. We know from mathematics that the number of real numbers in ε must be infinity, and in physics none of these real numbers are excluded. Therefore, within the infinitesimal interval ε, the number N of values of a variable (as, for example **r**) is also infinite in the interval $\Delta_\mathbf{r}$. Thus, also the number density $\rho = N/\Delta_\mathbf{r}$ of **r**-values is infinite in the interval $\Delta_\mathbf{r}$, where the interval $\Delta_\mathbf{r}$ can take any value but must be different from infinity.

In conclusion, the possible values defined by $\Psi^*(\mathbf{r}, t)\Psi(\mathbf{r}, t)$ and $\Psi^*(\mathbf{p}, E)\Psi(\mathbf{p}, E)$ are realized (with an infinite number density) already within an infinitesimal time-interval ε of almost zero, although the ranges $\Delta_\mathbf{r}, \Delta_t, \Delta_\mathbf{p}, \Delta_E$ can take any value (however, it must be different from infinity). This is of particular importance for time t and will be analyzed in more detail below.

This property (let us call it ε-property) is remarkable, but also in classical mechanics we work with an infinite number of space-points within an infinitesimal interval $\Delta\tau = \varepsilon$. A classical particle, which moves with velocity **v** from one space-position to another defining the space-distance of $\Delta_\mathbf{r}$, and if this process takes place within an infinitesimal time interval $\Delta\tau = \varepsilon$, the space-interval $\Delta_\mathbf{r}$ must also be infinitesimal, otherwise we were not be able to define a reasonable classical velocity **v** (classical mechanics is a local theory). Nevertheless, the interval $\Delta_\mathbf{r}$ contains an infinite number of real numbers, that is, the particle runs monotonically over an infinite number of real numbers within the infinitesimal time interval $\Delta\tau = \varepsilon$.

However, in quantum theory the world is non-local, and the laws are non-deterministic. In particular, the definition of the velocity **v** in the classical sense of the word is not possible. In contrast to classical mechanics, within projection theory the space-time intervals $\Delta_\mathbf{r}$ and Δ_t can take any values (different from infinity) in the case of an infinitesimal time interval $\Delta\tau = \varepsilon$. Both intervals ($\Delta_\mathbf{r}$ and Δ_t) are *statistically* occupied in the course of time τ. Within the infinitesimal time interval $\Delta\tau = \varepsilon$ the system jumps statistically from one space-time point $\mathbf{r}_i$, t_i to another point $\mathbf{r}_j$, t_j where $\mathbf{r}_{ij} = \mathbf{r}_i - \mathbf{r}_j$ and $t_{ij} = t_i - t_j$ can take arbitrary values within the intervals $\Delta_\mathbf{r}$, Δ_t: $\mathbf{r}_{ij} \leq \Delta_\mathbf{r}$, $t_{ij} \leq \Delta_t$. Although the time interval $\Delta\tau = \varepsilon$ is infinitesimal, the number density in both intervals is infinity because the number of events within $\Delta\tau = \varepsilon$ is infinite. Note, that this property is independent on $\Delta_\mathbf{r}, \Delta_t, \Delta_\mathbf{p}, \Delta_E$; the ranges $\Delta_\mathbf{r}, \Delta_t, \Delta_\mathbf{p}, \Delta_E$ can take any value (however, it must be different from infinity).

The situation in connection with the number of events (ε-property) within a small time interval $\Delta\tau = \varepsilon$ is close to Born's probability interpretation[85] and, as we have outlined above, also in classical mechanics. While classical mechanics is a local theory, the theoretical picture given here led to an inherently non-local representation with respect to space and time.

---

[85] See Chapter 1.

There are essential differences between Born's theory and the investigation given here. 1. Within Born's probability interpretation point-like particles are needed. Within the theoretical picture given here such point-like particles are not defined. 2. Within Born's theory only the positions **r** behaves statistically. Within projection theory we have an extension due to the system-specific time t which is not defined in usual quantum theory. Not only the space positions **r** behave statistically (Born's theory) but each space-time point **r**, t jumps statistically.

2.3.8.5. *Does god play dice?*

Whether or not the statistical behaviour is also a property of basic reality can principally not be said from the observer's point of view. As we have discussed in detail in Section 2.1, a human being cannot be make any statement about basic (objective) reality. What we only know with certainty within projection theory is that basic (objective) reality must exist. In this connection, Figure 2.7 is instructive.

Albert Einstein asserted that "God does not play dice". But how does Einstein know what God is able or willing to do? What can people say about the abilities of God? Nothing! Such kind of questions are inappropriate, and this assessment is supported by the foundations of projection theory. From the point of view of projection theory, we cannot state that "God plays dice" but also not that "God does not play dice". Both statements are not allowed within projection theory when we identify God with basic reality.

2.3.9. *Time*

Concerning time t we have the following situation. At time τ only one value of the system-specific time t is realized with a certain probability. However, if we consider an infinitesimal time interval $\Delta\tau = \varepsilon$ an infinite number of t-values are occupied, that is, the whole history (the complete past and future) of the system, described by the range Δ_t (life-time) of the distribution $\Psi^*(\mathbf{r},t)\Psi(\mathbf{r},t)$, is given within the infinitesimal time interval $\Delta\tau = \varepsilon$ of our clocks.[86]

---

[86] We have discussed this point in detail in Section 2.3.8.4.

This is the case for any $\tau_i \pm \Delta\tau$, that is, the law $\Psi^*(\mathbf{r},t)\Psi(\mathbf{r},t)$ is independent of τ. It is stationary with respect to time τ. Thus, we may say the following. Despite the statistical fluctuations, the whole of time — past, present and future — is laid out frozen before us. In other words, there is no connection between τ and the system-specific time t, that is, the introduction of τ within projection theory, outlined so far, would give no sense. Therefore, we have to construct a connection between τ and the system-specific time t, because the existence of τ is a matter of fact. This connection between the reference time τ and the system-specific time t will be constructed in the next section.

Let us state once more that the time τ has nothing to do with the system under investigation, but merely plays the role of a reference time and is defined by our clocks which we use in everyday life.

2.3.9.1. Reference time and selection processes

Principal remarks

So far there seems to be no connection between τ and the system-specific time t. Despite the statistical fluctuations, at time τ the whole of time t — past, present and future — is laid out frozen before us:

$$\Psi^*(\mathbf{r},t)\Psi(\mathbf{r},t), \quad -\infty < \mathbf{r} < \infty, \quad -\infty < t < \infty. \quad (2.102)$$

This is valid for all times τ. In other words, the reference time τ is not correlated to specific t-values for such a situation. From this point of view it would give no sense to introduce τ within projection theory.

Such a process defined by equation (2.102) could not produce a certain time-feeling because t jumps statistically between the various t-values without giving time t a certain direction; there is no past, present and future definable.

We know however from our observations that we always observe only certain configurations of reality at time τ:

$$\Psi^*(\mathbf{r},t_0)\Psi(\mathbf{r},t_0), \quad -\infty < \mathbf{r} < \infty, \quad (2.103)$$

for $t_0 = \tau$. Each photography represents such a configuration in space, at a certain time $\tau = t_0$. Why does nature work in this way? Why such selection processes from $\Psi^*(\mathbf{r}, t)\Psi(\mathbf{r}, t)$ to $\Psi^*(\mathbf{r}, t_0)\Psi(\mathbf{r}, t_0)$[87]? The answer is probably given by evolution. There is an important basic principle in connection with evolution.[88] As little outside world as possible. This guarantees optimal chances for survival. This principle is clearly reflected in the transition from $\Psi^*(\mathbf{r}, t)\Psi(\mathbf{r}, t)$ to $\Psi^*(\mathbf{r}, t_0)\Psi(\mathbf{r}, t_0)$ and reflects a certain kind of selection. The occurrence of the reference time τ is obviously one of the features for that.

Introduction of the reference system

How does nature organize that, i.e., the transition from the situation defined by equation (2.102) to that given by equation (2.103)? This cannot be due to an internal transformation within the system under investigation alone, that is, without the influence of another process. Besides the system under investigation only the observer's function appear within the frame of our analysis and, therefore, the transition from $\Psi^*(\mathbf{r}, t)\Psi(\mathbf{r}, t)$ to $\Psi^*(\mathbf{r}, t_0)\Psi(\mathbf{r}, t_0)$[87] must be due to an interplay between the system under investigation, described by $\Psi(\mathbf{r}, t)$, and the observer's function. However, the observer's function has been characterized so far by one parameter only: it is the reference time τ that is measured by our clocks in everyday life. Clearly, only one parameter τ is not sufficient for the description of the interplay between the system characterized by $\Psi(\mathbf{r}, t)$ and the observer's function. How can we characterize the observation process more realistically?

For this purpose let us define a reference system that is produced inside the observer, and let us formally describe it by the wave function $\Psi_{ref}(t)$ and the probability distribution $\Psi^*_{ref}(t)\Psi_{ref}(t)$, respectively, and we would like to characterize the time variable for characterizing the reference system by $\gamma = t$. Clearly, also $\gamma = t$ is a system-specific quantity.[89]

The source for the existence of the wave function $\Psi_{ref}(t)$ are energy fluctuations in reality $((\mathbf{p}, E)$-space) leading to $\Psi_{ref}(E)$ and

---

[87] See equations (2.102) and (2.103).
[88] Interested readers might refer to Schommers, W., (1994) Space and Time, Matter and Mind: World Scientific.
[89] For simplicity, we would like to assume that the reference system is not dependent on any position $\mathbf{r}$.

$\Psi^*_{ref}(E)\Psi_{ref}(E)$, respectively. $\Psi_{ref}(E)$ and $\Psi_{ref}(t)$ are connected by[90]

$$\Psi_{ref}(E) = \frac{1}{(2\pi\hbar)^{1/2}} \int_{-\infty}^{\infty} \Psi_{ref}(t) \exp\left\{i\frac{E}{\hbar}t\right\} dt. \qquad (2.104)$$

In conclusion, at time τ we have two probability distributions, $\Psi^*_{ref}(t)\Psi_{ref}(t)$ and $\Psi^*(\mathbf{r},t)\Psi(\mathbf{r},t)$; one for the description of time $\gamma = t$ of the reference system and the other for the description of time t of the system under investigation.

The reference system, described by $\Psi_{ref}(t)$, has two functions:

1. To describe the nature of the reference time τ more specifically, and
2. To select $\Psi^*(\mathbf{r},t_0)\Psi(\mathbf{r},t_0)$ from $\Psi^*(\mathbf{r},t)\Psi(\mathbf{r},t)$.[91]

We will recognize below that both functions are interconnected: selection is not possible without the existence of a systematically varying reference time. First, we will discuss the reference time in more detail, and after that the selection process.

2.3.9.2. Structure of reference time

We have stated so far that time τ runs monotonically from the past to the future. However, this time-feeling must also be due to a process (inside the brain of the observer) and is therefore also a system-specific time.

This process might differ considerably from those treated within the frame of usual quantum theory. The reason is simple: within projection theory the equation for the determination of $\Psi(\mathbf{r},t)$[92] is more general than Schrödinger's equation of usual quantum theory because the function $V(x,y,z,t)$ is in general more complex than the classical potential $U(x,y,z)$ which is used in usual quantum theory.[93] In principle, $V(x,y,z,t)$ could have an imaginary part. The

---

[90] See also equation (2.43).
[91] See equations (2.102) and (2.103).
[92] See equation (2.35).
[93] See Section 2.3.7.3.

appearance of the function $V(x, y, z, t)$ is a logical consequence within the theoretical structures of projection theory.[94]

We would like to describe the distribution $\Psi^*_{ref}(t)\Psi_{ref}(t)$ introduced in the last section. The probability of finding a certain value $\gamma = t$ for the reference time in the interval $\Delta\gamma = \Delta t$ around $\gamma = t$ is given by $\Psi^*_{ref}(t)\Psi_{ref}(t)\Delta t$. In other words, the reference time $\gamma = t$ becomes uncertain because $\Psi^*_{ref}(t)\Psi_{ref}(t)$ has a certain width Δ_τ and, therefore, no longer runs strictly from the past to the future as is suggested by our clocks used in everyday life. However, the probability distribution $\Psi^*_{ref}(t)\Psi_{ref}(t)$ for the reference time $\gamma = t$ should be a relatively sharp function and, furthermore, because the time of our clocks τ runs monotonically from the past to the future, also the distribution $\Psi^*_{ref}(t)\Psi_{ref}(t)$ must run monotonically from the past to the future and we have

$$\Psi^*_{ref}(t)\Psi_{ref}(t) \to \Psi^*_{ref}(\tau - t)\Psi_{ref}(\tau - t). \quad (2.105)$$

We never measure the time τ but $\gamma = t$, and $\gamma = t$ is uncertain due to $\Psi^*_{ref}(\tau - t)\Psi_{ref}(\tau - t)$. However, because $\Psi^*_{ref}(\tau - t)\Psi_{ref}(\tau - t)$ can be assumed to be a relatively sharp function, $\gamma = t$ should be close to τ. Due to τ the whole curve $\Psi^*_{ref}(\tau - t)\Psi_{ref}(\tau - t)$ moves strictly from the past to the future, but the values $\gamma = t$ for the reference time fluctuates around τ.

Since we experience, for example, a moving car with velocity **v** as a strict sequence of configurations $A(\mathbf{r} - \mathbf{v}\tau)$, that is, we have for τ the strict sequence $\tau_1 < \tau_2 \cdots$. But this feature is based on our macroscopic observations and, therefore, in general we have to admit a certain small width Δ_τ for $\Psi^*_{ref}(\tau - t)\Psi_{ref}(\tau - t)$.

We have already outlined in Section 2.3.9.1 that the source for the existence of the wave function $\Psi_{ref}(t)$ are energy fluctuations in reality (($\mathbf{p}, E$)-space) leading to $\Psi_{ref}(E)$ and $\Psi^*_{ref}(E)\Psi_{ref}(E)$, respectively. $\Psi_{ref}(E)$ and $\Psi_{ref}(t)$ are connected by equation (2.104). However, the law for the energy fluctuations $\Psi^*_{ref}(E)\Psi_{ref}(E)$ does not change when

---

[94] See also Section 2.3.7.

the function $\Psi_{ref}(t)$ is shifted by τ. With

$$\phi_{ref}(E) = \frac{1}{(2\pi\hbar)^{1/2}} \int_{-\infty}^{\infty} \Psi_{ref}(\tau - t) \exp\left\{-i\frac{E}{\hbar}t\right\} dt, \quad (2.106)$$

$$\phi_{ref}(E) = \Psi_{ref}(E) \exp\left\{-i\frac{E}{\hbar}\tau\right\}, \quad (2.107)$$

we get

$$\phi^*_{ref}(E)\phi_{ref}(E) = \Psi^*_{ref}(E)\Psi_{ref}(E). \quad (2.108)$$

That is, the expression $\Psi^*_{ref}(E)\Psi_{ref}(E)$ is independent on τ. In Section 2.3.8.2, we have outlined that not $\Psi_{ref}(E)$ is the relevant function but $\Psi^*_{ref}(E)\Psi_{ref}(E)$.

In summary, at time τ we have a probability distribution $\Psi^*_{ref}(\tau - t)\Psi_{ref}(\tau - t)$ for the description of the reference time. In principle, the parameter τ could also be a statistical quantity, that is, the whole curve $\Psi^*_{ref}(\tau - t)\Psi_{ref}(\tau - t)$ could jump statistically without any systematic sequence for τ. However, this would lead to problems with observations and we do not want to analyze this case here. We would like to assume here that τ is the time of our clocks and therefore runs systematically and monotonically from the past to the future. However, at the quantum level this time τ does not exist with certainty. We only observe $\gamma = t$ which is inherently uncertain, and this uncertainty is described by $\Psi^*_{ref}(\tau - t)\Psi_{ref}(\tau - t)$. However, because $\Psi^*_{ref}(\tau - t)\Psi_{ref}(\tau - t)$ can be assumed to be a relatively sharp function, $\gamma = t$ should be close to τ. Due to τ the whole curve $\Psi^*_{ref}(\tau - t)\Psi_{ref}(\tau - t)$ moves strictly from the past to the future and, simultaneously, the values $\gamma = t$ for the reference time fluctuate around τ.

2.3.9.3. Selections

Convolution integral

We have outlined in Section 2.3.9.1 that the existence of the wave function $\Psi_{ref}(t)$ for the reference system should also be responsible for selection processes. In fact, the transition from $\Psi^*(\mathbf{r}, t)\Psi(\mathbf{r}, t)$ to $\Psi^*(\mathbf{r}, t_0)\Psi(\mathbf{r}, t_0)$ can be explained on the basis of this kind of function, i.e., by $\Psi_{ref}(t)$. In other words, the interplay between the two systems (the reference system described by $\Psi_{ref}(t), \Psi_{ref}(E)$ and, on the other hand, the system under investigation described by $\Psi(\mathbf{r}, t), \Psi(\mathbf{r}, E)$) should lead to the selection process. This process obviously filters out the configuration $\Psi^*(\mathbf{r}, t_0)\Psi(\mathbf{r}, t_0)$ from $\Psi^*(\mathbf{r}, t)\Psi(\mathbf{r}, t)$.

How can we model that? We have two processes, characterized in $(\mathbf{p}, E)$-space by $\Psi_{ref}(E)$ and $\Psi(\mathbf{r}, E)$, that are independent of each other and can therefore be coupled in the form of

$$\Psi_O(\mathbf{r}, E) = \Psi_{ref}(E)\Psi(\mathbf{r}, E), \tag{2.109}$$

leading to the effect of observation by the registration of $\Psi_O(\mathbf{r}, E)$ and $\Psi_O^*(\mathbf{r}, E)\Psi_O(\mathbf{r}, E)$, respectively. How are these observations transformed into $(\mathbf{r}, t)$-space where our observations take place and where the filtering process from $\Psi^*(\mathbf{r}, t)\Psi(\mathbf{r}, t)$ to $\Psi^*(\mathbf{r}, t_0)\Psi(\mathbf{r}, t_0)$ becomes evident? Using the Fourier transform for $\Psi_O(\mathbf{r}, E)$, we have

$$\Psi_O(\mathbf{r}, E) = \int_{-\infty}^{\infty} \Psi_{CON}(\mathbf{r}, \tau) \exp\left\{i\frac{E}{\hbar}\tau\right\} \frac{d\tau}{(2\pi\hbar)^{1/2}}, \tag{2.110}$$

immediately obtaining the following convolution integral:

$$\Psi_{CON}(\mathbf{r}, \tau) = \int_{-\infty}^{\infty} \Psi_{ref}(\tau - t)\Psi(\mathbf{r}, t) \frac{dt}{(2\pi\hbar)^{1/2}}. \tag{2.111}$$

Equations (2.110) and (2.111) directly lead to equation (2.109): $\Psi_O(\mathbf{r}, E) = \Psi_{ref}(E)\Psi(\mathbf{r}, E)$.

PROOF.

With

$$\exp\left\{i\frac{E}{\hbar}\tau\right\} = \exp\left\{i\frac{E}{\hbar}(\tau - t)\right\} \exp\left\{i\frac{E}{\hbar}t\right\}, \tag{2.112}$$

and

$$\Psi_O(\mathbf{r}, E) = \int_{-\infty}^{\infty}\int_{-\infty}^{\infty} \Psi_{ref}(\tau - t)\Psi(\mathbf{r}, t)$$
$$\times \exp\left\{i\frac{E}{\hbar}\tau\right\} \frac{dt}{(2\pi\hbar)^{1/2}} \frac{d\tau}{(2\pi\hbar)^{1/2}}, \tag{2.113}$$

where equations (2.110) and (2.111) have been used, we immediately obtain equation (2.109) using the substitution $z = \tau - t$:

$$\Psi_O(\mathbf{r}, E) = \left\{ \int_{-\infty}^{\infty} \Psi_{ref}(z) \exp\left\{ i\frac{E}{\hbar}z \right\} \frac{dz}{(2\pi\hbar)^{1/2}} \right\}$$
$$\times \left\{ \int_{-\infty}^{\infty} \Psi(\mathbf{r}, t) \exp\left\{ i\frac{E}{\hbar}t \right\} \frac{dt}{(2\pi\hbar)^{1/2}} \right\} \quad (2.114)$$
$$= \Psi_{ref}(E)\Psi(\mathbf{r}, E).$$

Let us assume that $\Psi_{ref}(\tau - t)$ in (2.111) is a very sharp function, approximately described by a delta function:

$$\Psi_{ref}(\tau - t) = C\delta(\tau - t), \quad (2.115)$$

where C is a constant. The width Δ_τ of the function $\Psi_{ref}(\tau - t)$ should principally not be zero but can take any value with $\Delta_\tau \neq 0$. Therefore, we can use the delta function with any degree of accuracy.

Using equation (2.115) we obtain.

$$\Psi_{CON}(\mathbf{r}, \tau) = \frac{C}{(2\pi\hbar)^{1/2}} \Psi(\mathbf{r}, \tau), \quad (2.116)$$

and with equation (2.110)

$$\Psi_O(\mathbf{r}, E) = \frac{C}{2\pi\hbar} \int_{-\infty}^{\infty} \Psi(\mathbf{r}, \tau) \exp\left\{ i\frac{E}{\hbar}\tau \right\} d\tau. \quad (2.117)$$

In other words, from all the configurations $\Psi(\mathbf{r}, t)$, the configuration $\Psi(\mathbf{r}, \tau)$ for time $t = \tau$ is filtered out. In particular, from equation (2.116) it directly follows that there is just the filtering process required above, that is, the transition from $\Psi^*(\mathbf{r}, t)\Psi(\mathbf{r}, t)$ to $\Psi^*(\mathbf{r}, t_0)\Psi(\mathbf{r}, t_0)$[95] where $t_0 = \tau$. That is just what we observe: a certain configuration in space (for example, a tree) is observed at time τ (time of our clocks).

---

[95] See equations (2.102) and (2.103).

Two types of time variables

Using equations (2.114) and (2.115), we get

$$\Psi_O(\mathbf{r}, E) = \frac{C}{2\pi \hbar} \int_{-\infty}^{\infty} \Psi(\mathbf{r}, t) \exp\left\{i \frac{E}{\hbar} t\right\} dt, \qquad (2.118)$$

with

$$\Psi_O(\mathbf{r}, E) = \frac{C}{(2\pi \hbar)^{1/2}} \Psi(\mathbf{r}, E) \cdot \exp\left\{\frac{i}{\hbar} E\tau\right\}. \qquad (2.119)$$

Equations (2.117) and (2.118) define two representations for $\Psi_O(\mathbf{r}, E)$ by using two types of time variables, namely t and τ, that are completely different in character. Time t is a statistical quantity and jumps arbitrarily without any direction from one point to another in accordance with the probability distribution $\Psi^*(\mathbf{r}, t)\Psi(\mathbf{r}, t)$. The reference time τ behaves non-statistically and is defined by our clocks that we use in everyday life going monotonically from the past to the future. However, the physical character of t and τ is not relevant in the determination of $\Psi_O(\mathbf{r}, E)$ and does not influence the integrals. Since the ranges of t and τ are identical, the value for the integrals (2.117) and (2.118) must also be identical.

Rectangular form for the reference time distribution

We used for the wave function $\Psi_{ref}(\tau - t)$ a delta function,[96] that is, we have chosen a distribution with the width Δ_τ of zero. However, in reality $\Psi_{ref}(\tau - t)$ should have a non-vanishing width, i.e., $\Delta_\tau \neq 0$. Let us therefore use for $\Psi_{ref}(\tau - t)$ a rectangular form:

$$\Psi_{ref}(\tau - t) = B \quad \text{for } \tau - \frac{\Delta_\tau}{2} \leq t \leq \tau + \frac{\Delta_\tau}{2},$$
$$\Psi_{ref}(\tau - t) = 0 \quad \text{for } t < \tau - \frac{\Delta_\tau}{2} \text{ and } t > \tau + \frac{\Delta_\tau}{2}, \qquad (2.120)$$

---

[96] See equation (2.115).

where B is a constant. With equation (2.104), we obtain the following expression for $\Psi_{ref}(E)$:

$$\Psi_{ref}(E) = \sqrt{\frac{2\hbar}{\pi} \frac{B}{E}} \exp\left\{i\frac{E}{\hbar}\tau\right\} \sin\left[\frac{E\,\Delta_\tau}{2\hbar}\right]. \tag{2.121}$$

What can say about the convolution integral (2.111) in the case of equation (2.120)? We immediately get

$$\Psi_{CON}(\mathbf{r},\tau) = \int_{-\infty}^{\infty} \Psi_{ref}(\tau - t)\Psi(\mathbf{r},t)\frac{dt}{(2\pi\hbar)^{1/2}}$$

$$= B \int_{\tau-\frac{\Delta_\tau}{2}}^{\tau+\frac{\Delta_\tau}{2}} \Psi(\mathbf{r},t)\frac{dt}{(2\pi\hbar)^{1/2}}. \tag{2.122}$$

Here the variable t is defined between $\tau - \Delta_\tau/2$ and $\tau + \Delta_\tau/2$ where τ defines the centre of the rectangle.

On the other hand, the wave function $\Psi(\mathbf{r},t)$ of the system under investigation, which appears in equation (2.122), is generally given by[97]

$$i\hbar\frac{\partial}{\partial t}\Psi(\mathbf{r},t) = -\frac{\hbar^2}{2m_0}\Delta\Psi(\mathbf{r},t) + V(x,y,z,t)\Psi(\mathbf{r},t).$$

If the width Δ_τ is sufficiently small, we may assume that the function $V(x,y,z,t)$ is in a good approximation independent on t, that is, $V(x,y,z,t)$ is a constant in the interval

$$\tau - \Delta_\tau/2 \leq t \leq \tau + \Delta_\tau/2, \tag{2.123}$$

and we would like to assume that $V(x,y,z,t)$ is given by the value at position τ that just defines the middle of the rectangular:

$$V(x,y,z,t) = V(x,y,z,\tau). \tag{2.124}$$

Then, the system behaves within the interval (2.123) like a stationary system, and instead of equation (2.35), we have

$$i\hbar\frac{\partial}{\partial t}\Psi(\mathbf{r},t) = -\frac{\hbar^2}{2m_0}\Delta\Psi(\mathbf{r},t) + V(x,y,z,\tau)\Psi(\mathbf{r},t). \tag{2.125}$$

---

[97] See equation (2.35).

The corresponding expression in $(\mathbf{r}, E)$-space is given by

$$E \Psi(\mathbf{r}, E) = -\frac{\hbar^2}{2m_0} \Delta \Psi(\mathbf{r}, E) + V(x, y, z, \tau) \Psi(\mathbf{r}, E), \quad (2.126)$$

instead of equation (2.50). As can easily be verified, the solutions of equations (2.125) and (2.126) are connected by

$$\Psi(\mathbf{r}, t) = C \exp\left\{-i\frac{E}{\hbar}t\right\} \Psi(\mathbf{r}, E). \quad (2.127)$$

Using this expression the convolution integral (2.122) takes the form[98]

$$\Psi_{CON}(\mathbf{r}, \tau) = \sqrt{\frac{2\hbar}{\pi} \frac{CB}{E_0}} \sin\left[\frac{E_0}{\hbar} \frac{\Delta_\tau}{2}\right] \exp\left\{-i\frac{E_0}{\hbar}\tau\right\} \Psi(\mathbf{r}, E_0)$$

$$= \sqrt{\frac{2\hbar}{\pi} \frac{B}{E_0}} \sin\left[\frac{E_0}{\hbar} \frac{\Delta_\tau}{2}\right] \Psi(\mathbf{r}, \tau), \quad (2.128)$$

where we assumed that the system is in the state with the energy $E = E_0$. We observe at time τ the space configuration $\Psi^*(\mathbf{r}, E_0)\Psi(\mathbf{r}, E_0)$ of the system under investigation which is in the energy state $E = E_0$. Again, τ runs monotonically from the past to the future and because the approximation $V(x, y, z, t) = V(x, y, z, \tau)$[99] varies with τ, also the energies and wave functions $\Psi(\mathbf{r}, t)$ and $\Psi(\mathbf{r}, E)$ vary with τ. There is a continuous succession of stationary cases.

Effect of motion

The law (2.116) means the following: $\Psi^*(\mathbf{r}, t)\Psi(\mathbf{r}, t)$ will be systematically scanned by $\Psi^*_{ref}(\tau-t)\Psi_{ref}(\tau-t)$ and only those values of t can be observed which correspond with the reference time τ.[100] Therefore, the sense of time τ is to select a certain configuration of $\Psi^*(\mathbf{r}, t_k)\Psi(\mathbf{r}, t_k)$ with $t_k = \tau$. Clearly, $\Psi^*(\mathbf{r}, t)\Psi(\mathbf{r}, t)$ is a static function[101] and does not change with the course of time τ, and the effect of motion we

---

[98] See also Chapter 3, Appendix 3.D, equation (3.D.5).
[99] See equation (2.124).
[100] See also Figure 2.10
[101] See Section 2.3.8.4, ε-property.

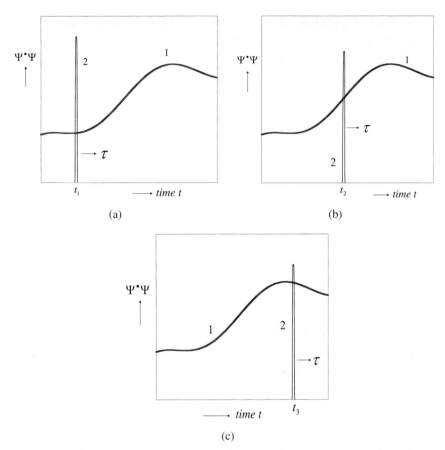

Fig. 2.10. The system under investigation, characterized by curve 1, fluctuates statistically between all possible configurations t, defined by the range of $\Psi^*(\mathbf{r},t)\Psi(\mathbf{r},t)$. On the other hand, curve 2, described by $\Psi^*_{ref}(\tau-t)\Psi_{ref}(\tau-t)$, characterizes the observers time feeling and this must also be due to a process. Δ_τ is the width of $\Psi^*_{ref}(\tau-t)\Psi_{ref}(\tau-t)$ and must be relatively small. (It is a delta function in the case of equation (2.115), that is, $\Delta_\tau = 0$.) At reference time t_B, with $\tau - \Delta_\tau/2 \leq t_B \leq \tau + \Delta_\tau/2$, all the configurations t are realized simultaneously, each with a certain probability $\Psi^*(\mathbf{r},t)\Psi(\mathbf{r},t)$. However, only those configurations can be observed which correspond with the reference time t_B since the human observer can only measure within the frame of curve 2 which is described by $\Psi^*_{ref}(\tau-t_B)\Psi_{ref}(\tau-t_B)$. In other words, there is a selection process. At times t_{B1}, t_{B2}, t_{B3} etc., the configurations (a) $\Psi^*(\mathbf{r},t_1)\Psi(\mathbf{r},t_1)$, (b) $\Psi^*(\mathbf{r},t_2)\Psi(\mathbf{r},t_2)$, (c) $\Psi^*(\mathbf{r},t_3)\Psi(\mathbf{r},t_3)$ etc., are selected.

experience in connection with $\Psi^*(\mathbf{r},t)\Psi(\mathbf{r},t)$ is entirely due to the "motion" of the reference time τ.

It should be emphasized that this concept of time is compatible with the fact that we know a lot about our past but nothing about our future.

2.3.9.4. *Information inside, information outside*

The transition from $\Psi^*(\mathbf{r},t)\Psi(\mathbf{r},t)$ to $\Psi^*(\mathbf{r},\tau)\Psi(\mathbf{r},\tau)$ means that at time τ only a small fraction of the whole information $\Psi^*(\mathbf{r},t)\Psi(\mathbf{r},t)$ is accessible to the observer. The picture of reality at time τ is given by $\Psi^*(\mathbf{r},\tau)\Psi(\mathbf{r},\tau)$. All the other information $\Psi^*(\mathbf{r},t)\Psi(\mathbf{r},t)$, $t \neq \tau$, is ignored (filtered out) and is not realized in $(\mathbf{r},t)$-space. Therefore, it cannot be experienced (measured, seen or felt). Clearly, the information $\Psi^*(\mathbf{r},t)\Psi(\mathbf{r},t)$, $t \neq \tau$ exists together with $\Psi^*(\mathbf{r},\tau)\Psi(\mathbf{r},\tau)$ in $(\mathbf{p},E)$-space in the form of $(\mathbf{p},E)$-fluctuations described by $\Psi^*(\mathbf{p},E)\Psi(\mathbf{p},E)$.

Measurements and the experience of other physical effects, in particular, in connection with the human body can only come into existence if there is a relation to $(\mathbf{r},t)$-space[102] Real effects are not observable if they have no connection to $(\mathbf{r},t)$-space as, for example, without space and time.

A picture in space and time only exists at $t = \tau$ where we have of course to consider that the reference time τ is uncertain because the function $\Psi^*_{ref}(\tau-t)\Psi_{ref}(\tau-t)$ has the small width Δ_τ which should be much smaller than the width Δ_t of $\Psi^*(\mathbf{r},t)\Psi(\mathbf{r},t)$: $\Delta_\tau \ll \Delta_t$.

We have seen in Section 2.3.8.4 that within the interval $\Delta\tau \neq 0$ (where $\Delta\tau$ can be infinitesimal) an infinite number of t-values of $\Psi^*(\mathbf{r},t)\Psi(\mathbf{r},t)$ are occupied within the range Δ_t of $\Psi^*(\mathbf{r},t)\Psi(\mathbf{r},t)$, that is, the entire history (the complete past and future) of the system, described by $\Psi^*(\mathbf{r},t)\Psi(\mathbf{r},t)$ having the width Δ_t, is given within the very small time interval $\Delta\tau \neq 0$ which can in principle be distinctly smaller than the width Δ_τ of the sharp function $\Psi^*_{ref}(\tau-t)\Psi_{ref}(\tau-t)$: $\Delta\tau \ll \Delta_\tau$. But this means that we have also an infinite number of

---
[102] See Section 2.3.8.3.

t-values within the section described by

$$\Psi^*(\mathbf{r}, t)\Psi(\mathbf{r}, t)\big|_{\Delta_t}. \quad (2.129)$$

(If the interval Δ_t contains an infinite number of events, also the fractions Δ_τ/Δ_t and $\Delta\tau/\Delta_t$ must contain an infinite number of events.)

2.3.9.5. Reality outside

The selected structure $\Psi(\mathbf{r}, \tau)$[103] and $\Psi^*(\mathbf{r}, \tau)\Psi(\mathbf{r}, \tau)$, respectively, appears at time τ spontaneously (unconsciously) in the head of the observer. We principally do not know what is outside; basic reality is not accessible to the observer.[104] But we are able to construct a "fictitious reality" which we have called $(\mathbf{p}, E)$-space, that is, space outside is different from that in the head of the observer $((\mathbf{r}, t)$-space). This has drastic consequences. For example, geometrical optics describes how a certain geometrical structure (a tree etc.) is transferred from reality outside into the observers head. However, in this case it is assumed that reality outside is also embedded in $(\mathbf{r}, t)$-space, and this is not true from the point of view of projection theory since reality is embedded in $(\mathbf{p}, E)$-space and not in $(\mathbf{r}, t)$-space. Reality outside, described in $(\mathbf{p}, E)$-space, is projected onto $(\mathbf{r}, t)$-space. From this point of view, geometrical optics can merely be considered as an auxiliary method.

We have stated above that the time structure $\Psi^*(\mathbf{r}, t)\Psi(\mathbf{r}, t)$ is systematically scanned by $\Psi_{ref}^*(\tau - t)\Psi_{ref}(\tau - t)$. This means that there must be a connection between both time structures, and this is because both systems interact with the same environment.

2.3.9.6. Constancy phenomena

The constancy of the shape of the function $\Psi_{ref}^*(\tau - t)\Psi_{ref}(\tau - t)$ is needed for a reliable observation of the world outside. This has to be managed by the physiological apparatus of the observer. It protects the perceptual functions against perturbations, in particular those from

---

[103] See equation (2.116).
[104] See also the discussion in Section 2.3.4.3.

the outside world (environment). But this kind of protection is not a new phenomenon in connection with the perceptual functions of the observer's physiological apparatus. Various constancy phenomena are known in this respect. Konrad Lorenz states:

> *Of special interest to the scientist striving for objectivation is the study of those perceptual functions which convey to us the experience of qualities constantly inherent in certain things in our environment. If, of course, we perceive a certain object (say a sheet of paper) as "white", even when different coloured lights, reflecting different wavelengths, are thrown on it. This so-called constancy phenomenon is achieved by the function of a highly complex physiological apparatus which computes, from the colour of the illumination and the colour reflected, the object's constantly inherent property which we call its colour. ... Other neural mechanisms enable us to see that an object which we observe from various sides retains one and the same shape even though the image on our retina assumes a great variety of forms. Other mechanisms make it possible for us to apprehend that an object we observe from various distances remains the same size, although the size of the retina image decreases with distance.*[105]

2.3.9.7. Schrödinger's equation and its limitations

Derivation of Schrödinger's equation from the principles of projection theory

From the general equation,[106]

$$i\hbar \frac{\partial}{\partial t}\Psi(\mathbf{r},t) = -\frac{\hbar^2}{2m_0}\Delta\Psi(\mathbf{r},t) + V(x,y,z,t)\Psi(\mathbf{r},t),$$

we can easily deduce Schrödinger's equations, and we can learn something about its validity. Let Δ be a small interval between two observations at $\tau = \tau_1$ and $\tau = \tau_2$ with $\Delta = \tau_2 - \tau_1$. In accordance

---
[105] Refer to Lorenz, K., (1973) Die Rückseite des Spiegels: Piper.
[106] See also equation (2.35).

with equation (2.116), we then observe the two selected space configurations $\Psi^*(\mathbf{r}, \tau_1)\Psi(\mathbf{r}, \tau_1)$ and $\Psi^*(\mathbf{r}, \tau_2)\Psi(\mathbf{r}, \tau_2)$ which are identical with two space configurations $\Psi^*(\mathbf{r}, t_1)\Psi(\mathbf{r}, t_1)$ and $\Psi^*(\mathbf{r}, t_2)\Psi(\mathbf{r}, t_2)$ with $t_1 = \tau_1$ and $t_2 = \tau_2$ because the events in connection with the system under investigation and the reference system are synchronized. In other words, we have

$$\Psi(\mathbf{r}, t_1) = \Psi(\mathbf{r}, \tau_1), \tag{2.130}$$

and

$$\Psi(\mathbf{r}, t_2) = \Psi(\mathbf{r}, \tau_2). \tag{2.131}$$

Clearly, also the potentials[107] are connected by

$$V(\mathbf{r}, t_1) = V(\mathbf{r}, \tau_1),$$
$$V(\mathbf{r}, t_2) = V(\mathbf{r}, \tau_2). \tag{2.132}$$

In particular, we have

$$\frac{\partial \Psi(\mathbf{r}, t)}{\partial t} = \lim_{\Delta \to 0} \frac{\Psi(\mathbf{r}, t_2) - \Psi(\mathbf{r}, t_1)}{t_2 - t_1}$$
$$= \lim_{\Delta \to 0} \frac{\Psi(\mathbf{r}, \tau_2) - \Psi(\mathbf{r}, \tau_1)}{\tau_2 - \tau_1} = \frac{\partial \Psi(\mathbf{r}, \tau)}{\partial \tau}. \tag{2.133}$$

When the interval Δ tends to zero, we have $t_1 = t_2 = t$ and $\tau_1 = \tau_2 = \tau$ and we immediately obtain with equation (2.35) the following equation:

$$i\hbar \frac{\partial}{\partial \tau} \Psi(\mathbf{r}, \tau) = -\frac{\hbar^2}{2m_0} \Delta \Psi(\mathbf{r}, \tau) + V(x.y.z, \tau)\Psi(\mathbf{r}, \tau). \tag{2.134}$$

This equation takes the form of Schrödinger's equation of usual quantum theory which we have deduced here from the general case described by equation (2.35). In equation (2.134), the system-specific time t does not appear as it must be in the case of usual quantum theory. However, equation (2.134) represents a modified version of Schrödinger's equation[108] because the "classical" potential $U(x.y.z, t = \tau)$ is replaced by $V(x.y.z, t = \tau)$ of projection theory.

---

[107] See equation (2.35).
[108] See equation (1.3). Here time t has to be identified with the reference time τ.

The essential point in connection with equation (2.134) is that there is no longer a system-specific time t since τ is an external parameter defined by our clocks that we use in everyday life. Therefore, equation (2.134) has exactly the form of Schrödinger's equation of usual quantum theory, and we would like to call equation (2.134) simply "Schrödinger's equation". The difference between $U(x.y.z, t = \tau)$ and $V(x.y.z, t = \tau)$ is not of relevance in this connection. From the point of view of projection theory, Schrödinger's equation is a total stranger. Usual quantum theory feigns a reality quite different from that of projection theory.

Let us briefly repeat the main differences between Schrödinger's equation (2.134) and the general equation (2.35). For certain observations, the reference time expressed by $\Psi^*_{ref}(\tau - t)\Psi_{ref}(\tau - t)$ is relevant,[109] and selection processes only give sense when this function has a certain width Δ_τ different from zero. If the width Δ_τ is exactly zero, the value for the reference time τ is sharp without any uncertainty and we can only observe one value, say t_a, of the system-specific time t and, therefore, only one configuration $\Psi^*(\mathbf{r}, t_a)\Psi(\mathbf{r}, t_a)$ with $t_a = \tau$. t_a takes one of the possible values $-\infty < t < \infty$ for which $\Psi^*(\mathbf{r}, t)\Psi(\mathbf{r}, t)$ is different from zero. In the case of $\Delta_\tau \neq 0$, the situation is different. Let Δ_t be the range of t for which the distribution $\Psi^*(\mathbf{r}, t)\Psi(\mathbf{r}, t)$ is not zero, and let us assume that the range Δ_t is different from infinity. How many values t can come into existence in the time interval

$$\Delta_\tau = \varepsilon \ll \Delta_t,$$

where ε is infinitesimal but different from zero? This number of values for t and configurations $\Psi^*(\mathbf{r}, t)\Psi(\mathbf{r}, t)$, respectively, is identical with the number of τ-values in the interval $\Delta_\tau = \varepsilon$, and this number is given by the number of all real numbers in $\Delta_\tau = \varepsilon$ and must be infinity.[110] Therefore, within the infinitesimal interval $\Delta_\tau = \varepsilon$, the number N of values of the variable t is also infinite in the interval

$$\Delta_t \gg \Delta_\tau = \varepsilon.$$

---

[109] See Sections 2.3.9.2 and 2.3.9.3.
[110] See Section 2.3.8.4.

Thus, the number density $\rho = N/\Delta_t$ of t-values is also infinite in the interval Δ_t, where the interval Δ_t can take any value but must be different from infinity. Hence, because $\rho = N/\Delta_t$ is infinite, the number of configurations $\Psi^*(\mathbf{r}, t)\Psi(\mathbf{r}, t) = \Psi^*(\mathbf{r}, \tau)\Psi(\mathbf{r}, \tau)$ in the small interval $\Delta_\tau = \varepsilon \ll \Delta_t$ also must be infinity. However, since $\Delta_\tau = \varepsilon$ is infinitesimal all these configurations $\Psi^*(\mathbf{r}, \tau)\Psi(\mathbf{r}, \tau)$ in $\Delta_\tau = \varepsilon$ are practically not different from each other. In other words, at the reference time τ (within $\Delta_\tau = \varepsilon$) only the information about one configuration is given ($\Psi^*(\mathbf{r}, \tau)\Psi(\mathbf{r}, \tau)$), and this restricted information is described by Schrödinger's equation (2.134) which we have derived above from the general equation (2.35). In contrast to Schrödinger's equation, equation (2.35) describes at the reference time τ (within $\Delta_\tau = \varepsilon$), the whole scenario expressed by $\Psi^*(\mathbf{r}, t)\Psi(\mathbf{r}, t)$ within the entire t-range of $-\infty < t < \infty$. In the next section, we will continue this discussion.

Space-time information

The system-specific time t has been eliminated in the procedure for the derivation of Schrödinger's equation[111] from the general equation (2.35). The reason is obvious: we only used the selected information about reality. In other words, Schrödinger's equation only formulates the selected case, that is, the configuration $\Psi^*(\mathbf{r}, t_0 = \tau)\Psi(\mathbf{r}, t_0 = \tau)$ which has been selected from $\Psi^*(\mathbf{r}, t)\Psi(\mathbf{r}, t)$.[112] All the other configurations for $t \neq \tau$ are neglected. We observe at time τ only a certain configuration belonging to t_0: $\Psi^*(\mathbf{r}, t_0)\Psi(\mathbf{r}, t_0)$ for $-\infty < \mathbf{r} < \infty$. However, projection theory makes at time τ a statement about the whole reality and not only about the selected information: Despite the statistical fluctuations, at time τ the whole of time t, that is, the past, present and future, is laid out frozen before us[113]:

$$\Psi^*(\mathbf{r}, t)\Psi(\mathbf{r}, t), \quad -\infty < \mathbf{r} < \infty, \quad -\infty < t < \infty.$$

This is valid for all times τ, that is, nothing is changed and we have always the same situation if the wave function $\Psi(\mathbf{r}, t)$, described by the

---

[111] See equation (2.134).
[112] See equations (2.102) and (2.103).
[113] See equation (2.102).

general equation (2.35), does not vary with time τ. In other words, in this case the configuration $\Psi^*(\mathbf{r}, t)\Psi(\mathbf{r}, t)$ remains constant when we go from τ_1 to τ_2 and so on:

$$\begin{aligned}\tau_1 &: \quad \Psi_1^*(\mathbf{r}, t)\Psi_1(\mathbf{r}, t); \quad -\infty < \mathbf{r} < \infty, -\infty < t < \infty \\ \tau_2 &: \quad \Psi_2^*(\mathbf{r}, t)\Psi_2(\mathbf{r}, t); \quad -\infty < \mathbf{r} < \infty, -\infty < t < \infty \\ &\vdots \qquad \vdots \qquad \qquad \vdots \end{aligned} \quad (2.135)$$

with

$$\Psi_1^*(\mathbf{r}, t)\Psi_1(\mathbf{r}, t) = \Psi_2^*(\mathbf{r}, t)\Psi_2(\mathbf{r}, t) = \cdots = \Psi^*(\mathbf{r}, t)\Psi(\mathbf{r}, t). \quad (2.136)$$

However, $\Psi^*(\mathbf{r}, t)\Psi(\mathbf{r}, t)$ is not fully realized in $(\mathbf{r}, t)$-space but only in $(\mathbf{p}, E)$-space in the form of $\Psi^*(\mathbf{p}, E)\Psi(\mathbf{p}, E)$. In $(\mathbf{r}, t)$-space only the selected configuration $\Psi^*(\mathbf{r}, \tau = t_k)\Psi(\mathbf{r}, \tau = t_k)$ is present at time τ.

Furthermore, we can state the following.[114] Each space-time point $\mathbf{r}, t$ contains, in a certain sense, the total information about the system. This is because we need, as a consequence of the Fourier transform (2.4), for the determination of $\Psi \mathbf{r}_k, t_k$ at space-time point $\mathbf{r}_k, t_k$ the whole information $\Psi(\mathbf{p}, E)$ about the system in $(\mathbf{p}, E)$-space. In principle, we could measure at any space-time point $\mathbf{r}_k, t_k$ the complete distribution $\Psi^*(\mathbf{p}, E)\Psi(\mathbf{p}, E)$ within a certain time interval $\Delta\tau$, and $\Delta\tau$ can be infinitesimal but has to be different from zero. If we would be able to determine from these data the wave function $\Psi(\mathbf{p}, E)$ we could, on the other hand, determine the complete wave function $\Psi(\mathbf{r}, t)$ for $-\infty \leq \mathbf{r} \leq \infty$, $-\infty \leq t \leq \infty$. That is, any space-time point $(\mathbf{r}_k, t_k)$ contains (within a certain infinitesimal time interval $\Delta\tau$) also the information of all the other space-time points for which $\Psi(\mathbf{r}, t)$ is defined. We come to the following scheme:

$$\Psi^*(\mathbf{r}_k, t_k)\Psi(\mathbf{r}_k, t_k) \to \Psi^*(\mathbf{p}, E)\Psi(\mathbf{p}, E)$$
$$\downarrow$$
$$\Psi(\mathbf{p}, E) \to \Psi(\mathbf{r}, t), \quad -\infty \leq \mathbf{r} \leq \infty,$$
$$-\infty \leq t \leq \infty.$$

---

[114] See also Section 2.3.6.

Let us analyze this point in greater detail[115]

1. At a certain space-time point, say $\mathbf{r}_k$ and t_k, the measurement of one of the possible values for $\mathbf{p}$ and for E is done in the intervals $\mathbf{r}_k, \mathbf{r}_k + \Delta \mathbf{r}_k$ and t_k, $t_k + \Delta t_k$ with the probability density of $\Psi^*(\mathbf{r}_k, t_k)\Psi(\mathbf{r}_k, t_k)$, where $\Delta V_k = \Delta \mathbf{r}_k \Delta t_k$ can be infinitesimal but different from zero.
2. Let $\Delta_\mathbf{r}, \Delta_t, \Delta_\mathbf{p}, \Delta_E$ be the ranges of $\mathbf{r}, t, \mathbf{p}$ and E for which the distributions $\Psi^*(\mathbf{r}, t)\Psi(\mathbf{r}, t)$ and $\Psi^*(\mathbf{p}, E)\Psi(\mathbf{p}, E)$ are not zero, and let us assume that the ranges $\Delta_\mathbf{r}, \Delta_t, \Delta_\mathbf{p}, \Delta_E$ are different from infinity. How many values $\mathbf{r}, t, \mathbf{p}$ and E can come into existence in the time interval $\Delta \tau = \varepsilon$, where ε is infinitesimal but different from zero? This number of values for each variable is identical with the number of τ-values in the interval ε, and this number is given by the number of all real numbers in ε. We know from mathematics that the number of real numbers within ε must be infinity, and in physics none of these real numbers are excluded. Therefore, within the infinitesimal interval ε, the number N of realized values of a variable (as, for example, $\mathbf{r}$) is also infinite in the interval $\Delta_\mathbf{r}$. Thus, the number density $\rho = N/\Delta_\mathbf{r}$ of $\mathbf{r}$-values is also infinite in the interval $\Delta_\mathbf{r}$, where the interval $\Delta_\mathbf{r}$ can take any value but must be different from infinity.
3. Thus, also within the interval $\Delta V_k = \Delta \mathbf{r}_k \Delta t_k$ in connection with $\Psi^*(\mathbf{r}_k, t_k)\Psi(\mathbf{r}_k, t_k)$, we have an infinite number of events because $\Delta_\mathbf{r} \Delta_t / \Delta V_k$ remains finite and is not infinity.
4. Conclusion: Within the infinitesimal interval $\Delta V_k = \Delta \mathbf{r}_k \Delta t_k$ around any space-time position, say $\mathbf{r}_k$ and t_k, the complete information about the system $\Psi^*(\mathbf{p}, E)\Psi(\mathbf{p}, E)$ with $-\infty < \mathbf{p} < \infty$, $-\infty < E < \infty$ is given on the basis of $\Psi^*(\mathbf{r}_k, t_k)\Psi(\mathbf{r}_k, t_k)$ to any degree of accuracy within the infinitesimal intervals $\Delta V_k = \Delta \mathbf{r}_k \Delta t_k$ and $\Delta \tau = \varepsilon$. If we are able to determine the wave function $\Psi(\mathbf{p}, E)$ from $\Psi^*(\mathbf{p}, E)\Psi(\mathbf{p}, E)$ (eventually, with the help of models) we could determine the complete wave function $\Psi(\mathbf{r}, t)$ for the complete region $-\infty \leq \mathbf{r} \leq \infty$, $-\infty \leq t \leq \infty$.

---

[115] See also Sections 2.3.8.3 and 2.3.8.4.

In summary, at any space-time point $\mathbf{r}_k, t_k$ the complete information about the system is determined. This is because projection theory is inherently non-local in character.[116] Within projection theory the entire information about a certain system (world) is enclosed in each space-time point $\mathbf{r}, t$. Such a statement cannot be made on the basis of Schrödinger's equation.[117] But what can we say about usual quantum theory?

Information in connection with usual quantum theory

Within usual quantum theory, Schrödinger's equation cannot be deduced, that is, the roots of this equation remain unknown, and it is assumed in usual quantum theory that this equation describes reality completely and not only certain selected configurations. Selection in the above sense is not defined in usual quantum theory but is exclusively a feature of projection theory. Therefore, it is not surprising that the property (2.135) with (2.136) does not hold in usual quantum theory. In connection with Schrödinger's equation[118] the space structure changes when we go from τ_1 to τ_2 and so on, and, instead of equations (2.135) with (2.136), we get

$$\begin{aligned} \tau_1: & \quad \Psi_1^*(\mathbf{r}, \tau_1)\Psi_1(\mathbf{r}, \tau_1); \quad -\infty \leq \mathbf{r} \leq \infty \\ \tau_2: & \quad \Psi_2^*(\mathbf{r}, \tau_2)\Psi_2(\mathbf{r}, \tau_2); \quad -\infty \leq \mathbf{r} \leq \infty, \\ & \quad \vdots \qquad\qquad \vdots \qquad\qquad \vdots \end{aligned} \quad (2.137)$$

with

$$\Psi_1^*(\mathbf{r}, \tau_1)\Psi_1(\mathbf{r}, \tau_1) \neq \Psi_2^*(\mathbf{r}, \tau_2)\Psi_2(\mathbf{r}, \tau_2) \neq \ldots \quad (2.138)$$

This is also valid for the energies. Within projection theory the energies in $(\mathbf{r}, E)$-space are formulated by[119]

$$E\Psi(\mathbf{r}, E) = -\frac{\hbar^2}{2m_0}\Delta\Psi(\mathbf{r}, E) + V\left(x, y, z, -i\hbar\frac{\partial}{\partial E}\right)\Psi(\mathbf{r}, E),$$

---

[116] See also Section 2.3.6.
[117] See equation (2.134).
[118] See equation (2.134).
[119] See equation (2.50).

and the energy distribution $\{E\}$ resulting from this equation is exactly the same for all times τ. In usual quantum theory we have only one space and this is the $(\mathbf{r}, \tau)$-space. The system-specific time t is not defined here. Reality (masses, energies etc.) are embedded in $(\mathbf{r}, \tau)$-space. It is well-known that in usual quantum theory the energies are determined on the basis on Schrödinger's equation by time-dependent perturbation theory, and, in contrast to projection theory, we obtain for each time τ a certain energy spectrum $\{E\}$ which are different from each other:

$$\tau_1 : \{E_1\}, \quad \tau_2 : \{E_2\}, \cdots \tag{2.139}$$

with

$$\{E_1\} \neq \{E_2\} \neq \cdots \tag{2.140}$$

This feature is different from that we obtain of projection theory. Here we have $\{E_1\} = \{E_2\} = \cdots$ instead of equation (2.140). The above equation (2.50) for the determination of $\Psi(\mathbf{r}, E)$ does not exist in usual quantum theory. In particular, there is no operator

$$\hat{t} = -i\hbar \frac{\partial}{\partial E},$$

for the time variable t defined in usual quantum theory.[120] Conventional quantum theory is only a restricted representation of quantum phenomena and this can lead to relevant changes in the results.

Summary

In conclusion, there is a big difference between usual quantum theory and projection theory. From the point of view of projection theory, the results of usual quantum theory are strongly restricted by the fact that the system-specific time t does not appear here.

The features discussed above are entirely due to the fact that Schrödinger's equation[121] is obtained if in the general equation (2.35) the system-specific time t is replaced by the reference time τ. The

---

[120] See Section 2.3.5 and equation (2.11).
[121] See equation (2.134).

τ-independent case of projection theory[122] appears in usual quantum theory as time-dependent process[123] with respect to time τ (t is not defined here).

2.3.9.8. Real situation

General remarks

The measurement of one of the possible values for **p** and for E is done in the space-time interval $\mathbf{r}, \mathbf{r} + \Delta\mathbf{r}$ and $t, t + \Delta t$ with the probability density of $\Psi^*(\mathbf{r}, t)\Psi(\mathbf{r}, t)$. One of the possible values for **p** and for E is given with the probability density $\Psi^*(\mathbf{p}, E)\Psi(\mathbf{p}, E)$. In this way we can determine the probability density $\Psi^*(\mathbf{p}, E)\Psi(\mathbf{p}, E)$ for all possible values **p** and E ($-\infty < \mathbf{p}, E < \infty$) at a certain space-time point $\mathbf{r}_k, t_k$ (within the space-time interval $\mathbf{r}_k, \mathbf{r}_k + \Delta\mathbf{r}_k$ and $t_k, t_k + \Delta t_k$). This is particularly valid for a certain selected configuration $\Psi^*(\mathbf{r}, t_k)\Psi(\mathbf{r}, t_k)$. We have discussed all these points in Sections 2.3.8.3 and 2.3.9.7.

Aspects

In other words, we measure for each selected space-configuration $\Psi^*(\mathbf{r}, \tau = t_k)\Psi(\mathbf{r}, \tau = t_k)$ the same energy spectrum $\{E\}$, and $\{E\}$ is independent on τ and is valid for each space-configuration $\Psi^*(\mathbf{r}, t_m)\Psi(\mathbf{r}, t_m)$, with $t_m \neq t_k$. This statement is true in connection with the τ-independent case of projection theory.[124]

Let *observer A* be the advocate of projection theory (the general equation (2.35) is valid), and let *observer B* be the advocate of usual quantum theory (Schrödinger's equation (2.134) is valid). We would like to suppose here that projection theory is true but not Schrödinger's equation[121] of usual quantum theory.

Both observers perform measurements for the determination of the energy spectra $\{E\}$ for various times τ (using identical measuring instruments). Result of the measurements: all energy spectra $\{E\}$ are exactly the same, that is, they are not dependent on time τ.

---

[122] See equations (2.135) and (2.136).
[123] See equations (2.137) and (2.138).
[124] See equations (2.135) and (2.136).

Observer A concludes that he is concerned with the τ-independent case of projection theory.[125] On the other hand, *observer B* believes that these measurements reflect the stationary case of usual quantum theory, that is, he misjudges the situation and assumes that the potential $V(x, y, z, \tau)$ in equation (2.134) is independent on time τ and is given by $V(x, y, z)$ leading to the stationary case of usual quantum theory

$$i\hbar\frac{\partial}{\partial \tau}\Psi(\mathbf{r},\tau) = -\frac{\hbar^2}{2m_0}\Delta\Psi(\mathbf{r},\tau) + V(x,y,z)\Psi(\mathbf{r},\tau) \qquad (2.141)$$

having the same energy spectrum $\{E\}$ for all times τ and this is in accordance with the measurements. Then, $V(x, y, z)$ can be chosen by *observer B* in such a way that it describes the measured energy spectrum $\{E\}$ sufficiently well.

However, this analysis is wrong from the point of view of projection theory. Here the energy distribution $\{E\}$ resulting from equation (2.50)[126] is exactly the same for all times τ and, therefore, all possible energies are given at time τ. This is of course also the case for the measurements. Therefore, from the point of view of *observer A*, equation (2.141) represents at time τ all possible energy spectra $\{E\}$ at once which can appear in accordance with equation (2.134) in the time interval $-\infty \leq \tau \leq \infty$. Therefore, equation (2.141) represents a compressed form from the point of view of projection theory, and the result of the experiments discussed above are explained by an effective potential $V_{eff}(x, y, z)$ and not by $V(x, y, z)$ and also not by equation (2.141) since the measurements have led to energy spectra $\{E\}$ which are independent on τ. Therefore, instead of equation (2.141), we get

$$i\hbar\frac{\partial}{\partial \tau}\Psi(\mathbf{r},\tau) = -\frac{\hbar^2}{2m_0}\Delta\Psi(\mathbf{r},\tau) + V_{eff}(x,y,z)\Psi(\mathbf{r},\tau). \qquad (2.142)$$

In conclusion, *observer B* believes that we have

$$V(x, y, z, \tau) \to V(x, y, z), \qquad (2.143)$$

---

[125] See also Sections 2.3.9.2 and 2.3.9.3.
[126] See also Section 2.3.9.3.

and *observer A* judges the situation in connection with Schrödinger's equation by

$$V(x,y,z,\tau) \to V_{eff}(x,y,z). \qquad (2.144)$$

We would like to illustrate the situation by a simple example. Let the potential $V(x,y,z)^{127}$ lead to an energy spectrum $\{E\}$ with two components which are exactly the same for all times τ, that is, we have $\tau_1 : E_{11}, E_{12}; \tau_2 : E_{11}, E_{12}; \ldots$ On the other hand, in the case of equation (2.134) these two energy values vary with time and we get $\tau_1 : E_{11}, E_{12}; \tau_2 : E_{21}, E_{22}; \ldots$ with $E_{11} \neq E_{21}$ and $E_{12} \neq E_{22}$. In the case of equations (2.144) and (2.142) we have the effective potential $V_{eff}(x,y,z)$ which reflects the compressed form of this τ-dependent situation, described by equation (2.134), and we obtain in the case of the effective potential $V_{eff}(x,y,z)$ for all times τ the τ-independent energy spectrum $\{E\}$ having the components $E_{11}, E_{12}, E_{21}, E_{22}, \ldots$, which consists not only of two components E_{11}, E_{12} as in the case of $V(x,y,z)$.

Clearly, for *observer A* Schrödinger' equation (2.134) is wrong and, therefore, also the transition (2.144) must be wrong from the point of view of projection theory. However, transition (2.144) explains how the situation is modelled within the frame of Schrödinger's equation (2.134). *Observer B* does not know projection theory and assumes that Schrödinger's equation (2.134) reflects the most basic level of reality. However, usual quantum theory feigns a reality which is quite different from that of projection theory.

2.3.9.9. τ-Dependent systems

If however $\Psi(\mathbf{r},t)$ varies with time τ the whole system (reality) changes with respect to the variables $\mathbf{r}$ and t. Let be $\Psi_1^*(\mathbf{r},t)\Psi_1(\mathbf{r},t)$ the configuration at time τ_1 and $\Psi_2^*(\mathbf{r},t)\Psi_2(\mathbf{r},t)$ the configuration at time τ_2, then the whole space-time structure ($-\infty \leq \mathbf{r} \leq \infty$, $-\infty \leq t \leq \infty$) is changed by the transition from τ_1 to τ_2. We immediately recognize that it gives no sense to transfer this situation to the Schrödinger case

---

[127] See equations (2.143) and (2.141).

because the dependence on τ in the case of Schrödinger's equation[128] only comes into play when $\Psi(\mathbf{r}, t)$ does not vary with respect to τ (see also the remarks above). However, since equation (2.142) describes stationary systems we can formally define non-stationary systems within the frame of Schrödinger's equation (2.142) by $V_{eff}(x, y, z, \tau)$, that is, $V_{eff}(x, y, z)$ in equation (2.142) is simply replaced by $V_{eff}(x, y, z, \tau)$. In other words, due to the shortcomings in usual quantum theory, we have to introduce a third transformation and instead of equation (2.144), we obtain

$$V(x, y, z, \tau) \to V_{eff}(x, y, z) \to V_{eff}(x, y, z, \tau). \qquad (2.145)$$

Again, all these kinds of transformations are fundamentally wrong from the point of view of projection theory.

Clearly, also in the case of the general equation (2.35) of projection theory, we can formally define a stationary case by $V(x, y, z, t) = V(x, y, z)$. However, we will outline in Chapter 3 (Appendix 3.D) that stationary systems (stationary with respect to time t) cannot exist within projection theory.

2.3.9.10. *Some additional remarks*

Equation (2.134)

$$i\hbar \frac{\partial}{\partial \tau} \Psi(\mathbf{r}, \tau) = -\frac{\hbar^2}{2m_0} \Delta \Psi(\mathbf{r}, \tau) + V(x, y, z, \tau) \Psi(\mathbf{r}, \tau)$$

represents a modified form of Schrödinger's equation (1.3) of usual quantum theory:

$$i\hbar \frac{\partial}{\partial \tau} \Psi(\mathbf{r}, \tau) = -\frac{\hbar^2}{2m_0} \Delta \Psi(\mathbf{r}, \tau) + U(x, y, z, \tau) \Psi(\mathbf{r}, \tau).$$

The difference between $U(x, y, z, \tau)$ and $V(x, y, z, \tau)$ does not influence the principal structure of equation (2.134) which has exactly the same structure as Schrödinger's equation of usual quantum theory,

---

[128] See equation (2.134).

and this is because the system-specific time t does not appear in this equation.

The starting point for the derivation of the general equation (2.35)

$$i\hbar \frac{\partial}{\partial t} \Psi(\mathbf{r}, t) = -\frac{\hbar^2}{2m_0} \Delta \Psi(\mathbf{r}, t) + V(x, y, z, t) \Psi(\mathbf{r}, t)$$

is the classical equation (2.20)

$$E = \frac{\mathbf{p}^2}{2m_0} + U(x, y, z),$$

and the corresponding equation of usual quantum theory is expressed by the stationary Schrödinger equation (2.42),

$$i\hbar \frac{\partial}{\partial \tau} \Psi(\mathbf{r}, \tau) = -\frac{\hbar^2}{2m_0} \Delta \Psi(\mathbf{r}, \tau) + U(x, y, z) \Psi(\mathbf{r}, \tau).$$

Equation (2.134) is based on equation (2.35) and not on $E = \mathbf{p}^2/2m + U(x, y, z)$.

From the non-stationary classical case,

$$E = \frac{\mathbf{p}^2}{2m_0} + U(x, y, z, \tau) \qquad (2.146)$$

follows the non-stationary Schrödinger equation (2.42) of usual quantum theory[129] which has exactly the same structure as equation (2.134), as already outlined above.

In the case of equation (2.146) our general equation (2.35), which is stationary with respect to τ, becomes non-stationary and in the case of a sufficiently sharp function for the reference time structure $\Psi^*_{ref}(\tau - t) \Psi_{ref}(\tau - t)$,[130] we get

$$i\hbar \frac{\partial}{\partial t} \psi(\mathbf{r}, t)_\tau = -\frac{\hbar^2}{2m_0} \Delta \psi(\mathbf{r}, t)_\tau + V(x, y, z, t)_\tau \psi(\mathbf{r}, t)_\tau, \qquad (2.147)$$

with $\psi(\mathbf{r}, t)_\tau \equiv \psi(\mathbf{r}, t, \tau)$ and $V(x, y, z, t)_\tau \equiv V(x, y, z, t, \tau)$.

---

[129] See also Chapter 1.
[130] See, in particular, Section 2.3.9.2.

2.3.9.11. *Uncertainty relation for time and energy*

The fact that there is no system-specific time t in usual quantum theory is the reason why there is no uncertainty relation for energy and time which would agree in its physical content with the uncertainty relation for the momentum **p** and position **r**.[131] As already discussed in Section 1.2, the significance of the well-known relation (1.17) is entirely different from that of equation (1.16). This difference is symbolically expressed by the use of Δ instead of δ.

It is easy to show[132] that within projection theory a relation for E and t (system-specific time) must exist in analogy to the uncertainty relation for the momentum **p** and position **r**.[131] Therefore, instead of[133]

$$\Delta E \Delta \tau \geq \frac{\hbar}{2},$$

we get in projection theory the relation,

$$\delta E \, \delta t \geq \frac{\hbar}{2}. \quad (2.148)$$

We measure at instant τ the values E and t of the system under investigation with a certain probability.[134] In other words, the quantities E and t are inherently uncertain.

2.3.9.12. *Time within special theory of relativity*

Block universe

John Wheeler's statement[135] "Time is nature's way of keeping everything from happening at once" is very close to that we obtained here[136] about the nature of time. Wheeler's statement is an interpretation of

---

[131] See equation (1.16).
[132] Interested readers might refer to Schommers, W., (2008) Advanced Science Letters **1**, 59.
[133] See conventional quantum theory, equation (1.17).
[134] See Scheme (2.101).
[135] Refer to Al-Khalili, J., (1999) Black Holes, Worm Holes & Time Machines: Institute of Physics.
[136] See Sections 2.3.9.1 to 2.3.9.10.

the situation given within Special Theory of Relativity. This is well-summarized by Jim Al-Khalili[135]

> *Minkowski's four-dimensional space-time is often referred to as the block universe model. Once time is treated like a fourth dimension of space we can imagine the whole of space and time modelled as a four-dimensional block.[...] Here we have a view of the totality of existence in which the whole of time — past, present and future — is laid out frozen before us. Many physicists, including Einstein later in his life, pushed this model to its logical conclusion: in four-dimensional space-time, nothing ever moves. All events which have ever happened or ever will happen exist together in the block universe and there is no distinction between past and future. This implies that nothing unexpected can ever happen. Not only is the future preordained but it is already out there and is as unalterably fixed as the past.[...] Is this picture really necessary? After all, we can just as easily imagine a Newtonian spacetime modelled as a four-dimensional block. The difference is that in that case space and time are independent of each other, whereas in relativity the two are linked. One of the consequences of relativity is that no two observers will be able to agree on when "now" is. By abandoning absolute time we must also admit that the notion of a universal present moment does not exist. For one observer, all events in the Universe that appear to be simultaneous can be linked together to form a certain cross-sectional slice through space-time which that observer calls "now". But another observer, moving relative to the first, will have a different slice that will cross the first. Some events that lie on the first observer's "now" slice will be in the second observer's past while others will be in his future. This mind-boggling result is known as the relativity of simultaneity, and is the reason why many physicists have argued that since there is no absolute division between past and future then there can be no passage of time, since we cannot agree on where the present should be.[...] Worse than that, if one observer sees an event A occur before an event B, then it is possible for another observer to witness B before A. If two observers cannot even agree on the order that*

things happen, how can we ever define an objective passage of time as a sequence of events?[137]

This behaviour is also reflected in $\Psi^*(\mathbf{r}, t)\Psi(\mathbf{r}, t)$ because the range of the time interval Δ_t is *statistically* occupied in the course of time τ and we cannot agree on where the present should be, just as in the case of the block universe of Special Theory of Relativity. Without the reference time structure $\Psi^*_{ref}(\tau - t)\Psi_{ref}(\tau - t)$,[138] which does not exist in Special Theory of Relativity, we would not be able to distinguish between past, present and future. Also the statement "in four-dimensional space-time nothing ever moves" is in accordance with that which we have pointed out above in connection with projection theory: $\Psi^*(\mathbf{r}, t)\Psi(\mathbf{r}, t)$ is a static function and does not change. The general equation (2.35) just describes this situation. (However, in principle $\Psi^*(\mathbf{r}, t)\Psi(\mathbf{r}, t)$ could also vary with time τ.[139]) Furthermore, the statement "Not only is the future preordained but is already out there and is as unalterably fixed in the past" is also in complete accordance with that we have outlined above about the nature of time.

There is however an essential difference between the result of Special Theory of Relativity and the investigation given here. While the block universe of Special Theory of Relativity is valid for the whole universe and cannot be influenced by the observer, $\Psi^*(\mathbf{r}, t)\Psi(\mathbf{r}, t)$ reflects only a certain system and can definitely be influenced by an observer (for example, to put the system at time τ_1 into another environment). Such an influence means that the interaction between the system and its environment is changed and this is described by a new wave function, say $\Phi(\mathbf{r}, t)$. In other words, instead of $\Psi^*(\mathbf{r}, t)\Psi(\mathbf{r}, t)$ we have now $\Phi^*(\mathbf{r}, t)\Phi(\mathbf{r}, t)$ for times $\tau > \tau_1$.

As in the case of $\Psi^*(\mathbf{r}, t)\Psi(\mathbf{r}, t)$, also in connection with $\Phi^*(\mathbf{r}, t)\Phi(\mathbf{r}, t)$, the following is valid. For certain times $\tau > \tau_1$, only one value of the system-specific time is realized with a certain probability. However, if we consider an infinitesimal time interval $\Delta \tau = \varepsilon$ an infinite

---

[137] Refer to Al-Khalili, J., (1999) Black Holes, Worm Holes & Time Machines: Institute of Physics.
[138] See Section 2.3.9.2.
[139] See Section 2.3.9.10.

number of t-values are occupied. In other words, the whole history (the complete past and future) of the system, described by the range Δ_t (life-time) of the distribution $\Phi^*(\mathbf{r},t)\Phi(\mathbf{r},t)$, is given within the infinitesimal time interval $\Delta\tau = \varepsilon$ of our clocks. This is the case for any $\tau_i \pm \Delta\tau$, that is, the law $\Phi^*(\mathbf{r},t)\Phi(\mathbf{r},t)$ is independent of $\tau > \tau_1$. It is stationary with respect to time $\tau > \tau_1$.[140] Thus, also here we may say the following. Despite the statistical fluctuations, the whole of time t — its past, present and future — is laid out frozen before us. But the past and the future of $\Phi^*(\mathbf{r},t)\Phi(\mathbf{r},t)$ have nothing to do the past and the future of $\Psi^*(\mathbf{r},t)\Psi(\mathbf{r},t)$.

Feynman diagrams

The need of a reference time also within the block universe of Special Theory of Relativity has been discussed very clearly by Kenneth Ford within the frame of elementary particle physics. Here certain particles can go backward in time and we obtain time-reversed trajectories. Ford wrote:

> *Unfortunately, no one knows how to test the theory of time-reversed trajectories experimentally. It must be accepted (if at all) for the symmetry it introduces into our picture of the world and for the simplicity it introduces into the description of antiparticles. To see that the direction of motion through time cannot be measured, imagine yourself a sub-microscopic observer of the scene depicted in Figure 2.11. If you put a ruler horizontally across the bottom of the diagram and then push it slowly upward across the diagram, the intersections of the world line with the moving ruler edge give a rough history of your observations. The point is that it is the ruler edge of your observation which is moving in time, not the world lines themselves. The particle world lines may be regarded as perfectly static, simply there, painted in space-time like lines on a map. To a creature capable of comprehending the whole span of time as we comprehend the span of space, the activity of annihilation and creation represented in this diagram*

---

[140] See also Section 2.2.9.7.

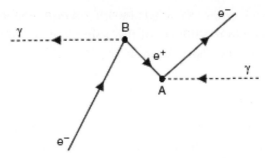

Fig. 2.11. Feynman diagram for photon-electron scattering. In connection with this figure Kenneth Ford wrote: *According to the normal view of time unrolling one direction, we start at the bottom of the diagram and read up. First, an electron and a photon are approaching each other. At vertex A, the photon creates an electron–positron pair. The new electron flies away, while the positron collides with the first electron at vertex B. There they undergo mutual annihilation and a new photon is born. The alternative view, which Feynman showed to be also consistent, is to picture the first electron proceeding to point B, where it emits a photon and reverses its path through time. It "then" travels to point A where it absorbs the incoming photon and once again reverses its course through time, flying off in the "right" direction. Either view is permissible and logically consistent.*[141]

> is not activity at all. It is a stationary display, a picture painted in space and time. It is the fact that man, the observer, can comprehend only one instant of time at each moment that converts the stationary display into motion and activity. At one time, the human observer sees an electron in one place. At a later time, he sees it at a different place. It is natural for him to believe that the electron, like the man, has moved forward in time to get from the first place to the second place. But there is no essential reason to believe this. We know only that the electron world line has traced out a certain path in space-time, but we have no possible way of knowing in what order the points of that path were traced out, nor even whether it makes sense to speak of an order or a direction of tracing the world line.[142]

Here the role of the distribution $\Psi^*(\mathbf{r}, t)\Psi(\mathbf{r}, t)$ must be identified with the role of the Feynman diagram.[143] Also the Feynman diagram

---

[141] Refer to Ford, K.W., (1963) Elementary particles: Blaisdell Publishing Company.
[142] Refer to Ford, K.W., (1963) Elementary particles: Blaisdell Publishing Company.
[143] See Figure 2.11.

should be considered as static situation in analogy to the distribution $\Psi^*(\mathbf{r},t)\Psi(\mathbf{r},t)$. On the other hand, the ruler edge exactly plays the role of the reference time τ described by $\Psi^*_{ref}(\tau-t)\Psi_{ref}(\tau-t)$.

2.4. Summary

Within the memory of man, all essential things are represented within the frame of pictures. This is the most fundamental statement within projection theory. These pictures are direct impressions from basic reality (outside world) which is principally not accessible to human observers. In particular, the pictures represent only a restricted part of basic reality. Only that information is observable which is useful for a human, and that what is useful is dictated by the biological evolution.

The descriptions of these fundamental pictures are done on the basis of fictitious realities which are constructed on the space-time elements $\mathbf{r}$ and t. The variables of fictitious realities are the momentum $\mathbf{p}$ and the energy E. Thus, we have two spaces in projection theory: $(\mathbf{r},t)$-space and $(\mathbf{p},E)$-space. The physical processes take place in $(\mathbf{p},E)$-space and this information is projected onto $(\mathbf{r},t)$-space. $(\mathbf{r},t)$-space does not contain real material objects but only geometrical structures. In particular, $(\mathbf{r},t)$-space cannot be an element of the outside world but is exclusively positioned in the brain of the observer and appears simultaneously with the objects (or more precisely with the geometrical structures of them). This viewpoint is close to that which the philosopher Immanuel Kant (1724–1804) proposed.

This so-called projection principle is compatible with the fact the elements x,y,z,t of space and time can in principle not be observed and, therefore, we have to assume that they do not exist as specific physical objects in the outside world. We came to the conclusion that the material objects which we observe in everyday life are not embedded in space (space-time). Space and time have to be considered as auxiliary elements for the representation of the selected information about the outside world (basic reality). The experiments with other biological systems support the view that these pictures must be species-dependent. In other words, there are projections onto space

(space-time). The projection principle introduced in connection with $(\mathbf{r}, t)$-space and $(\mathbf{p}, E)$-space is compatible with this general finding.

Both spaces, i.e., $(\mathbf{r}, t)$-space and $(\mathbf{p}, E)$-space, are connected by a Fourier transform, and we were able to deduce the basic equations for the determination of the wave functions: $\Psi(\mathbf{p}, E)$ for $(\mathbf{p}, E)$-space and $\Psi(\mathbf{r}, t)$ for $(\mathbf{r}, t)$-space. It turns out that the functions $\Psi^{\bullet}(\mathbf{r}, t)\Psi(\mathbf{r}, t)$ and $\Psi^{\bullet}(\mathbf{p}, E)\Psi(\mathbf{p}, E)$ have to be interpreted as probability densities. Within conventional quantum theory, it is not possible to deduce the basic equation (Schrödinger's equation) for the determination of the wave function. As in classical physics, within usual quantum theory everything is embedded in space (space-time).

Furthermore, projection theory is able to extend conventional quantum theory. The formalism leads in a most natural way to the existence of a system-specific time which is not known in usual quantum theory. Within usual quantum theory, the time is still an external parameter as in classical mechanics and has nothing to do with the system under investigation.

3

Free, Non-interacting Systems (Particles)

3.1. General Remarks

In the treatment of free systems (particles) we would like to start with the general equations with non-vanishing interaction. The analysis will exclusively be done within $(\mathbf{r}, t)$-space and $(\mathbf{p}, E)$-space. Then, the following two equations are of relevance[1]:

$(\mathbf{r}, t)$-space

$$i\hbar \frac{\partial}{\partial t} \Psi(\mathbf{r}, t) = -\frac{\hbar^2}{2m_0} \Delta \Psi(\mathbf{r}, t) + V(x, y, z, t) \Psi(\mathbf{r}, t). \quad (3.1)$$

$(\mathbf{p}, E)$-space

$$E\Psi(\mathbf{p}, E) = \frac{\mathbf{p}^2}{2m_0} \Psi(\mathbf{p}, E)$$
$$+ V\left(i\hbar \frac{\partial}{\partial p_x}, i\hbar \frac{\partial}{\partial p_y}, i\hbar \frac{\partial}{\partial p_z}, -i\hbar \frac{\partial}{\partial E}\right) \Psi(\mathbf{p}, E). \quad (3.2)$$

Again, both spaces are equivalent and, therefore, also equations (3.1) and (3.2) are equivalent concerning their information content. Free, non-interacting systems are defined on the basis of equations (3.1) and (3.2) if the interaction terms are zero, that is, if we have for $(\mathbf{r}, t)$-space:

$$V(x, y, z, t) \Psi(\mathbf{r}, t) = 0, \quad (3.3)$$

---

[1] See Section 2.3.7.

and for $(\mathbf{p}, E)$-space:

$$V\left(i\hbar\frac{\partial}{\partial p_x}, i\hbar\frac{\partial}{\partial p_y}, i\hbar\frac{\partial}{\partial p_z}, -i\hbar\frac{\partial}{\partial E}\right)\Psi(\mathbf{p}, E) = 0. \qquad (3.4)$$

Then, the equations for the determination of the wave functions for $(\mathbf{r}, t)$-space as well as for $(\mathbf{p}, E)$-space are given by
$(\mathbf{r}, t)$-space:

$$i\hbar\frac{\partial}{\partial t}\Psi(\mathbf{r}, t) = -\frac{\hbar^2}{2m_0}\Delta\Psi(\mathbf{r}, t). \qquad (3.5)$$

$(\mathbf{p}, E)$-space:

$$E\Psi(\mathbf{p}, E) = \frac{\mathbf{p}^2}{2m_0}\Psi(\mathbf{p}, E). \qquad (3.6)$$

In the case of a free system, which does not interact with other systems, $\mathbf{p}, E$-fluctuations are zero:

$$\Delta\mathbf{p} = 0, \ \Delta E = 0, \qquad (3.7)$$

and the values $\mathbf{p}$ and E remain constant. We would like to call these constant values $\mathbf{p}_0$ and E_0. Then, we have

$$\Psi(\mathbf{p}, E) \to \Psi(\mathbf{p}_0, E_0), \qquad (3.8)$$

and equation (3.6) takes the form

$$E_0\Psi(\mathbf{p}_0, E_0) = \frac{\mathbf{p}_0^2}{2m_0}\Psi(\mathbf{p}_0, E_0). \qquad (3.9)$$

Mathematically, for $\Psi(\mathbf{p}_0, E_0) = \infty$ and $\Psi(\mathbf{p}_0, E_0) = 0$, the momentum $\mathbf{p}_0$ and the energy E_0 may take values independent from each other, i.e., $\mathbf{p}_0$ and E_0 can be chosen arbitrarily in this case, and in general we have

$$E_0 \neq \frac{\mathbf{p}_0^2}{2m_0}. \qquad (3.10)$$

This is in principle possible but we will see below that equation (3.10) leads to contradictions. Only in the case of

$0 < \Psi(\mathbf{p}_0, E_0) < \infty$, we have

$$E_0 = \frac{\mathbf{p}_0^2}{2m_0}, \tag{3.11}$$

as in the classical case. For a more detailed discussion we need to investigate the behaviour of the system in both spaces, $(\mathbf{r}, t)$-space and $(\mathbf{p}, E)$-space, simultaneously.

Remark

The case $\Psi(\mathbf{p}_0, E_0) = 0$ is possible if the values for $\mathbf{p}_0$ and E_0 are simultaneously zero: $\mathbf{p}_0 = 0$ and $E_0 = 0$, i.e., $\mathbf{p}_0$ and E_0 can no longer be chosen arbitrarily in this specific case. We will discuss this feature below and will recognize that the case $\Psi(\mathbf{p}_0, E_0) = 0$ in connection with $\mathbf{p}_0 = 0$ and $E_0 = 0$ means non-existence of the free system. The situation for free systems within usual quantum theory is discussed in Appendix 3.A.

3.2. The Behaviour of the Basic Equations

When we use equation (2.20) as starting point for $U(x, y, z) = 0$, the simultaneous treatment of the equations (3.5) and (3.6) for the determination of $\Psi(\mathbf{r}, t)$ and $\Psi(\mathbf{p}, E)$ leads to the following results. The general relationship between these functions is[2]

$$\Psi(\mathbf{r}, t) = \frac{1}{(2\pi\hbar)^2} \int_{-\infty}^{\infty} \Psi(\mathbf{p}, E) \exp\left\{i\left[\frac{\mathbf{p}}{\hbar} \cdot \mathbf{r} - \frac{E}{\hbar}t\right]\right\} d\mathbf{p}\, dE. \tag{3.12}$$

When we apply the procedure introduced in Section 2.3.7, we get

$$i\hbar \frac{\partial}{\partial t} \Psi(\mathbf{r}, t) = \frac{1}{(2\pi\hbar)^2} \int_{-\infty}^{\infty} E\Psi(\mathbf{p}, E) \exp\left\{i\left[\frac{\mathbf{p}}{\hbar} \cdot \mathbf{r} - \frac{E}{\hbar}t\right]\right\} d\mathbf{p}\, dE, \tag{3.13}$$

---

[2] See Section 2.3.4.3.

and

$$-\frac{\hbar^2}{2m_0}\Delta\Psi(\mathbf{r},t) = \frac{1}{(2\pi\hbar)^2}\int_{-\infty}^{\infty}\frac{\mathbf{p}^2}{2m_0}\Psi(\mathbf{p},E)$$
$$\times \exp\left\{i\left[\frac{\mathbf{p}}{\hbar}\cdot\mathbf{r} - \frac{E}{\hbar}t\right]\right\}d\mathbf{p}\,dE. \quad (3.14)$$

This directly leads to

$$\left\{i\hbar\frac{\partial}{\partial t} + \frac{\hbar^2}{2m_0}\Delta\right\}\Psi(\mathbf{r},t) = \frac{1}{(2\pi\hbar)^2}\int_{-\infty}^{\infty}\left\{E - \frac{\mathbf{p}^2}{2m_0}\right\}\Psi(\mathbf{p},E)$$
$$\times \exp\left[\frac{i}{\hbar}(\mathbf{p}\cdot\mathbf{r} - Et)\right]d\mathbf{p}\,dE, \quad (3.15)$$

with

$$\left\{i\hbar\frac{\partial}{\partial t} + \frac{\hbar^2}{2m_0}\Delta\right\}\Psi(\mathbf{r},t) = f(\mathbf{r},t)_{free}, \quad (3.16)$$

and

$$\left\{E - \frac{\mathbf{p}^2}{2m_0}\right\}\Psi(\mathbf{p},E) = f(\mathbf{p},E)_{free}, \quad (3.17)$$

where

$$f(\mathbf{r},t)_{free} = \frac{1}{(2\pi\hbar)^2}\int_{-\infty}^{\infty}f(\mathbf{p},E)_{free}\exp\left[\frac{i}{\hbar}(\mathbf{p}\cdot\mathbf{r} - Et)\right]d\mathbf{p}\,dE. \quad (3.18)$$

Since the values $\mathbf{p}$ and E remain constant, we would like to call again these constant values $\mathbf{p}_0$ and E_0, as in Section 3.1: $\mathbf{p}, E \to \mathbf{p}_0, E_0$. Then we get

$$\left\{E_0 - \frac{\mathbf{p}_0^2}{2m_0}\right\}\Psi(\mathbf{p}_0,E_0) = f(\mathbf{p}_0,E_0)_{free}, \quad (3.19)$$

and instead of equation (3.18)

$$
\begin{aligned}
f(\mathbf{r}, t)_{free} &= \frac{1}{(2\pi\hbar)^2} \int_{-\infty}^{\infty} f(\mathbf{p}, E)_{free} \exp\left[\frac{i}{\hbar}(\mathbf{p}\cdot\mathbf{r} - Et)\right] d\mathbf{p}\, dE \\
&= \frac{1}{(2\pi\hbar)^2} f(\mathbf{p}_0, E_0)_{free} \exp\left[\frac{i}{\hbar}(\mathbf{p}_0\cdot\mathbf{r} - E_0 t)\right] \\
&\quad \times \lim_{\Delta\mathbf{p},\Delta E \to 0} \int_{\mathbf{p}_0, E_0}^{\mathbf{p}_0+\Delta\mathbf{p}, E+\Delta E} d\mathbf{p}\, dE \\
&= \frac{1}{(2\pi\hbar)^2} \left\{ E_0 - \frac{\mathbf{p}_0^2}{2m_0} \right\} \Psi(\mathbf{p}_0, E_0) \exp\left[\frac{i}{\hbar}(\mathbf{p}_0\cdot\mathbf{r} - E_0 t)\right] \\
&\quad \times \lim_{\Delta\mathbf{p},\Delta E \to 0} \int_{\mathbf{p}_0, E_0}^{\mathbf{p}_0+\Delta\mathbf{p}, E+\Delta E} d\mathbf{p}\, dE, \quad (3.20)
\end{aligned}
$$

with

$$
\lim_{\Delta\mathbf{p},\Delta E \to 0} \int_{\mathbf{p}_0, E_0}^{\mathbf{p}_0+\Delta\mathbf{p}, E+\Delta E} d\mathbf{p}\, dE = 0. \quad (3.21)
$$

Let us now discuss several cases in order to find a relationship between $\mathbf{p}_0$ and E_0 but also to be able to make definite statements about possible values for $\Psi(\mathbf{p}_0, E_0)$.

3.2.1. The Case $f(\mathbf{p}_0, E_0)_{free} = \infty$

On the basis of equation (3.19), the following relations must be valid in the case of $f(\mathbf{p}_0, E_0)_{free} = \infty$: $\Psi(\mathbf{p}_0, E_0) = \infty$ and in general we have $E_0 \neq \mathbf{p}_0^2/2m_0{}^3$ with $\mathbf{p}_0 \neq \infty$ and $E_0 \neq \infty$.[4] With equation (3.21), we obtain for $f(\mathbf{r}, t)_{free}$ an uncertain relation:

$$
f(\mathbf{r}, t)_{free} = \infty \cdot 0, \quad (3.22)
$$

for all $\mathbf{r}$ and t. Then, equation (3.16) is no longer defined but becomes uncertain as well. Thus, we have to exclude the case $f(\mathbf{p}_0, E_0)_{free} = \infty$. In particular, equation (3.22) would be in contradiction to equation (3.5).

---
[3] See equation (3.10).
[4] In nature, the possible values for $\mathbf{p}_0$ and E_0 must be finite.

Equation (3.22) reflects the situation with respect to $(\mathbf{r}, t)$-space. But what about the situation in $(\mathbf{p}, E)$-space? Since both spaces are equivalent we should obtain a similar behaviour in $(\mathbf{p}, E)$-space. The following discussion confirms that.

Due to $\Psi(\mathbf{p}_0, E_0) = \infty$, the values for $\mathbf{p}_0$ and E_0 may take values that are independent of each other, so that we have $E_0 \neq \mathbf{p}_0^2/2m_0$, that is, $\mathbf{p}_0$ and E_0 can be chosen arbitrarily and must not fulfil the equation $E = \mathbf{p}^2/2m_0$. Thus, we have

$$\Psi(\mathbf{p}_0, E_0) = \infty,$$
$$E_0 \neq \mathbf{p}_0^2/2m_0, \tag{3.23}$$

in the case of $f(\mathbf{p}_0, E_0)_{free} = \infty$, as we have already quoted above.

We may also argue as follows. If there is a relationship or connection between $\mathbf{p}_0$ and E_0, this connection must be uncertain or undefined. It can be expressed formally by

$$E_0 = f_1(\mathbf{p}_0) \neq \frac{\mathbf{p}_0^2}{2m_0},$$
$$E_0 = f_2(\mathbf{p}_0) \neq \frac{\mathbf{p}_0^2}{2m_0}, \tag{3.24}$$
$$\vdots$$

with

$$f_1(\mathbf{p}_0) \neq f_2(\mathbf{p}_0) \neq \cdots \tag{3.25}$$

In conclusion, the analysis in $(\mathbf{p}, E)$-space also leads to an undefined situation. Therefore, also from this point of view, we have to exclude the case $f(\mathbf{p}_0, E_0)_{free} = \infty$ leading to $\Psi(\mathbf{p}_0, E_0) = \infty$.

3.2.2. On the Relationship Between $\mathbf{p}_0$ and E_0

In the case of a classical system, there are by definition no $\mathbf{p}, E$-fluctuations, i.e., we have $\Delta \mathbf{p} = 0$ and $\Delta E = 0$. This automatically means that the values for $\mathbf{p}$ and E are not uncertain and take at each time τ constant values which we would like to designate $\mathbf{p}_0$ and E_0, as

Free, Non-interacting Systems (Particles)

we did for the free quantum-theoretical system. Thus, for a free classical system we get with[5]

$$E = \frac{\mathbf{p}^2}{2m_0} + U(x,y,z), \tag{3.26}$$

and for $U(\mathbf{r}) = 0$ and $\mathbf{p} = \mathbf{p}_0$ and $E = E_0$ the expression

$$\left\{ E_0 = \frac{\mathbf{p}_0^2}{2m_0} \right\}_{classical\ system}. \tag{3.27}$$

Since the variables $\mathbf{p}$ and E of our free quantum-theoretical system also do not fluctuate we have, as in the classical case, $\Delta \mathbf{p} = 0$ and $\Delta E = 0$. Also, here $\mathbf{p}$ and E are not uncertain but take at each time τ the constant values $\mathbf{p}_0$ and E_0. In other words, the momentum and the energy of a free quantum-theoretical system behave strictly classically, and the relationship between $\mathbf{p}_0$ and E_0 must be identical with equation (3.27):

$$\left\{ E_0 = \frac{\mathbf{p}_0^2}{2m_0} \right\}_{classical\ system} = \left\{ E_0 = \frac{\mathbf{p}_0^2}{2m_0} \right\}_{free\ quantum-theoretical\ system}, \tag{3.28}$$

and we may write for the free quantum-theoretical system

$$E_0 = \frac{\mathbf{p}_0^2}{2m_0}. \tag{3.29}$$

In the case of $\Psi(\mathbf{p}_0, E_0) = \infty$,[6] we do not have $E_0 = \mathbf{p}_0^2/2m_0$ but the relationship between $\mathbf{p}_0$ and E_0 becomes uncertain, i.e., $\mathbf{p}_0$ and E_0 may take values independent of each other. (Note that the values for $\mathbf{p}_0$ and E_0 are *not* uncertain but the relationship between $\mathbf{p}_0$ and E_0 is.) As pointed out in Section 3.2.1, we may have functions $E_0 = f_1(\mathbf{p}_0)$, $E_0 = f_2(\mathbf{p}_0)$, etc.[7] in the case of $\Psi(\mathbf{p}_0, E_0) = \infty$, and these functions are in general different from $E_0 = \mathbf{p}_0^2/2m_0$. For example,

---

[5] See Section 2.3.7.1.
[6] See Section 3.2.1, equation (3.23).
[7] See equation (3.24).

the values of $\mathbf{p}_0$ and E_0 might lead to $E_0 = 5\mathbf{p}_0^2/2m_0$ and we could define $f_1(\mathbf{p}_0)$ by $f_1(\mathbf{p}_0) = 5\mathbf{p}_0^2/2m_0$. In other words, if the accidental quantum-theoretical solution for $\mathbf{p}_0$ and E_0 lead to $E_0 = 5\mathbf{p}_0^2/2m_0$, we have no possibility to reach the correct relation $E_0 = \mathbf{p}_0^2/2m_0$, since both values for $\mathbf{p}_0$ and E_0 remain constant after their creation and, therefore, relation $E_0 = 5\mathbf{p}_0^2/2m_0$ remains unchanged. There is no hidden mechanism defined that could transform $E_0 = 5\mathbf{p}_0^2/2m_0$ to $E_0 = \mathbf{p}_0^2/2m_0$. Therefore, relation $E_0 = 5\mathbf{p}_0^2/2m_0$ must be wrong. Only equation (3.29) can be correct.

In conclusion, the situation described by equation (3.24) would lead to a contradiction because our basic initial condition would be changed. For example, in the case of $E_0 = f_1(\mathbf{p}_0) \neq \mathbf{p}_0^2/2m_0$ instead of [8]

$$E_0 = \frac{\mathbf{p}_0^2}{2m_0} + U(x,y,z), \quad E = \frac{\mathbf{p}^2}{2m_0} + U(x,y,z), \qquad (3.30)$$

we would have

$$E_0 = f_1(\mathbf{p}_0) + U(\mathbf{r}), \quad E = f_1(\mathbf{p}) + U(x,y,z), \qquad (3.31)$$

leading to

$$i\hbar \frac{\partial}{\partial t} \Psi(\mathbf{r},t) = f_1 \left\{ -i\hbar \frac{\partial}{\partial x}, -i\hbar \frac{\partial}{\partial y}, -i\hbar \frac{\partial}{\partial z} \right\} \Psi(\mathbf{r},t)$$

$$\neq -\frac{\hbar^2}{2m_0} \Delta \Psi(\mathbf{r},t), \qquad (3.32)$$

$$E_0 \Psi(\mathbf{p}_0, E_0) = f_1(\mathbf{p}_0) \Psi(\mathbf{p}_0, E_0) \neq \frac{\mathbf{p}_0^2}{2m_0} \Psi(\mathbf{p}_0, E_0), \qquad (3.33)$$

or

$$E \Psi(\mathbf{p}, E) = f_1(\mathbf{p}) \Psi(\mathbf{p}, E) \neq \frac{\mathbf{p}^2}{2m_0} \Psi(\mathbf{p}, E). \qquad (3.34)$$

This is however in contradiction to the initial condition (3.26) on which the equations (3.5) and (3.6) are based. Also from this point of view, the case $\Psi(\mathbf{p}_0, E_0) = \infty$ has to be excluded because it violates our input condition (3.26).

---

[8] See equation (3.26).

3.2.3. The Case $f(\mathbf{p}_0, E_0)_{free} \neq \infty$

Using equation (3.19), we obtain with $f(\mathbf{p}_0, E_0)_{free} \neq \infty$

$$\left\{ E_0 - \frac{\mathbf{p}_0^2}{2m_0} \right\} \Psi(\mathbf{p}_0, E_0) \neq \infty. \tag{3.35}$$

Since we always have $\mathbf{p}_0 \neq \infty$ and $E_0 \neq \infty$, in nature the term

$$\left\{ E_0 - \frac{\mathbf{p}_0^2}{2m_0} \right\}$$

remains finite but we may have

$$E_0 - \frac{\mathbf{p}_0^2}{2m_0} \neq 0. \tag{3.36}$$

Then, the function $\Psi(\mathbf{p}_0, E_0)$ may not be infinite, that is, we must have

$$\Psi(\mathbf{p}_0, E_0) \neq \infty, \tag{3.37}$$

and we get with equation (3.20)

$$f(\mathbf{r}, t)_{free} = 0. \tag{3.38}$$

In other words, equation (3.5) is fulfilled now.

In the case of equation (3.35), we also have to exclude $\Psi(\mathbf{p}_0, E_0) = 0$, otherwise we have $E_0 \neq \mathbf{p}_0^2/2m_0$ and the momentum $\mathbf{p}_0$ and the energy E_0 may take values independent from each other, i.e., $\mathbf{p}_0$ and E_0 can be chosen arbitrarily as in the case of $\Psi(\mathbf{p}_0, E_0) = \infty$. However, $\Psi(\mathbf{p}_0, E_0) = 0$ is a trivial condition and makes no sense for $\mathbf{p}_0 \neq 0$ and $E_0 \neq 0$ but is only possible if $\mathbf{p}_0$ and E_0 are simultaneously zero: $\mathbf{p}_0 = 0$ and $E_0 = 0$, i.e., $\mathbf{p}_0$ and E_0 can no longer be chosen arbitrarily in this specific case. We have already mentioned this point above and we will give more details below.

What about the function $f(\mathbf{p}_0, E_0)_{free}$? What values can this function take for $\mathbf{p} = \mathbf{p}_0$ and $E = E_0$? Let us discuss this point by means of equation (3.19):

$$\left\{E_0 - \frac{\mathbf{p}_0^2}{2m_0}\right\} \Psi(\mathbf{p}_0, E_0) = f(\mathbf{p}_0, E_0)_{free} \neq \infty. \quad (3.39)$$

With

$$g(\mathbf{p}_0, E_0) = \frac{f(\mathbf{p}_0, E_0)_{free}}{\Psi(\mathbf{p}_0, E_0)}, \quad (3.40)$$

and

$$\Psi(\mathbf{p}_0, E_0) \neq 0, \quad (3.41)$$

we may reformulate equation (3.39) and we get

$$\left\{E_0 - \frac{\mathbf{p}_0^2}{2m_0}\right\} = g(\mathbf{p}_0, E_0). \quad (3.42)$$

$\Psi(\mathbf{p}_0, E_0) \neq 0$ however makes sense only if $\mathbf{p}_0 \neq 0$ and $E_0 \neq 0$.
Then, we obtain in analogy to equation (3.24):

$$E_0 = f_a(\mathbf{p}_0, E_0), \quad (3.43)$$

with

$$f_a(\mathbf{p}_0, E_0) = g(\mathbf{p}_0, E_0) + \frac{\mathbf{p}_0^2}{2m_0}. \quad (3.44)$$

Therefore, the same physical arguments must hold as we applied in Section 3.2.2. Since the variables $\mathbf{p}$ and E do not fluctuate in the case of a free system, we have $\Delta\mathbf{p} = 0$ and $\Delta E = 0$, as in the classical case. The values $\mathbf{p} = \mathbf{p}_0$ and $E = E_0$ are not uncertain but remain constant in the course of time τ. In other words, the quantities $\mathbf{p}_0$ and E_0 of our free quantum-theoretical system behave strictly classically, and the relationship between $\mathbf{p}_0$ and E_0 must be identical with equation (3.29): $E_0 = \mathbf{p}_0^2/2m_0$. The violation of the initial condition (3.26) would be a

Free, Non-interacting Systems (Particles)

contradiction. Then, we must have[9]

$$f(\mathbf{p}_0, E_0)_{free} = 0, \qquad (3.45)$$

and because of $\Psi(\mathbf{p}_0, E_0) \neq \infty$,[10] we have $E_0 = \mathbf{p}_0^2/2m_0$, that is, equation (3.29) is fulfilled. In conclusion, we have not only $f(\mathbf{p}_0, E_0)_{free} \neq \infty$ but due to the behaviour of $\mathbf{p}_0$ and E_0, we must have $f(\mathbf{p}_0, E_0)_{free} = 0$.

The result given by equation (3.45) can also be obtained for $\mathbf{p}_0 \neq 0$ and $E_0 \neq 0$ by applying the inverse transformation of equation (3.18) which is expressed by

$$f(\mathbf{p}, E)_{free} = \frac{1}{(2\pi\hbar)^2} \int_{-\infty}^{\infty} f(\mathbf{r}, t)_{free} \exp\left[-\frac{i}{\hbar}(\mathbf{p}\cdot\mathbf{r} - Et)\right] d\mathbf{r}\, dt. \qquad (3.46)$$

Because the function $f(\mathbf{r}, t)_{free}$ must always be zero, we get with equation (3.46) $f(\mathbf{p}_0, E_0)_{free} = 0$ for $\mathbf{p} = \mathbf{p}_0 \neq 0$ and $E = E_0 \neq 0$.[11] However, also in the case of $\mathbf{p} = \mathbf{p}_0 = 0$ and $E = E_0 = 0$ the function $f(\mathbf{p}, E)_{free}$ must be zero because we have to consider the following point: Both spaces, i.e. $(\mathbf{r}, t)$-space and $(\mathbf{p}, E)$-space, are equivalent and, therefore, the interaction situation should exactly be the same for both spaces. As we have outlined in Chapter 2, functions line $f(\mathbf{r}, t)_{free}$ and $f(\mathbf{p}_0, E_0)_{free}$ play the role of interaction terms. Since equation (3.38) is valid, we have $f(\mathbf{r}, t)_{free} = 0$ in $(\mathbf{r}, t)$-space and, therefore, not only $V(x, y, z)$ is zero, as we have required in Section 3.2, but also the quantum-theoretical part of the potential, that is, $V(x, y, z)$ must be zero too, where equations (3.1) and (3.16) have been considered. Thus, also the quantum-theoretical part of the potential in $(\mathbf{p}, E)$-space must be zero leading to equation (3.4):

$$v\left(i\hbar\frac{\partial}{\partial f_x}, i\hbar\frac{\partial}{\partial f_y}, i\hbar\frac{\partial}{\partial f_z}, -i\hbar\frac{\partial}{\partial E}\right)\Psi(f, E) = 0$$

---

[9] See equation (3.39).
[10] See equation (3.37).
[11] See equation (3.38).

This directly leads to equation (3.6) and with equation (3.19) we obtain $f(\mathbf{p}, E)_{free} = 0$ for all values $\mathbf{p} = \mathbf{p}_0$ and $E = E_0$, in particular for $\mathbf{p}_0 = 0$ and $E_0 = 0$.

3.2.4. Conclusion

In summary, in the case of $f(\mathbf{p}_0, E_0)_{free} \neq \infty$, we obtain relations for $\Psi(\mathbf{p}_0, E_0)$ and $\Psi(\mathbf{r}, t)$ that are well-defined and consistent with the equations (3.5) and (3.6). Furthermore, instead of

$$\Psi(\mathbf{p}_0, E_0) = \infty,$$
$$E_0 \neq \mathbf{p}_0^2/2m_0,$$

for $f(\mathbf{p}_0, E_0)_{free} = \infty,$[12] we get

$$\Psi(\mathbf{p}_0, E_0) \neq \infty,$$
$$E_0 = \mathbf{p}_0^2/2m_0,$$

for $f(\mathbf{p}_0, E_0)_{free} \neq \infty$. In fact, the value for $f(\mathbf{p}_0, E_0)_{free}$ must be zero.[13]

Let us continue our discussion on the basis of relation (3.3). In particular, we would like to discuss the meaning of $E_0 = \mathbf{p}_0^2/2m_0$ and the physical picture behind this relation in the quantum-theoretical case.

3.3. Classical and Quantum-Theoretical Elements

Equation (3.29) is identical with the classical case: $E_0 = \mathbf{p}_0^2/2m_0$. The reason is obvious. The system behaves classical in $(\mathbf{p}, E)$-space. The quantum-mechanical states in $(\mathbf{p}, E)$-space are characterized by the appearance of the operators[14]

$$i\hbar \frac{\partial}{\partial p_x}, \ i\hbar \frac{\partial}{\partial p_y}, \ i\hbar \frac{\partial}{\partial p_z}, \ -i\hbar \frac{\partial}{\partial E}, \qquad (3.47)$$

---

[12] See equation (3.23).
[13] See equation (3.45).
[14] See Section 2.3.7.

and these operators exclusively come into play by the interaction V.[15] Because the interaction term becomes zero in the case of a free system, it is understandable that the variables $\mathbf{p}_0$ and E_0 behave classically, satisfying the classical relation $E_0 = \mathbf{p}_0^2/2m_0$.

Although we obtain the classical relation $E_0 = \mathbf{p}_0^2/2m_0$ in $(\mathbf{p}, E)$-space, the free system discussed here is nevertheless a quantum-mechanical system, i.e., the usual picture of classical physics cannot be applied. For example, a velocity $\mathbf{v} = d\mathbf{r}/dt$ in the classical sense of the word in not definable in $(\mathbf{p}, E)$-space because the variables $\mathbf{r}$ and t do not appear in $(\mathbf{p}, E)$-space. Therefore, the momentum $\mathbf{p}$ and the energy E cannot be expressed by the relations of usual classical mechanics:

$$\mathbf{p} = m_0 \mathbf{v},$$
$$E = \frac{1}{2} m_0 \mathbf{v}^2. \qquad (3.48)$$

This follows directly from the fact that (fictitious) "reality" is not embedded in space and time, i.e., $((\mathbf{r}, t)$-space), but in $(\mathbf{p}, E)$-space.[16]

Within projection theory we have a (fictitious) "reality", which is embedded in $(\mathbf{p}, E)$-space, and we have a "picture of reality", that is represented in $(\mathbf{r}, t)$-space, and therefore, we work simultaneously in $(\mathbf{r}, t)$-space and $(\mathbf{p}, E)$-space. In order to obtain more information about the behaviour and features of free systems, we should discuss the functions $\Psi(\mathbf{p}, E) = \Psi(\mathbf{p}_0, E_0)$ and $\Psi(\mathbf{r}, t)$ simultaneously and not only the $(\mathbf{p}_0, E_0)$-behaviour as we did so far.

3.4. Behaviour of the Wave Function in (r, t)-Space and (p, E)-Space

In the case of a free system, there is the following relationship between the functions $\Psi(\mathbf{r}, t)$ and $\Psi(\mathbf{p}, E)$:

$$\Psi(\mathbf{r}, t) = C\Psi(\mathbf{p}, E) \exp\left\{\frac{i}{\hbar}(\mathbf{p} \cdot \mathbf{r} - Et)\right\}. \qquad (3.49)$$

---

[15] See equation (3.2).
[16] Clearly, within usual classical mechanics the system-specific time t is not defined but only the external time τ, and we have here $\mathbf{v} = d\mathbf{r}/d\tau$.

Because the variables **p** and E of the free system are constants which we have designated as $\mathbf{p}_0$ and E_0, we have again $\Psi(\mathbf{p}, E) = \Psi(\mathbf{p}_0, E_0)$ and, instead of equation (3.49) we may write[17]

$$\Psi(\mathbf{r}, t) = C\Psi(\mathbf{p}_0, E_0) \exp\left\{\frac{i}{\hbar}(\mathbf{p}_0 \cdot \mathbf{r} - E_0 t)\right\}. \tag{3.50}$$

When we use equation (3.5) we get with equation (3.49) expression (3.6), and with equation (3.50) expression (3.9). With $\Psi(\mathbf{p}_0, E_0) < \infty$,[18] equation (3.29) is valid: $E_0 = \mathbf{p}_0^2/2m_0$.

The constant C used in equation (3.50) can be determined as follows: With equation (2.5) and equation (3.37), we get with $\mathbf{r}_p = (x_p, y_p, z_p)$,

$$\begin{aligned}
\Psi(\mathbf{p}, E) &= \frac{1}{(2\pi\hbar)^2} \int_{-\infty}^{\infty} \Psi(\mathbf{r}, t) \exp\left\{-i\left[\frac{\mathbf{p}}{\hbar} \cdot \mathbf{r} - \frac{E}{\hbar}t\right]\right\} dt\, d\mathbf{r} \\
&= \lim_{\substack{\mathbf{p}_0 \to \mathbf{p} \\ E_0 \to E}} \Psi(\mathbf{p}_0, E_0) C \frac{1}{(2\pi\hbar)^2} \lim_{\mathbf{r}_p \to \infty} \lim_{t_E \to \infty} \int_{-\mathbf{r}_p}^{\mathbf{r}_p} \int_{-t_E}^{t_E} \\
&\quad \times \exp\left\{\frac{i}{\hbar}[(\mathbf{p}_0 - \mathbf{p}) \cdot \mathbf{r} - (E_0 - E)t]\right\} d\mathbf{r}\, dt \\
&= \Psi(\mathbf{p}, E) C \frac{1}{(2\pi\hbar)^2} \lim_{\substack{p_{0x} \to p_x \\ p_{0y} \to p_y \\ p_{0z} \to p_z}} \lim_{E_0 \to E} \lim_{\substack{x_p \to \infty \\ y_p \to \infty \\ z_p \to \infty}} \\
&\quad \times \lim_{t_E \to \infty} (2\hbar)^4 \frac{\sin(p_x - p_{0x})x_p/\hbar}{(p_x - p_{0x})} \\
&\quad \times \frac{\sin(p_y - p_{0y})y_p/\hbar}{(p_y - p_{0y})} \frac{\sin(p_z - p_{0z})z_p/\hbar}{(p_z - p_{0z})} \frac{\sin(E_0 - E)t_E/\hbar}{(E_0 - E)}.
\end{aligned} \tag{3.51}$$

In particular, we have

$$\lim_{\substack{x_p \to \infty \\ y_p \to \infty \\ z_p \to \infty}} \lim_{t_E \to \infty} (2\hbar)^4 \frac{\sin(p_x - p_{0x})x_p/\hbar}{(p_x - p_{0x})}$$

---

[17] Other relations for $\Psi(\mathbf{r}, t)$ are treated in Appendix 3.B and Appendix 3.C.
[18] The reason is discussed above.

$$\times \frac{\sin(p_y - p_{0y})y_p/\hbar}{(p_y - p_{0y})} \frac{\sin(p_z - p_{0z})z_p/\hbar}{(p_z - p_{0z})} \frac{\sin(E_0 - E)t_E/\hbar}{(E_0 - E)}$$

$$= (2\pi)^4 \delta(p_x/\hbar - p_{0x}/\hbar)\delta(p_y/\hbar - p_{0y}/\hbar)$$
$$\times \delta(p_z/\hbar - p_{0z}/\hbar)\delta(E_0/\hbar - E/\hbar), \tag{3.52}$$

where the well-known relation

$$\delta(a - b) = \frac{1}{\pi} \lim_{n \to \infty} \left[\frac{\sin n(a - b)}{a - b} \right] \tag{3.53}$$

has been used. Then, we obtain with $\Psi(\mathbf{p}_0, E_0) \neq \infty$,

$$C = \frac{\hbar^2}{\lim_{\substack{p_{0x} \to p_x \\ p_{0y} \to p_y \\ p_{0z} \to 0_z}} \lim_{E_0 \to E} (2\pi)^2 \delta(p_x/\hbar - p_{0x}/\hbar)}$$
$$\times \delta(p_y/\hbar - p_{0y}/\hbar)\delta(p_z/\hbar - p_{0z}/\hbar)\delta(E_0/\hbar - E/\hbar)$$

$$= \frac{\hbar^2}{(2\pi)^2 [\delta(0)]^4}, \tag{3.54}$$

and finally

$$C = 0. \tag{3.55}$$

Because of $\Psi(\mathbf{p}_0, E_0) < \infty$, we get with equation (3.50) the following result for $\Psi(\mathbf{r}, t)$:

$$\Psi(\mathbf{r}, t) = 0. \tag{3.56}$$

Remark

The values $\mathbf{p}$ and E remain constant and we have called these constant values $\mathbf{p}_0$ and E_0. Then we get with equation (3.12)

$$\Psi(\mathbf{r}, t) = \frac{1}{(2\pi\hbar)^2} \int_{-\infty}^{\infty} \Psi(\mathbf{p}, E) \exp\left\{i\left[\frac{\mathbf{p}}{\hbar} \cdot \mathbf{r} - \frac{E}{\hbar}t\right]\right\} d\mathbf{p}\, dE$$

$$= \frac{1}{(2\pi\hbar)^2} \Psi(\mathbf{p}_0, E_0) \exp\left[\frac{i}{\hbar}(\mathbf{p}_0 \cdot \mathbf{r} - E_0 t)\right]$$

$$\times \lim_{\Delta_\mathbf{p}, \Delta_E \to 0} \int_{\mathbf{p}_0, E_0}^{\mathbf{p}_0 + \Delta_\mathbf{p}, E + \Delta_E} d\mathbf{p}\, dE. \tag{3.57}$$

Equation (3.57) has exactly the form of equation (3.50), that is, we have

$$\Psi(\mathbf{r}, t) = C\Psi(\mathbf{p}_0, E_0) \exp\left[\frac{i}{\hbar}(\mathbf{p}_0 \cdot \mathbf{r} - E_0 t)\right],$$

with

$$C = \frac{1}{(2\pi\hbar)^2} \lim_{\Delta\mathbf{p}, \Delta E \to 0} \int_{\mathbf{p}_0, E_0}^{\mathbf{p}_0 + \Delta\mathbf{p}, E + \Delta E} d\mathbf{p}\, dE \to 0, \quad (3.58)$$

i.e., the constant C becomes zero as in the case of equation (3.55). Because equation (3.57) takes the form of equation (3.50), we may write in this case

$$C = \frac{1}{(2\pi\hbar)^2} \lim_{\Delta\mathbf{p}, \Delta E \to 0} \int_{\mathbf{p}_0, E_0}^{\mathbf{p}_0 + \Delta\mathbf{p}, E + \Delta E} d\mathbf{p}\, dE = \frac{\hbar^2}{(2\pi)^2 [\delta(0)]^4}, \quad (3.59)$$

where the equations (3.54) and (3.58) have been used.

Clearly, with the conditions $\Psi(\mathbf{p}_0, E_0) \neq \infty$,[19] and $C = 0$,[20] also the expression (3.57) directly leads to equation (3.56): $\Psi(\mathbf{r}, t) = 0$ for all values $\mathbf{r}$ and t.

Furthermore, when we calculate the expression[21]

$$\left\{i\hbar\frac{\partial}{\partial t} + \frac{\hbar^2}{2m_0}\Delta\right\} \Psi(\mathbf{r}, t) \quad (3.60)$$

by means of equation (3.57), we immediately get

$$\left\{i\hbar\frac{\partial}{\partial t} + \frac{\hbar^2}{2m_0}\Delta\right\} \Psi(\mathbf{r}, t) = \frac{1}{(2\pi\hbar)^2}\left\{E_0 - \frac{\mathbf{p}_0^2}{2m_0}\right\}\Psi(\mathbf{p}_0, E_0)$$

$$\times \exp\left[\frac{i}{\hbar}(\mathbf{p}_0 \cdot \mathbf{r} - E_0 t)\right]$$

$$\times \lim_{\Delta\mathbf{p}, \Delta E \to 0} \int_{\mathbf{p}_0, E_0}^{\mathbf{p}_0 + \Delta\mathbf{p}, E + \Delta E} d\mathbf{p}\, dE. \quad (3.61)$$

This is identical with the formula (3.16) when we use expression (3.20) for the function $f(\mathbf{r}, t)_{free}$.

---

[19] See equation (3.37).
[20] See equation (3.58).
[21] See equation (3.16).

3.5. Probability Considerations in Connection with $\Psi(\mathbf{p}_0, E_0)$

How have we to understand this result? The probability $\Delta W = \Psi^*(\mathbf{r}, t)\Psi(\mathbf{r}, t)\Delta \mathbf{r}\Delta t$ for an event within the space-time intervals $\Delta \mathbf{r}$ and $\Delta t (\Delta \mathbf{r} \neq \infty, \Delta t \neq \infty)$ must for $\Psi^*(\mathbf{r}, t)\Psi(\mathbf{r}, t) = 0\{\Psi(\mathbf{r}, t) = 0\}$ always be zero. In other words, a free system cannot be observed within $(\mathbf{r}, t)$-space.[22] At this stage of our discussion, we may say that the free system can exist in $(\mathbf{p}, E)$-space because the wave function $\Psi(\mathbf{p}_0, E_0)$ and the quantities $\mathbf{p}_0$ and E_0 may exist in principle, but nothing is observable. But we have to be careful. Let us examine this case in more detail.

The solution $\Psi(\mathbf{p}_0, E_0) = \infty$ is not allowed.[23] But is $\Psi(\mathbf{p}_0, E_0) \neq 0$ possible? It is not for the following reason. The probability

$$\Delta W = \Psi^*(\mathbf{p}_0, E_0)\Psi(\mathbf{p}_0, E_0)\Delta \mathbf{p}\Delta E \qquad (3.62)$$

of finding the system in the state $\mathbf{p}_0$ and E_0 must be zero because of $\Delta \mathbf{p} = 0$, $\Delta E = 0$ and because $\Psi^*(\mathbf{p}_0, E_0)\Psi(\mathbf{p}_0, E_0) \neq \infty$. The condition $\Delta \mathbf{p} = 0$, $\Delta E = 0$ follows directly from the fact that only the values $\mathbf{p}_0$ and E_0 are defined but not $\mathbf{p} \neq \mathbf{p}_0$ and $E \neq E_0$, i.e., we have no uncertainties with respect to the variables $\mathbf{p}$ and E in the case of a free system and, therefore, we have $\Delta \mathbf{p} = 0$ and $\Delta E = 0$.

It makes no sense at all to have $\Delta W = 0$ for the probability and, on the other hand, $\Psi^*(\mathbf{p}_0, E_0)\Psi(\mathbf{p}_0, E_0) \neq 0\{\Psi(\mathbf{p}_0, E_0) \neq 0\}$ for the probability density and the wave function, respectively? Such a behavior cannot be justified and is rather unphysical. It is particularly against the probability interpretation in connection with quantum phenomena. We may argue as follows. Generally, the property $\Psi^*(\mathbf{p}, E)\Psi(\mathbf{p}, E) \neq 0$ has to be connected to $\Delta W \neq 0$ but not to $\Delta W = 0$. Therefore, the property $\Delta W = 0$ has to be connected to

$$\Psi^*(\mathbf{p}_0, E_0)\Psi(\mathbf{p}_0, E_0) = 0\{\Psi(\mathbf{p}_0, E_0) = 0\},$$

and not to the case

$$\Psi^*(\mathbf{p}_0, E_0)\Psi(\mathbf{p}_0, E_0) \neq 0\{\Psi(\mathbf{p}_0, E_0) \neq 0\}.$$

---

[22] We perform our measurements exclusively in $(\mathbf{r}, t)$-space.
[23] See Section 3.1.

In other words, we should have the following links:

$$\Delta W \neq 0 \leftrightarrow \Psi^*(\mathbf{p}_0, E_0)\Psi(\mathbf{p}_0, E_0) \neq 0\{\Psi(\mathbf{p}_0, E_0) \neq 0\}, \quad (3.63)$$

$$\Delta W = 0 \leftrightarrow \Psi^*(\mathbf{p}_0, E_0)\Psi(\mathbf{p}_0, E_0) = 0\{\Psi(\mathbf{p}_0, E_0) = 0\}, \quad (3.64)$$

and nothing else. Since $\Delta W = 0$ for a free system, and also the wave function $\Psi(\mathbf{p}_0, E_0)$ must be zero in this case:

$$\Psi(\mathbf{p}_0, E_0) = 0. \quad (3.65)$$

We may also argue as follows. Both spaces, i.e., $(\mathbf{r}, t)$-space and $(\mathbf{p}, E)$-space, are completely equivalent concerning their contents (information). Both spaces only differ in the representation of the results.[24] The result $\Psi(\mathbf{r}, t) = 0$ means *non-existence* in $(\mathbf{r}, t)$-space.[25] Thus, we must also have *non-existence* in $(\mathbf{p}, E)$-space leading directly to $\Psi(\mathbf{p}_0, E_0) = 0$ which is identical with the result given by equation (3.65) that we have found on the basis of probability arguments. The property $\Psi^*(\mathbf{p}_0, E_0)\Psi(\mathbf{p}_0, E_0) = 0$ for all values of $\mathbf{p}_0$ and E_0 means that none of the values $\mathbf{p}_0, E_0$ can be realized and reflects non-existence.

3.6. Normalization Condition

With equation (3.50), we obtain for the normalization integral

$$\int_{-\infty}^{\infty} \Psi^*(\mathbf{r}, t)\Psi(\mathbf{r}, t) d\mathbf{r}\, dt.$$

The following expression

$$\int_{-\infty}^{\infty} \Psi^*(\mathbf{r}, t)\Psi(\mathbf{r}, t) d\mathbf{r}\, dt = C^2 \lim_{\substack{p_{0x} \to p'_{0x} \\ p_{0y} \to p'_{0y} \\ p_{0z} \to p'_{0z}}} \lim_{E_0 \to E'_0} \Psi^*(\mathbf{p}_0, E_0)\Psi(\mathbf{p}'_0, E'_0)$$

---

[24] See Section 2.3.4.3.
[25] See equation (3.56).

Free, Non-interacting Systems (Particles) 135

$$\times \int_{-\infty}^{\infty} \exp\left\{i\left[\frac{(\mathbf{p}_0 - \mathbf{p}_0')}{\hbar}\cdot \mathbf{r} - \frac{(E_0 - E_0')}{\hbar}t\right]\right\} dt\, d\mathbf{r}$$

$$= C^2 \lim_{\substack{p_{0x} \to p_{0x}' \\ p_{0y} \to p_{0y}' \\ p_{0z} \to p_{0z}'}} \lim_{E_0 \to E_0'} \Psi^*(\mathbf{p}_0, E_0)\Psi(\mathbf{p}_0', E_0') \lim_{\mathbf{r_p} \to \infty} \lim_{t_E \to \infty} \int_{-\mathbf{r_p}}^{\mathbf{r_p}} \int_{-t_E}^{t_E}$$

$$\times \exp\left\{\frac{i}{\hbar}\left[(\mathbf{p}_0 - \mathbf{p}_0')\cdot \mathbf{r} - (E_0 - E_0')t\right]\right\} d\mathbf{r}\, dt$$

$$= C^2 \Psi^*(\mathbf{p}_0, E_0)\Psi(\mathbf{p}_0, E_0) \lim_{\substack{p_{0x} \to p_{0x}' \\ p_{0y} \to p_{0y}' \\ p_{0z} \to p_{0z}'}} \lim_{E_0 \to E_0'} \lim_{\substack{x_p \to \infty \\ y_p \to \infty \\ z_p \to \infty}}$$

$$\times \lim_{t_E \to \infty} (2\hbar)^4 \frac{\sin(p_{0x}' - p_{0x})x_p/\hbar}{(p_{0x}' - p_{0x})}$$

$$\times \frac{\sin(p_{0y}' - p_{0y})y_p/\hbar}{(p_{0y}' - p_{0y})} \frac{\sin(p_{0z}' - p_{0z})z_p/\hbar}{(p_{0z}' - p_{0z})} \frac{\sin(E_0 - E_0')t_E/\hbar}{(E_0 - E_0')},$$

(3.66)

where the function $\Psi(\mathbf{p}_0, E_0)$ is a constant, and C takes the form[26]

$$C = \left\{ \begin{array}{l} \dfrac{1}{(2\pi\hbar)^2} \lim_{\substack{p_{0x} \to p_{0x}' \\ p_{0y} \to p_{0y}' \\ p_{0z} \to p_{0z}'}} \lim_{E_0 \to E_0'} \lim_{\substack{x_p \to \infty \\ y_p \to \infty \\ z_p \to \infty}} \\ \times \lim_{t_E \to \infty} (2\hbar)^4 \dfrac{\sin(p_{0x}' - p_{0x})x_p/\hbar}{(p_{0x}' - p_{0x})} \dfrac{\sin(p_{0y}' - p_{0y})y_p/\hbar}{(p_{0y}' - p_{0y})} \\ \times \dfrac{\sin(p_{0z}' - p_{0z})z_p/\hbar}{(p_{0z}' - p_{0z})} \dfrac{\sin(E_0 - E_0')t_E/\hbar}{(E_0 - E_0')} \end{array} \right\}^{-1}.$$

(3.67)

Using

$$\int_{-\infty}^{\infty} \Psi^*(\mathbf{r}, t)\Psi(\mathbf{r}, t) d\mathbf{r}\, dt = \Psi^*(\mathbf{p}_0, E_0)\Psi(\mathbf{p}_0, E_0) B, \quad (3.68)$$

---

[26] See Section 3.4.

with

$$B = C^2 \lim_{\substack{p_{0x} \to p'_{0x} \\ p_{0y} \to p'_{0y} \\ p_{0z} \to p'_{0z}}} \lim_{E_0 \to E'_0} \lim_{\substack{x_p \to \infty \\ y_p \to \infty \\ z_p \to \infty}} \lim_{t_E \to \infty} A, \qquad (3.69)$$

and

$$A = (2\hbar)^4 \frac{\sin(p'_{0x} - p_{0x})x_p/\hbar}{(p'_{0x} - p_{0x})} \frac{\sin(p'_{0y} - p_{0y})y_p/\hbar}{(p'_{0y} - p_{0y})}$$
$$\times \frac{\sin(p'_{0z} - p_{0z})z_p/\hbar}{(p'_{0z} - p_{0z})} \frac{\sin(E_0 - E'_0)t_E/\hbar}{(E_0 - E'_0)}, \qquad (3.70)$$

we obtain with the expression for C given by equation (3.67):

$$B = (2\pi\hbar)^4 \frac{\lim_{\substack{p_{0x} \to p'_{0x} \\ p_{0y} \to p'_{0y} \\ p_{0z} \to p'_{0z}}} \lim_{E_0 \to E'_0} \lim_{\substack{x_p \to \infty \\ y_p \to \infty \\ z_p \to \infty}} \lim_{t_E \to \infty} A}{\lim_{\substack{p_{0x} \to p'_{0x} \\ p_{0y} \to p'_{0y} \\ p_{0z} \to p'_{0z}}} \lim_{E_0 \to E'_0} \lim_{\substack{x_p \to \infty \\ y_p \to \infty \\ z_p \to \infty}} \lim_{t_E \to \infty} A^2}$$

$$= (2\pi\hbar)^4 \lim_{\substack{p_{0x} \to p'_{0x} \\ p_{0y} \to p'_{0y} \\ p_{0z} \to p'_{0z}}} \lim_{E_0 \to E'_0} \lim_{\substack{x_p \to \infty \\ y_p \to \infty \\ z_p \to \infty}} \lim_{t_E \to \infty} \frac{A}{A^2}$$

$$= (2\pi\hbar)^4 \frac{1}{\lim_{\substack{p_{0x} \to p'_{0x} \\ p_{0y} \to p'_{0y} \\ p_{0z} \to p'_{0z}}} \lim_{E_0 \to E'_0} \lim_{\substack{x_p \to \infty \\ y_p \to \infty \\ z_p \to \infty}} \lim_{t_E \to \infty} A}, \qquad (3.71)$$

where A has the properties $A \neq 0$ and $A \neq \infty$. With equation (3.53), we finally get

$$B = \frac{\hbar^4}{\lim_{\substack{p_{0x} \to p_x \\ p_{0y} \to p_y \\ p_{0z} \to 0_z}} \lim_{E_0 \to E} \delta(p_x/\hbar - p_{0x}/\hbar)\delta(p_y/\hbar - p_{0y}/\hbar)}$$
$$\times \delta(p_z/\hbar - p_{0z}/\hbar)\delta(E_0 - E)$$

$$= \frac{\hbar^4}{[\delta(0)]^4}, \qquad (3.72)$$

and

$$\int_{-\infty}^{\infty} \Psi^*(\mathbf{r},t)\Psi(\mathbf{r},t)d\mathbf{r}\,dt = \Psi^*(\mathbf{p}_0,E_0)\Psi(\mathbf{p}_0,E_0)\frac{\hbar^4}{[\delta(0)]^4} = 0. \tag{3.73}$$

Because of

$$\int_{-\infty}^{\infty} \Psi^*(\mathbf{r},t)\Psi(\mathbf{r},t)d\mathbf{r}\,dt = \int_{-\infty}^{\infty} \Psi^*(\mathbf{p},E)\Psi(\mathbf{p},E)d\mathbf{p}\,dE,$$

also the integral

$$\int_{-\infty}^{\infty} \Psi^*(\mathbf{p},E)\Psi(\mathbf{p},E)d\mathbf{p}\,dE$$

must be zero. Independent from this argument, we immediately find

$$\int_{-\infty}^{\infty} \Psi^*(\mathbf{p},E)\Psi(\mathbf{p},E)d\mathbf{p}\,dE$$
$$= \Psi^*(\mathbf{p}_0,E_0)\Psi(\mathbf{p}_0,E_0)\lim_{\Delta_{Bp}\Delta_E \to 0}\int_{\mathbf{p}_0,E_0}^{\mathbf{p}_0+\Delta_\mathbf{p},E+\Delta_E} d\mathbf{p}\,dE$$
$$= 0, \tag{3.74}$$

where $\Psi^*(\mathbf{p}_0,E_0)\Psi(\mathbf{p}_0,E_0)$ is zero[27] and also the integral.[28]

3.7. Mean Values for the Momentum and the Energy

What can we say about the values $\mathbf{p}_0$ and E_0? We have stated the following in Section 3.5. Because of $\Psi^\bullet(\mathbf{p}_0,E_0)\Psi(\mathbf{p}_0,E_0) = 0$, none of the values $\mathbf{p}_0$ and E_0 can be realized and this property reflects non-existence. By the investigation of the mean values for $\bar{\mathbf{p}}$ and $\bar{E}$, we are able to define the notion "non-existence" in greater detail.

---

[27] See equation (3.65).
[28] See equation (3.21).

Because there are no **p**, E-fluctuations[29] the mean values $\bar{\mathbf{p}}$ and $\bar{E}$ must be identical with $\mathbf{p}_0$ and E_0:

$$\bar{\mathbf{p}} = \mathbf{p}_0 = (p_{0x}, p_{0y}, p_{0z}),$$
$$\bar{E} = E_0. \qquad (3.75)$$

These mean values are expressed by

$$\bar{p}_\alpha = \int_{-\infty}^{\infty} \Psi^*(\mathbf{r}, t) \left[-i\hbar \frac{\partial}{\partial \alpha} \right] \Psi(\mathbf{r}, t) d\mathbf{r}\, dt$$
$$= \int_{-\infty}^{\infty} p_{0\alpha} \Psi^*(\mathbf{p}, E) \Psi(\mathbf{p}, E) d\mathbf{p}\, dE, \quad \alpha = x, y, z, \quad (3.76)$$

and

$$\bar{E} = \int_{-\infty}^{\infty} \Psi^*(\mathbf{r}, t) \left[i\hbar \frac{\partial}{\partial t} \right] \Psi(\mathbf{r}, t) d\mathbf{r}\, dt$$
$$= \int_{-\infty}^{\infty} E \Psi^*(\mathbf{p}, E) \Psi(\mathbf{p}, E) d\mathbf{p}\, dE. \qquad (3.77)$$

Using $\Psi(\mathbf{r}, t)$, the expression given by equation (3.50), we obtain for $(\mathbf{r}, t)$-space

$$\bar{p}_\alpha = \int_{-\infty}^{\infty} \Psi^*(\mathbf{r}, t) \left[-i\hbar \frac{\partial}{\partial \alpha} \right] \Psi(\mathbf{r}, t) d\mathbf{r}\, dt$$
$$= p_{0\alpha} \int_{-\infty}^{\infty} \Psi^*(\mathbf{r}, t) \Psi(\mathbf{r}, t) d\mathbf{r}\, dt = 0, \quad \alpha = x, y, z, \quad (3.78)$$

and

$$\bar{E} = \int_{-\infty}^{\infty} \Psi^*(\mathbf{r}, t) \left[i\hbar \frac{\partial}{\partial t} \right] \Psi(\mathbf{r}, t) d\mathbf{r}\, dt$$
$$= E_0 \int_{-\infty}^{\infty} \Psi^*(\mathbf{r}, t) \Psi(\mathbf{r}, t) d\mathbf{r}\, dt = 0, \qquad (3.79)$$

---

[29] $\Delta \mathbf{p} = 0, \Delta E = 0$, see equation (3.7).

Free, Non-interacting Systems (Particles)

with[30]

$$\int_{-\infty}^{\infty} \Psi^*(\mathbf{r}, t)\Psi(\mathbf{r}, t)d\mathbf{r}\, dt = 0,$$

where $\Psi(\mathbf{r}, t)$ is zero for all values of $\mathbf{r}$ and t.

We get the same results when we work within $(\mathbf{p}, E)$-space. Because there are no $\mathbf{p}, E$-fluctuations[31] we simply have

$$p_\alpha = p_{0\alpha}, \quad E = E_0 \quad \text{and} \quad \Psi(\mathbf{p}, E) = \Psi(\mathbf{p}_0, E_0) = \text{const.}, \quad (3.80)$$

$$\bar{p}_\alpha = \int_{-\infty}^{\infty} p_\alpha \Psi^*(\mathbf{p}, E)\Psi(\mathbf{p}, E)d\mathbf{p}\, dE$$

$$= p_{0\alpha}\Psi^*(\mathbf{p}_0, E_0)\Psi(\mathbf{p}_0, E_0) \lim_{\Delta_{Bp}\Delta_E \to 0} \int_{\mathbf{p}_0, E_0}^{\mathbf{p}_0+\Delta_\mathbf{p}, E+\Delta_E} d\mathbf{p}\, dE$$

$$= 0, \quad \alpha = x, y, z, \qquad (3.81)$$

$$\bar{E} = \int_{-\infty}^{\infty} E_0 \Psi^*(\mathbf{p}, E)\Psi(\mathbf{p}, E)d\mathbf{p}\, dE$$

$$= E_0 \Psi^*(\mathbf{p}_0, E_0)\Psi(\mathbf{p}_0, E_0) \lim_{\Delta_{Bp}\Delta_E \to 0} \int_{\mathbf{p}_0, E_0}^{\mathbf{p}_0+\Delta_\mathbf{p}, E+\Delta_E} d\mathbf{p}\, dE$$

$$= 0, \qquad (3.82)$$

with[32]

$$\Psi^*(\mathbf{p}_0, E_0)\Psi(\mathbf{p}_0, E_0) = 0, \qquad (3.83)$$

and[33]

$$\lim_{\Delta_{Bp}\Delta_E \to 0} \int_{\mathbf{p}_0, E_0}^{\mathbf{p}_0+\Delta_\mathbf{p}, E+\Delta_E} d\mathbf{p}\, dE = 0.$$

Thus, the $(\mathbf{p}, E)$-space analysis is identical with the analysis performed in $(\mathbf{r}, t)$-space, and this must be the case because both spaces are completely equivalent concerning their information content.

---

[30] See equation (3.73).
[31] $\Delta \mathbf{p} = 0, \Delta E = 0$.
[32] See equation (3.65).
[33] See equation (3.21).

Conclusion

In other words, the mean momentum and the mean energy must be zero: $\bar{\mathbf{p}} = 0, \bar{E} = 0$. With equation (3.75), we therefore have

$$\mathbf{p}_0 = 0, E_0 = 0. \tag{3.84}$$

This means that only solutions for $\Psi(\mathbf{p}_0, E_0)$ can exist for $\mathbf{p}_0 = 0$ and $E_0 = 0$. Then, we get with equation (3.65)

$$\Psi(\mathbf{p}_0 = 0, E_0 = 0) = 0. \tag{3.85}$$

Solutions for $\mathbf{p}_0 \neq 0, E_0 \neq 0$ are forbidden within projection theory.

On the other hand, the property $\Psi^\bullet(\mathbf{p}_0, E_0)\Psi(\mathbf{p}_0, E_0) = 0$ for all values of $\mathbf{p}_0$ and E_0,[34] means that none of the values $\mathbf{p}_0, E_0$ can be realized and reflects non-existence. Thus, the result $\mathbf{p}_0 = 0, E_0 = 0$ defines non-existence,[35] that is, a system which is characterized by $\mathbf{p}_0$ and E_0 does not exist if $\mathbf{p}_0 = 0, E_0 = 0$.

3.8. The (p, E)-Pool

What does the result $\mathbf{p}_0 = 0, E_0 = 0$ mean? Let us assume that the entire reality (cosmos) consists of a certain number of systems which are interacting with each other and let us denote its total momentum by $\mathbf{p}_{\cos}$ and its total energy by $E_{\cos}$. Then, we have a $(\mathbf{p}, E)$-pool which we would like to denote by $(\mathbf{p}_{\cos}, E_{\cos})$. When a new system, having at time τ the momentum $\mathbf{p}$ and the energy E, is created, we have

$$(\mathbf{p}_{\cos}, E_{\cos}) \to \tau : (\mathbf{p}'_{\cos}, E'_{\cos}) + (\mathbf{p}', E'). \tag{3.86}$$

In connection with equation (3.86), all have to be arranged in the course of time τ in accordance with the conservation law for momentum and energy. The new system $(\mathbf{p}', E')$ interacts with its surrounding $(\mathbf{p}'_{\cos}, E'_{\cos})$ which is defined by the remaining part of the cosmos where $\mathbf{p}'_{\cos}$ is the total momentum at time τ of this remaining part of the cosmos and $E'_{\cos}$ is its total energy at the same time τ. The new system interacts with the surrounding through $\mathbf{p}, E$-fluctuations

---

[34] See Section 3.5.
[35] See equation (3.84).

Free, Non-interacting Systems (Particles)

$\Delta \mathbf{p}$ and ΔE in accordance with the conservation law for momentum and energy and we have

$$(\mathbf{p}'_{\cos}, E'_{\cos}) \xleftrightarrow{\Delta \mathbf{p}, \Delta E} (\mathbf{p}', E'), \qquad (3.87)$$

with

$$\tau : \mathbf{p}_{\cos} = \mathbf{p}'_{\cos} + \mathbf{p}', \quad E_{\cos} = E'_{\cos} + E'. \qquad (3.88)$$

In the case of a free, i.e., non-interacting system we have by definition $\Delta \mathbf{p} = 0$ and $\Delta E = 0$ and the values for $\mathbf{p}'$ and E' remain constant in the course of time τ:

$$\mathbf{p}' = \text{const.} = \mathbf{p}_0, \; E' = \text{const.} = E_0, \qquad (3.89)$$

and we have instead of equation (3.87), the relation

$$(\mathbf{p}'_{\cos}, E'_{\cos}) \xleftrightarrow{\Delta \mathbf{p}=0, \Delta E=0} (\mathbf{p}_0, E_0). \qquad (3.90)$$

In this connection, we may state the following. Our free system with $\mathbf{p}_0 = 0, E_0 = 0$, which we assume is not divisible into certain subsystems, cannot exist because a system with $\mathbf{p}_0 = 0$ and $E_0 = 0$ cannot be considered as created. Such a system does not exist because the cosmic $\mathbf{p}, E$-pool does not change in connection with $\mathbf{p}_0 = 0$, $E_0 = 0$ and instead of process (3.86), we have

$$(\mathbf{p}_{\cos}, E_{\cos}) = (\mathbf{p}'_{\cos}, E'_{\cos}). \qquad (3.91)$$

Such a non-elementary system $(\mathbf{p}_{\cos}, E_{\cos})$, consisting of interacting subsystems with $\mathbf{p}_1, E_1, \mathbf{p}_2, E_2, \ldots, \mathbf{p}_i, E_i, \ldots$, has a certain space-time extension $\Delta \mathbf{r}_{\mathbf{p}_0, E_0}, \Delta t_{\mathbf{p}_0, E_0}$ depending on the number of subsystems and the kind of interaction. If this interaction is assumed to be dependent on the space-time distances of the subsystems and attractive, the space-time extension $\Delta \mathbf{r}_{\mathbf{p}_0, E_0}, \Delta t_{\mathbf{p}_0, E_0}$ of the whole system is not infinity,[36] that is we have

$$\Delta \mathbf{r}_{\mathbf{p}_0, E_0}, \Delta t_{\mathbf{p}_0, E_0} < \infty. \qquad (3.92)$$

---

[36] That is, if the number of subsystems remains finite.

Again, the whole system ($\mathbf{p}_{\cos}, E_{\cos}$) with a certain space-time extension $\Delta \mathbf{r}_{\mathbf{p}_0, E_0}, \Delta t_{\mathbf{p}_0, E_0}$ does not interact with external systems but it exists because it is non-elementary.

Not only does the system ($\mathbf{p}_0, E_0$) behave like a free system with $\mathbf{p}_0 = 0$ and $E_0 = 0$, but also the cosmos ($\mathbf{p}_{\cos}, E_{\cos}$) itself. However, in contrast to the elementary case ($\mathbf{p}_0, E_0$), the total momentum and the total energy of a non-elementary system must not be zero[37]:

$$\mathbf{p}_{\cos} \neq 0, \quad E_{\cos} \neq 0. \tag{3.93}$$

We will recognize in Chapter 4 that the whole system ($\mathbf{p}_{\cos}, E_{\cos}$) moves arbitrarily through space and time, i.e., ($\mathbf{r}, t$)-space, and an observer who is resting relative to ($\mathbf{r}, t$)-space cannot perceive or detect this arbitrarily moving cosmos ($\mathbf{p}_{\cos}, E_{\cos}$). However, an observer who is resting relative to the cosmos is able to make measurements within this system, that is, he can perceive the details of the cosmos. An internal observer is coupled to the system (cosmos) and, therefore, he performs exactly the same statistical jumps.

Within the context of the arbitrarily moving cosmos ($\mathbf{p}_{\cos}, E_{\cos}$), the following should already be mentioned here. The space-time variables that describe the arbitrary motion are connected to a certain momentum and a certain energy, say $\mathbf{p}_\mathbf{r}, E_t$, with $\mathbf{p}_\mathbf{r} = 0$ and $E_t = 0$. However, in contrast to the elementary system ($\mathbf{p}_0, E_0$), the variables $\mathbf{p}_\mathbf{r}$ and E_t have nothing to do with the total momentum by $\mathbf{p}_{\cos}$ and the total energy by $E_{\cos}$ of the cosmos.

3.9. Free, Elementary Systems do not Exist

We have the following results for the free (non-interacting) case with the condition that the system is elementary, i.e., it is assumed that the system cannot be divided into subsystems. Free means that the potential function becomes zero: $V(x, y, z, t) = 0$. Then, also the operator $V(i\hbar \partial / \partial p_x, \ldots, -i\hbar \partial / \partial E)$[38] is not definable. The results can be summarized as follows[39]:

---

[37] We will outline this point in greater detail in Chapter 4.
[38] See Section 2.3.7.3.
[39] See also Figure 3.1.

1. The wave functions $\Psi(\mathbf{r},t)$ and $\Psi(\mathbf{p}_0, E_0)$ become zero.[40] Thus, also the probability densities are zero:

$$\Psi^*(\mathbf{r},t)\Psi(\mathbf{r},t) = 0, \qquad (3.94)$$

$$\Psi^*(\mathbf{p}_0, E_0)\Psi(\mathbf{p}_0, E_0) = 0. \qquad (3.95)$$

Consequently, the normalization integrals

$$\int_{-\infty}^{\infty} \Psi^*(\mathbf{r},t)\Psi(\mathbf{r},t) dt\, d\mathbf{r},$$

and

$$\int_{-\infty}^{\infty} \Psi^*(\mathbf{p}_0, E_0)\Psi(\mathbf{p}_0, E_0) dE\, d\mathbf{p},$$

are getting zero as well.[41]

2. It turned out that the systems momentum $\mathbf{p}_0$ and its energy E_0 become zero too: $\mathbf{p}_0 = 0, E_0 = 0$. That is, nothing has been taken

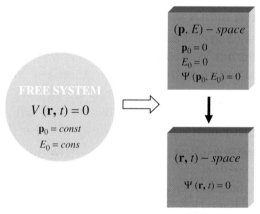

Fig. 3.1. In the case of non-interacting elementary system, we have $V(\mathbf{r},t) = 0$ and, therefore, the operator $V(i\hbar\partial/\partial p_x, \ldots, -i\hbar\partial/\partial E)$ is not definable. Then, the wave functions $\Psi(\mathbf{r},t)$ and $\Psi(\mathbf{p}_0, E_0)$ become zero and, furthermore, the systems momentum $\mathbf{p}_0$ and its energy E_0 become zero too: $\mathbf{p}_0 = 0, E_0 = 0$. That is, nothing has been taken away from the cosmic $\mathbf{p}, E$-pool. Therefore, from the point of view of projection theory, free elementary systems cannot be realized in nature.

---

[40] See equations (3.56) and (3.65).
[41] See equations (3.73) and (3.74).

away from the cosmic **p**, E-pool. Therefore, from the point of view of projection theory, free elementary systems cannot exist in nature. In this connection, it is interesting to note also that stationary systems[42] do not exist in projection theory.[43]

3.10. No Equation for the Determination of the Wave Function $\Psi(\mathbf{p}_0, E_0)$

3.10.1. Additional Physically Relevant Conditions?

In the case of an interacting system, the wave function $\Psi(\mathbf{p}, E)$, formulated in $(\mathbf{p}, E)$-space, is determined by the basic equation (3.2):

$$E\Psi(\mathbf{p}, E) = \frac{\mathbf{p}^2}{2m_0}\Psi(\mathbf{p}, E)$$
$$+ V\left(i\hbar\frac{\partial}{\partial p_x}, i\hbar\frac{\partial}{\partial p_y}, i\hbar\frac{\partial}{\partial p_z}, -i\hbar\frac{\partial}{\partial E}\right)\Psi(\mathbf{p}, E).$$

Free (non-interacting) systems are defined in $(\mathbf{p}, E)$-space on the basis of this equation[44] if the interaction term is zero and we get equation (3.9):

$$E_0\Psi(\mathbf{p}_0, E_0) = \frac{\mathbf{p}_0^2}{2m_0}\Psi(\mathbf{p}_0, E_0).$$

Because we have excluded the case $\Psi(\mathbf{p}_0, E_0) = \infty$, the wave function $\Psi(\mathbf{p}_0, E_0)$ vanishes in equation (3.9) and we have $E_0 = \mathbf{p}_0^2/m_0$.[45] That is, in the non-interacting case, we have no equation for the determination of the wave function, i.e., $\Psi(\mathbf{p}_0, E_0)$, in $(\mathbf{p}, E)$-space. $\Psi(\mathbf{p}_0, E_0)$ is not definable. An analytical expression for $\Psi(\mathbf{p}_0, E_0)$ is not deducible. In other words, the use of a certain analytical expression for $\Psi(\mathbf{p}_0, E_0)$ is not allowed and therefore unphysical.

---

[42] This is defined by $V(x, y, z, t) \to V(x, y, z)$.
[43] See Appendix 3.D.
[44] See equation (3.2).
[45] For details, see also Sections 3.2–3-4.

However, by the introduction of additional physically relevant conditions, the situation could be changed in principle. But also in this case, the basic equations (3.2) and (3.9) must be fulfilled. How could such additional physically relevant conditions be introduced in connection with an elementary system $(\mathbf{p}_0, E_0)$? This is obviously not possible if we restrict ourselves in the description on the variables $\mathbf{p}_0$ and E_0. There is no way for such an introduction of physically relevant conditions. In fact, the system $(\mathbf{p}_0, E_0)$ is very simple. We assumed that it is elementary in character and it cannot be divided into certain subsystems. Thus, the system only consists of four constant numbers $\mathbf{p}_0 = (p_x, p_y, p_z)$ and E_0 that do not change with time τ. If the variables $\mathbf{p}_0 = (p_x, p_y, p_z)$ and E_0 would change with time τ, the system would interact with other systems but, by definition, the system $(\mathbf{p}_0, E_0)$ does not interact with other systems.

In conclusion, no additional elements, conditions and assumptions can be introduced in order to obtain more information about the function $\Psi(\mathbf{p}_0, E_0)$, more than that we discussed above and which are dictated by the basic equations of projection theory. Therefore, we have to work without additional conditions and we have to continue our discussion on what we have said above.

In the non-interacting case, we have no equation for the determination of the wave function, i.e., $\Psi(\mathbf{p}_0, E_0)$ in $(\mathbf{p}, E)$-space. $\Psi(\mathbf{p}_0, E_0)$ is not definable; an analytical expression for $\Psi(\mathbf{p}_0, E_0)$ is not deducible. In other words, the use of a certain analytical expression for $\Psi(\mathbf{p}_0, E_0)$ is not allowed and would therefore be unphysical.

3.10.2. Multi-valuedness of the Wave Function $\Psi(\mathbf{p}_0, E_0)$

Then, we are confronted with the following situation. In principle, the function $\Psi(\mathbf{p}_0, E_0)$, only defined for $\mathbf{p}_0$ and E_0, may take any value without infinity: $\Psi(\mathbf{p}_0, E_0) = \infty$.[46] But this makes no sense because

---
[46] See Section 3.1.

only a one-valued probability density $\Psi^*(\mathbf{p}_0, E_0)\Psi(\mathbf{p}_0, E_0)$ is physically allowed.

What could the multi-valuedness of the probability density $\Psi^*(\mathbf{p}_0, E_0)\Psi(\mathbf{p}_0, E_0)$ mean? For example, it makes no sense to assume that $\Psi^*(\mathbf{p}_0, E_0)\Psi(\mathbf{p}_0, E_0)$ varies arbitrarily with time τ. Only in the case of a permanently changing surrounding and potential function, i.e., $V(x, y, z, t) = V(x, y, z, t, \tau)$, respectively, we could understand that $\Psi^*(\mathbf{p}_0, E_0)\Psi(\mathbf{p}_0, E_0)$ varies arbitrarily with time τ. However, also such an effect must be excluded because, by definition, the system $(\mathbf{p}_0, E_0)$ does interact with other external systems.

Therefore, the multi-valuedness of the wave function $\Psi(\mathbf{p}_0, E_0)$ reflects an unphysical situation and has therefore to be considered as strictly forbidden. Thus, not only the case $\Psi(\mathbf{p}_0, E_0) = \infty$ has to be excluded but also all the other possible values for $\Psi(\mathbf{p}_0, E_0)$, i.e., $\Psi^*(\mathbf{p}_0, E_0)\Psi(\mathbf{p}_0, E_0)$, apart from $\Psi(\mathbf{p}_0, E_0) = 0$, i.e., $\Psi^*(\mathbf{p}_0, E_0)\Psi(\mathbf{p}_0, E_0) = 0$, which reflects within projection theory *non-existence* of the system $(\mathbf{p}_0, E_0)$. In the next section, we would like to analyze this point in somewhat more detail.

3.10.3. *Existence and Non-Existence*

Within projection theory we have a (fictitious) "reality" which is embedded in $(\mathbf{p}, E)$-space, and we have a "picture of reality" which is represented in $(\mathbf{r}, t)$-space. Reality is projected onto space-time. As we have outlined in Section 2.3.4, the observer measures in $(\mathbf{r}, t)$-space the elements $(\mathbf{p}, E) = (\mathbf{p}_0, E_0)$ of reality with a probability density $\Psi^*(\mathbf{r}, t)\Psi(\mathbf{r}, t)$. The elements $(\mathbf{p}, E) = (\mathbf{p}_0, E_0)$ are given in reality, i.e., $(\mathbf{p}, E)$-space, with the probability density $\Psi^*(\mathbf{p}, E)\Psi(\mathbf{p}, E) = \Psi^*(\mathbf{p}_0, E_0)\Psi(\mathbf{p}_0, E_0)$. In the case of

$$\Psi^*(\mathbf{r}, t)\Psi(\mathbf{r}, t) = 0, \qquad (3.96)$$

with

$$\Psi^*(\mathbf{p}, E)\Psi(\mathbf{p}, E) = \Psi^*(\mathbf{p}_0, E_0)\Psi(\mathbf{p}_0, E_0) \neq 0, \qquad (3.97)$$

the system exists but is not observable. However, if

$$\Psi^*(\mathbf{p}, E)\Psi(\mathbf{p}, E) = \Psi^*(\mathbf{p}_0, E_0)\Psi(\mathbf{p}_0, E_0) = 0, \qquad (3.98)$$

the system does not exist. It cannot exist in $(\mathbf{p}, E)$-space because the probability density $\Psi^*(\mathbf{p}, E)\Psi(\mathbf{p}, E) = \Psi^*(\mathbf{p}_0, E_0)\Psi(\mathbf{p}_0, E_0)$, which defines its existence, is exactly zero and, therefore, any measurement at the space-time point $(\mathbf{r}, t)$ cannot lead to an event. In other words, a system which does not exist cannot be observed.

In the case of (3.96) and (3.97), the system exists but is not observable. However, the validity of the equations (3.96) and (3.98) means non-existence of the system.

In conclusion, in the case of $\Psi(\mathbf{p}_0, E_0) = 0$, i.e., $\Psi^*(\mathbf{p}_0, E_0)\Psi(\mathbf{p}_0, E_0) = 0$, the system $(\mathbf{p}_0, E_0)$ with mass m_0 is by definition non-existent.

3.10.4. Summary

We come to the conclusion that a free, elementary system cannot exist. There is obviously no other way to solve the problem in connection with the multi-valuedness of the wave function $\Psi(\mathbf{p}_0, E_0)$ and the probability density $\Psi^*(\mathbf{p}_0, E_0)\Psi(\mathbf{p}_0, E_0)$, respectively. We have discussed above that one cannot introduce new, physically acceptable conditions in connection with such simple, elementary systems. In particular, such systems cannot be changed because there is, by definition, no interaction possible with other systems.

3.11. Principle of Usefulness

Why does nature not admit non-interacting, free elementary systems without internal structure? Such free systems with constant momentum $\mathbf{p}_0$ and constant energy E_0 are in a certain sense "dead systems" because they are not involved in the processes in nature. There are no $(\mathbf{p}, E)$-fluctuations between a free system and its surroundings, and such systems would be useless. Such systems with

$$\mathbf{p}_0 = \text{const.} \neq 0, \tag{3.99}$$

and

$$E_0 = \text{const.} \neq 0, \tag{3.100}$$

have no place in nature because they would be completely detached from the scenario and this is against the principle of usefulness. Thus, from this point of view, free systems should not exist in nature. In fact, projection theory excludes such states explicitly because we have $\mathbf{p}_0 = 0$ and $E_0 = 0$ and $\Psi(\mathbf{r}, t) = 0$ and $\Psi(\mathbf{p}, E) = \Psi(\mathbf{p}_0, E_0) = 0$,[47] i.e., the properties given by the equations (3.99) and (3.100) do not exist within projection theory. In other words, projection theory supports the existence of the principle of usefulness.

In fact, the world of man is formed in accordance with the principle of usefulness, and a lot of examples, in particular in connection with evolution, support this view.[48] It is astonishing that this principle is obviously even reflected at the basic level of theoretical description (projection theory). However, this is not a description of the *objective reality* but it is "merely" man's view. Man is principally not able to recognize the properties of that which we have called "objective (basic) reality".

3.12. Further General Remarks

In the case of a non-interacting system, there are no $\mathbf{p}, E$-fluctuations, i.e., $\Delta \mathbf{p} = 0, \Delta E = 0$, and the constant values for the momentum $\mathbf{p}_0$ and the energy E_0 becomes zero within projection theory: $\mathbf{p}_0 = 0$ and $E_0 = 0$ with $\Psi(\mathbf{r}, t) = 0$ and $\Psi(\mathbf{p}, E) = \Psi(\mathbf{p}_0, E_0) = 0$. This behaviour is not restricted to $E_0 = \mathbf{p}_0^2/2m_0$,[49] but is of course also valid for the general relationship $E_0 = F(\mathbf{p}_0)$. The property $\mathbf{p}_0 = 0$ and $E_0 = 0$ comes exclusively into play by the fact that the $(\mathbf{p}, E)$-fluctuations for a free system vanish, i.e., we have $\Delta \mathbf{p} = 0, \Delta E = 0$. In other words, the transition

$$E_0 = \frac{\mathbf{p}_0^2}{2m_0} \to E_0 = F(\mathbf{p}_0), \qquad (3.101)$$

does not change the property $\mathbf{p}_0 = 0$ and $E_0 = 0$.

---

[47] See equations (3.56) and (3.65).
[48] See also Section 2.1.
[49] See Section 3.1.

Free, Non-interacting Systems (Particles) 149

Even in the case of non-vanishing $\mathbf{p}, E$-fluctuations, i.e., $\Delta \mathbf{p} \neq 0, \Delta E \neq 0$, our general statements are not restricted to the classical equation $E = \mathbf{p}^2/m_0 + U(x,y,z)$ as the starting point[50] but we could also formulate the background of the theory on the basis of the relationship $E = F(\mathbf{p}) + U(x,y,z)$ or $E = F(\mathbf{p}, \mathbf{r})$. That is, the transitions

$$E = \frac{\mathbf{p}^2}{2m_0} + U(x,y,z) \rightarrow$$
$$E = F(\mathbf{p}) + U(x,y,z) \rightarrow E = F(\mathbf{p}, x, y, z) \quad (3.102)$$

do not change the basic statements of projection theory. However, in the case of $E = \mathbf{p}^2/(2m_0) + U(x,y,z)$ we can work within usual mechanics in the classical limit.

3.13. Rest Mass Effect

The momentum $\mathbf{p}$ and the kinetic energy E of a particle with mass m_0 are given within the framework of classical mechanics by

$$\mathbf{p} = m_0 \mathbf{v},$$
$$E = \frac{1}{2} m_0 v^2, \quad (3.103)$$

where $\mathbf{v}$ is the velocity of the system. In the case of $\mathbf{v} = 0$, the energy E and the momentum $\mathbf{p}$ become zero but the particle nevertheless exists. It is resting and has the mass m_0.

In the relativistic case, its energy is not zero at $\mathbf{v} = 0$ but we have the rest energy $m_0 c^2$. The derivation of the expression $m_0 c^2$ is based on the postulates of Special Theory of Relativity. In particular, in connection with $m_0 c^2$ it is assumed that the Lorentz transformations are valid. However, we will recognize in Chapter 4 that the Lorentz transformations are not applicable within projection theory. Thus, the expression $m_0 c^2$ cannot be used to extend the classical, non-relativistic equation $E = \mathbf{p}^2/(2m_0)$ by rest mass effects. We showed however[51]

---

[50] See Section 2.2.7.1.
[51] Interested readers might refer to Schommers, W. (Ed.), (1989) Quantum theory and pictures of reality Springer-Verlag.

that the rest energy E_{m_0} can be expressed by

$$E_{m_0} = m_0 \gamma^{(1)} \tag{3.104}$$

in the first approximation without using the Lorentz transformations or other transformations.[52] $\gamma^{(1)}$ is a parameter and is in general not given by c^2.

For sufficiently small velocities **v** instead of equation (2.20), we may write with (3.104)

$$E = m_0 \gamma^{(1)} + \frac{\mathbf{p}^2}{2m_0} + U(x, y, z). \tag{3.105}$$

Applying the operator rules which we have found in Section 2.3.5, we obtain for both spaces the following equations:
($\mathbf{r}, t$)-space

$$i\hbar \frac{\partial}{\partial t} \Psi(\mathbf{r}, t) = m_0 \gamma^{(1)} \Psi(\mathbf{r}, t) - \frac{\hbar^2}{2m_0} \Delta \Psi(\mathbf{r}, t) + V(x, y, z, t) \Psi(\mathbf{r}, t), \tag{3.106}$$

($\mathbf{p}, E$)-space

$$E \Psi(\mathbf{p}, E) = m_0 \gamma^{(1)} \Psi(\mathbf{p}, E) + \frac{\mathbf{p}^2}{2m_0} \Psi(\mathbf{p}, E) + V \left(i\hbar \frac{\partial}{\partial p_x}, i\hbar \frac{\partial}{\partial p_y}, i\hbar \frac{\partial}{\partial p_z}, -i\hbar \frac{\partial}{\partial E} \right) \Psi(\mathbf{p}, E), \tag{3.107}$$

that is, the relativistic term is the same in both spaces. In the case of non-interacting systems, we get for
($\mathbf{r}, t$)-space:

$$V(x, y, z, t) \Psi(\mathbf{r}, t) = 0, \tag{3.108}$$

and for

---

[52] See also Appendix 3.E, equation (3.E.20).

Free, Non-interacting Systems (Particles)

$(\mathbf{p}, E)$-space:

$$V\left(i\hbar\frac{\partial}{\partial p_x}, i\hbar\frac{\partial}{\partial p_y}, i\hbar\frac{\partial}{\partial p_z}, -i\hbar\frac{\partial}{\partial E}\right)\Psi(\mathbf{p}, E) = 0. \quad (3.109)$$

Then, the equations for the determination of the wave functions for $(\mathbf{r}, t)$-space as well as for $(\mathbf{p}, E)$-space are given by

for $(\mathbf{r}, t)$-space:

$$i\hbar\frac{\partial}{\partial t}\Psi(\mathbf{r}, t) = m_0\gamma^{(1)}\Psi(\mathbf{r}, t) - \frac{\hbar^2}{2m_0}\Delta\Psi(\mathbf{r}, t), \quad (3.110)$$

for $(\mathbf{p}, E)$-space:

$$E\Psi(\mathbf{p}, E) = m_0\gamma^{(1)}\Psi(\mathbf{p}, E) + \frac{\mathbf{p}^2}{2m_0}\Psi(\mathbf{p}, E), \quad (3.111)$$

or

$$E_0\Psi(\mathbf{p}_0, E_0) = m_0\gamma^{(1)}\Psi(\mathbf{p}_0, E_0) + \frac{\mathbf{p}_0^2}{2m_0}\Psi(\mathbf{p}_0, E_0). \quad (3.112)$$

Applying the well-known function[53]

$$\Psi(\mathbf{r}, t) = C\Psi(\mathbf{p}_0, E_0)\exp\left\{\frac{i}{\hbar}(\mathbf{p}_0 \cdot \mathbf{r} - E_0 t)\right\}$$

to equation (3.110) we obtain equation (3.112). In other words, the term $m_0\gamma^{(1)}$ or similar constant expressions have no influence on the wave function $\Psi(\mathbf{r}, t)$ which is given in the case of non-interacting systems by equation (3.50). The mean values for the mean momentum $\bar{\mathbf{p}}$ and for the mean energy $\bar{E}$ become zero although we have considered the rest energy $m_0\gamma^{(1)}$. With $\bar{\mathbf{p}} = \mathbf{p}_0 = 0$ and $\bar{E} = E_0 = 0$,[54] we finally obtain with $2m_0 E_0 = 2m_0^2\gamma^{(1)} + \mathbf{p}_0^2$ for the mass m_0 a value of zero: $m_0 = 0$. Then, the properties $\mathbf{p}_0 = 0$, $E_0 = 0$ and $m_0 = 0$ define the non-existence of the system.

---

[53] See equation (3.50).
[54] See equation (3.89).

3.14. Summary

In the case of free systems the momentum $\mathbf{p} = \mathbf{p}_0$ and the energy $E = E_0$ remain constant in the course of time τ. There is no interaction between the system under investigation and other systems or environment, that is, the potential $V(x, y, z, t)$ is zero: $V(x, y, z, t) = 0$, and there are no $\mathbf{p}, E$-fluctuations in the case of free systems: $\Delta \mathbf{p} = 0$, $\Delta E = 0$. However, such non-interacting systems are completely useless and their existence would be against the principles of evolution.

Nevertheless, the world seems to be built up of subsystems which are normally assumed to be independent from the interaction between them. In other words, we suppose anyhow that there are really certain subsystems possible without mutual interactions. It is assumed that each of the subsystems can exist in an isolated state and that its existence is not dependent on certain external units and parameter, respectively. Our observations in everyday life suggest such a model. However, we have to be careful.

Such free non-interacting systems could be considered as useful, in the sense of evolution, if, at a certain time τ_{begin}, an interaction could be switched on and, in principle, could be switched off again at time τ_{end}. However, we do not observe such a switching-on/switching-off scenario in nature. In particular, it assumes a mechanism that is hostile against projection theory because it can obviously not be based on $\mathbf{p}, E$-fluctuations. In conclusion, free non-interacting systems should not exist in nature, and projection theory should be able to proof that if the projection principle contains the "principle of usefulness".[55]

The results of this chapter actually show that this is the case. We showed mathematically that free non-interacting systems cannot exist within projection theory. It turned out that the relevant quantities, the probability distributions $\Psi^*(\mathbf{r}, t)\Psi(\mathbf{r}, t)$ and $\Psi^*(\mathbf{p}_0, E_0)\Psi(\mathbf{p}_0, E_0)$, are exactly zero.

---

[55] See Section 2.1.

Appendix 3.A
Free System within Usual Quantum Theory
3.A.1. Superposition Ansatz

Within usual quantum theory, the superposition principle is assumed to be valid, that is, any wave function $\Psi(\mathbf{r})$ can be linearly superimposed with the help of a complete set of eigenfunctions $\Psi_f(\mathbf{r})$ and we have in the case of a continuous spectrum

$$\Psi(\mathbf{r}) = \int_{-\infty}^{\infty} a_f \Psi_f(\mathbf{r}) df, \qquad (3.A.1)$$

where a_f are constant coefficients and f are the eigenvalues. A variable for the system-specific time t does not exist here. Time within usual quantum theory is a simple external parameter[1] and we measure τ with our clocks.

If we use the eigenvalues $\mathbf{p}$ of the operator for the momentum, we have $f = \mathbf{p}$, and the eigenfunctions $\Psi_{\mathbf{p}}(\mathbf{r})$ are given by[2]

$$\Psi_{\mathbf{p}}(\mathbf{r}) = C \exp\left\{\frac{i}{\hbar}(\mathbf{p} \cdot \mathbf{r})\right\}. \qquad (3.A.2)$$

Clearly, the eigenvalues $\mathbf{p} = (p_x, p_y, p_z)$ form a continuous spectrum with a range from $-\infty$ to ∞. Instead of equation (3.A.1), we get

$$\Psi(\mathbf{r}) = \int_{-\infty}^{\infty} a_{\mathbf{p}} \Psi_{\mathbf{p}}(\mathbf{r}) d\mathbf{p}$$

$$= C \int_{-\infty}^{\infty} a_{\mathbf{p}} \exp\left\{\frac{i}{\hbar}(\mathbf{p} \cdot \mathbf{r})\right\} d\mathbf{p}. \qquad (3.A.3)$$

The superposition principle requires the condition[3]

$$\int_{-\infty}^{\infty} \Psi_{\mathbf{p}'}^{*}(\mathbf{r})\Psi_{\mathbf{p}}(\mathbf{r}) d\mathbf{r} = \delta(\mathbf{p}' - \mathbf{p}), \qquad (3.A.4)$$

---
[1] We have called it τ in all sections of this monograph; see in particular Section 2.3.9.
[2] Interested readers might refer to Landau, L.D., and Lifschitz, E.M., (1965) Quantum mechanics: Pergamon.
[3] Interested readers might refer to Landau, L.D., and Lifschitz, E.M., (1965) Quantum mechanics: Pergamon.

and this condition determines the constant C. With equation (3.A.2), we obtain

$$\int_{-\infty}^{\infty} \Psi_{\mathbf{p}'}^*(\mathbf{r})\Psi_{\mathbf{p}}(\mathbf{r})d\mathbf{r} = C^2 \int_{-\infty}^{\infty} \exp\left\{\frac{i}{\hbar}(\mathbf{p}\cdot\mathbf{r} - \mathbf{p}'\cdot\mathbf{r})\right\} d\mathbf{r}$$

$$= C^2(2\pi\hbar)^3\delta(\mathbf{p}' - \mathbf{p}), \quad (3.A.5)$$

leading to

$$C = \frac{1}{(2\pi\hbar)^{3/2}}. \quad (3.A.6)$$

In the case of a free non-interacting particle, the momentum $\mathbf{p}$ remains constant in the course of time τ and we have $\mathbf{p} = \mathbf{p}_0$ and the coefficients $a_\mathbf{p}$ must take the form

$$a_\mathbf{p} = \delta(p_x - p_{0x})\delta(p_y - p_{0y})\delta(p_z - p_{0z}) = \delta(\mathbf{p} - \mathbf{p}_0). \quad (3.A.7)$$

Then, we get with equations (3.A.3) and (3.A.6) for the wave function $\Psi(\mathbf{r})$ of a free system the form,

$$\Psi(\mathbf{r}) = \frac{1}{(2\pi\hbar)^{3/2}} \exp\left\{\frac{i}{\hbar}(\mathbf{p}_0\cdot\mathbf{r})\right\}. \quad (3.A.8)$$

The value $1/(2\pi\hbar)^{3/2}$ for the constant C has been determined on the basis of the superposition principle which is assumed to be valid within usual quantum theory.[4] Since we have no variable for the system-specific time t in usual quantum theory, a Fourier transform with respect to the variable t and E does not exist here. Therefore, a Fourier transform with respect to the variable t and E does not exist within usual quantum theory.

Equation (3.A.3), with $C = 1/(2\pi\hbar)^{3/2}$,

$$\Psi(\mathbf{r}) = \frac{1}{(2\pi\hbar)^{3/2}} \int_{-\infty}^{\infty} a_\mathbf{p} \exp\left\{\frac{i}{\hbar}(\mathbf{p}\cdot\mathbf{r})\right\} d\mathbf{p} \quad (3.A.9)$$

is nothing else than a Fourier transform, as in the case of projection theory, and we can identify the coefficient $a_\mathbf{p}$ with the wave function

---

[4] See equation (3.A.6).

Free, Non-interacting Systems (Particles)

$\Psi(\mathbf{p})$ of projection theory, and we get for the inverse Fourier transform

$$a_{\mathbf{p}} = \Psi(\mathbf{p}) = \frac{1}{(2\pi\hbar)^{3/2}} \int_{-\infty}^{\infty} \Psi(\mathbf{r}) \exp\left\{-\frac{i}{\hbar} \mathbf{p} \cdot \mathbf{r}\right\} d\mathbf{r}, \quad (3.A.10)$$

with

$$\int_{-\infty}^{\infty} \Psi^*(\mathbf{r})\Psi(\mathbf{r}) d\mathbf{r} = \int_{-\infty}^{\infty} a^*(\mathbf{p}) a(\mathbf{p}) d\mathbf{p}. \quad (3.A.11)$$

3.A.2. Mean Momentum for a Free System

The mean value $\bar{\mathbf{p}}$ of the momentum $\mathbf{p}$ is defined by[5]

$$\bar{\mathbf{p}} = \int_{-\infty}^{\infty} \Psi^*(\mathbf{r}) - i\hbar \frac{\partial}{\partial \mathbf{r}} \Psi(\mathbf{r}) d\mathbf{r}, \quad (3.A.12)$$

and must be

$$\bar{\mathbf{p}} = \mathbf{p}_0, \quad (3.A.13)$$

because the momentum for a free system remains constant in the course of time τ. However, condition (3.A.13) is not fulfilled in the case of equation (3.A.8) that is the result of usual quantum theory. This can easily be verified. With equation (3.A.12), we obtain

$$\bar{\mathbf{p}} = \int_{-\infty}^{\infty} \Psi^*(\mathbf{r}) - i\hbar \frac{\partial}{\partial \mathbf{r}} \Psi(\mathbf{r}) d\mathbf{r} = \frac{1}{(2\pi\hbar)^3} \mathbf{p}_0 \int_{-\infty}^{\infty} \exp\left\{-\frac{i}{\hbar} \mathbf{p}_0' \cdot \mathbf{r}\right\}$$
$$\times \exp\left\{\frac{i}{\hbar} \mathbf{p}_0 \cdot \mathbf{r}\right\} d\mathbf{r}$$
$$= \mathbf{p}_0 \lim_{\mathbf{p}_0 \to \mathbf{p}_0'} \delta(\mathbf{p}_0 - \mathbf{p}_0') = \mathbf{p}_0 [\delta(0)]^3 \to \infty. \quad (3.A.14)$$

That is, the mean value $\bar{\mathbf{p}}$ for the momentum is always infinity for each value $\mathbf{p}_0$, and this result is in contradiction to equation (3.A.13). Or we may say that in the case of $\bar{\mathbf{p}} = \infty$ only the value $\mathbf{p}_0 = \infty$ is a solution within usual quantum theory, and this must be excluded because we cannot have an infinite momentum in physical reality. The reason why $\bar{\mathbf{p}}$ becomes infinity is obvious: within usual quantum theory the wave

---

[5] Interested readers might refer to Landau, L.D., and Lifschitz, E.M., (1965) Quantum mechanics: Pergamon.

function $\Psi(\mathbf{r})$ cannot be normalized to unity in the case of a free non-interacting system but the integral

$$\int_{-\infty}^{\infty} \Psi^*(\mathbf{r})\Psi(\mathbf{r})d\mathbf{r} \qquad (3.A.15)$$

becomes infinity. We get

$$\begin{aligned}\int_{-\infty}^{\infty} \Psi^*(\mathbf{r})\Psi(\mathbf{r})d\mathbf{r} &= \frac{1}{(2\pi\hbar)^3} \int_{-\infty}^{\infty} \exp\left\{-\frac{i}{\hbar}\mathbf{p}_0' \cdot \mathbf{r}\right\} \\ &\quad \times \exp\left\{\frac{i}{\hbar}\mathbf{p}_0 \cdot \mathbf{r}\right\} d\mathbf{r} \\ &= \lim_{\mathbf{p}_0 \to \mathbf{p}_0'} \delta(\mathbf{p}_0 - \mathbf{p}_0') = \delta(0) \to \infty. \qquad (3.A.16)\end{aligned}$$

Within usual quantum theory the constant C is determined on the basis of the superposition principle and not as normalization constant.[6] The wave function $\Psi(\mathbf{r})$ of a free system cannot be normalized to unity within conventional quantum theory. This leads to a wrong value for the mean momentum $\bar{\mathbf{p}}$: Instead of $\bar{\mathbf{p}} = \mathbf{p}_0$,[7] we get $\bar{\mathbf{p}} = \mathbf{p}_0[\delta(0)]^3 \to \infty$.[8] Within projection theory, these problems do not appear.

3.A.3. Usual Quantum Theory and Projection Theory

Within usual quantum theory, the material world is embedded in space, i.e., $\mathbf{r}$-space. In connection with the superposition (3.A.3), the wave function $\Psi(\mathbf{r})$ and the coefficient $a_\mathbf{p}$ as well as the momentum $\mathbf{p}$ belong to $\mathbf{r}$-space.[9] Within usual quantum theory and classical physics, the $\mathbf{r}$-space comprises everything. Within conventional physics, the time always plays the role of an external parameter and is measured by our clocks. We have designated this time by τ.[10]

---

[6] See equation (3.A.2).
[7] See equation (3.A.13).
[8] See equation (3.A.14).
[9] See Figure 3.A.1(a).
[10] See Section 2.3.9.

Free, Non-interacting Systems (Particles)

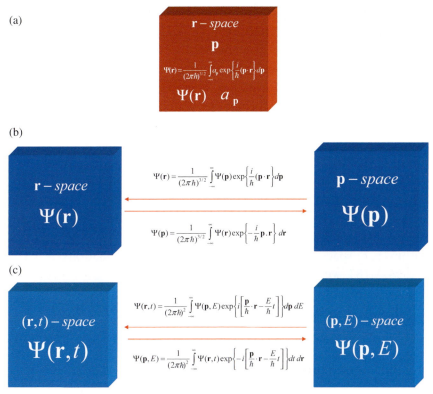

Fig. 3.A.1. On the superposition principle. (a) The situation within usual quantum theory. (b) The corresponding incomplete case within projection theory. (c) Projection theory, i.e., complete case.

The situation within projection theory is different: The coefficient $a_\mathbf{p}$ of usual quantum theory is identical with the wave function $\Psi(\mathbf{p})$ of projection theory and, instead of equation (3.A.9) we have

$$\Psi(\mathbf{r}) = \frac{1}{(2\pi\hbar)^{3/2}} \int_{-\infty}^{\infty} \Psi(\mathbf{p}) \exp\left\{\frac{i}{\hbar}(\mathbf{p}\cdot\mathbf{r})\right\} d\mathbf{p}. \qquad (3.A.17)$$

With the inverse transformation,

$$\Psi(\mathbf{p}) = \frac{1}{(2\pi\hbar)^{3/2}} \int_{-\infty}^{\infty} \Psi(\mathbf{r}) \exp\left\{-\frac{i}{\hbar}\mathbf{p}\cdot\mathbf{r}\right\} d\mathbf{r}. \qquad (3.A.18)$$

However, within projection theory the material world is not embedded in **r**-space but we have two spaces here: **r**-space and p-space. Both spaces are equivalent concerning their information content, and the Fourier transform (3.A.17) enables us to calculate the information $\Psi(\mathbf{r})$ of $\mathbf{r}$-space by means of the information $\Psi(\mathbf{p})$ of $\mathbf{p}$-space. On the other hand, the Fourier transform (3.A.18) enables the calculation of the information $\Psi(\mathbf{p})$, i.e., $\mathbf{p}$-space, by $\Psi(\mathbf{r})$, i.e., $\mathbf{r}$-space. The Fourier transformations connect the spaces. Therefore, equations (3.A.17) and (3.A.18) have nothing to do with a superposition in the sense of usual quantum theory. The situation is summarized in Fig. 3.A.1(b).

From the viewpoint of projection theory, a description within $\mathbf{r}$-space or $\mathbf{p}$-space must be considered as incomplete because the system-specific time t, which is not defined within usual quantum theory, has not been considered. Within projection theory, we work within $(\mathbf{r}, t)$-space or within $(\mathbf{p}, E)$-space. Instead of equations (3.A.17) and (3.A.18), we have

$$\Psi(\mathbf{r}, t) = \frac{1}{(2\pi\hbar)^2} \int_{-\infty}^{\infty} \Psi(\mathbf{p}, E) \exp\left\{i\left[\frac{\mathbf{p}}{\hbar} \cdot \mathbf{r} - \frac{E}{\hbar}t\right]\right\} d\mathbf{p}\, dE, \tag{3.A.19}$$

and the inverse transformation is given by

$$\Psi(\mathbf{p}, E) = \frac{1}{(2\pi\hbar)^2} \int_{-\infty}^{\infty} \Psi(\mathbf{r}, t) \exp\left\{-i\left[\frac{\mathbf{p}}{\hbar} \cdot \mathbf{r} - \frac{E}{\hbar}t\right]\right\} dt\, d\mathbf{r}. \tag{3.A.20}$$

As in the case of $\mathbf{r}$-space and $\mathbf{p}$-space, both spaces here are also equivalent concerning their information content. The equations enable us to transfer the information from $(\mathbf{p}, E)$-space to $(\mathbf{r}, t)$-space[11] and vice versa.[12] Also in this case the Fourier transformations connect the spaces. Thus, equations (3.A.19) and (3.A.20) have nothing to do with a superposition in the sense of usual quantum theory. The situation is summarized in Fig. 3.A.1(c).

Is the superposition principle valid in projection theory? Definitely not! We have investigated this point in Section 2.3.7.6. The

---
[11] See equation (3.A.19).
[12] See equation (3.A.20).

superposition ansatz within projection theory has to be formulated on the basis of a specific space as, for example, the $(\mathbf{r}, E)$-space with the wave function $\Psi(\mathbf{r}, E)$. In analogy to equation (3.A.1), we have to write

$$\varphi(\mathbf{r}) = \int_{-\infty}^{\infty} a_E \Psi(\mathbf{r}, E) dE. \qquad (3.A.21)$$

Using the equation for the determination of $\Psi(\mathbf{r}, E)$ in $(\mathbf{r}, E)$-space,[11]

$$E\Psi(\mathbf{r}, E) = -\frac{\hbar^2}{2m} \Delta \Psi(\mathbf{r}, E) + V\left(\mathbf{r}, -i\hbar \frac{\partial}{\partial E}\right) \Psi(\mathbf{r}, E), \qquad (3.A.22)$$

we can check the validity of the superposition ansatz (4.A.21). Due to the appearance of the operator

$$-i\hbar \frac{\partial}{\partial E},$$

for the time coordinate,[13] $\varphi(\mathbf{r})$ cannot be a solution for equation (3.A.22). Thus, the superposition ansatz (3.A.21) cannot be valid.

The same holds for $(\mathbf{r}, t)$-space and $(\mathbf{p}, E)$-space. Here the equations for the determination of $\Psi(\mathbf{r}, t)$ and $\Psi(\mathbf{p}, E)$ are expressed by equations (2.35) and (2.40). The linearly composed functions can be formulated by

$(\mathbf{r}, t)$-space:

$$\varphi(\mathbf{r}) = \int_{-\infty}^{\infty} b_t \Psi(\mathbf{r}, t) dt, \qquad (3.A.23)$$

and

$(\mathbf{p}, E)$-space:

$$\varphi(\mathbf{p}) = \int_{-\infty}^{\infty} a_E \Psi(\mathbf{p}, E) dE. \qquad (3.A.24)$$

As in the case of equation (3.A.21), these formulations[14] are not valid. More details concerning the superposition principle within projection theory are given in Section 2.2.7.6.

---

[13] This is not known within usual quantum theory.
[14] See equations (3.A.23) and (3.A.24).

Remark

Within usual quantum theory, the superposition principle is a basic property of all quantum-mechanical systems (independent on the interaction). The weakness of this principle has been discussed in Section 1.6. Let us repeat here the main points: Before a measurement, the properties of a quantum object are not independent realities, and it is meaningless to talk about the physical reality. They are not "either-or" alternative worlds, but there is a superposition of all possible properties or worlds. The effect of the measurement is to chop the overlapping worlds apart into disconnected worlds. In other words, the measurement process transforms the pure state, represented by the wave function Ψ or by the density matrix, into a mixture represented by the density matrix.

Within the Copenhagen interpretation of quantum theory, which is the most used interpretation, the collapse of the wave function and the assignment of statistical weights are not explained. Within this picture "*they are consequences of an external a priori metaphysics, which is allowed to intervene at this point and suspend the Schrödinger equation*".[15] In other words, although the Schrödinger equation[16] belongs to one of the basics of usual quantum theory, it is principally not able to explain the collapse of the wave function.

Appendix 3.B

If we use

$$\Psi(\mathbf{r}, t) = C \exp\left\{\frac{i}{\hbar}(\mathbf{p}_0 \cdot \mathbf{r} - E_0 t)\right\}, \qquad (3.B.1)$$

instead of

$$\Psi(\mathbf{r}, t) = C\Psi(\mathbf{p}_0, E_0) \exp\left\{\frac{i}{\hbar}(\mathbf{p}_0 \cdot \mathbf{r} - E_0 t)\right\}, \qquad (3.B.2)$$

we obtain solutions that are not acceptable.[1] Let us briefly discuss why.

---

[15] Interested readers might refer to de Witt, B., (1970) Physics today, September issue.
[16] Here the form of usual quantum theory is meant; see Chapter 1.
[1] See equation (3.49).

Free, Non-interacting Systems (Particles)

The wave function $\Psi(\mathbf{p}, E)$ is expressed by equation (2.5) and takes the following form in the case of equation (3.B.1):

$$\Psi(\mathbf{p}, E) = C(2\pi\hbar)^2 \delta(p_{0x} - p_x)\delta(p_{0y} - p_y)\delta(p_{0z} - p_z)\delta(E - E_0)$$
$$= C(2\pi\hbar)^2 \delta(\mathbf{p}_0 - \mathbf{p})\delta(E - E_0). \quad (3.B.3)$$

The constant C can be determined by normalizing the probability density $\Psi^*(\mathbf{r}, t)\Psi(\mathbf{r}, t)$ to unity:

$$\int_{-\infty}^{\infty} \Psi^*(\mathbf{r}, t)\Psi(\mathbf{r}, t) d\mathbf{r}\, dt = 1. \quad (3.B.4)$$

Because of

$$\int_{-\infty}^{\infty} \Psi^*(\mathbf{r}, t)\Psi(\mathbf{r}, t) d\mathbf{r}\, dt = \int_{-\infty}^{\infty} \Psi^*(\mathbf{p}, E)\Psi(\mathbf{p}, E) d\mathbf{p}\, dE, \quad (3.B.5)$$

also the probability density $\Psi^*(\mathbf{p}, E)\Psi(\mathbf{p}, E)$ in $(\mathbf{p}, E)$-space must be normalized to unity, we have

$$\int_{-\infty}^{\infty} \Psi^*(\mathbf{p}, E)\Psi(\mathbf{p}, E) d\mathbf{p}\, dE = 1. \quad (3.B.6)$$

Then, we get in the case of equations (3.B.1) and (3.B.4)

$$C^2 = \frac{1}{\lim_{\mathbf{p}_0 \to \mathbf{p}} \lim_{E_0 \to E} (2\pi\hbar)^4 \delta(p_{0x} - p_x)\delta(p_{0y} - p_y)\delta(p_{0z} - p_z)\delta(E - E_0)}$$

$$= \frac{1}{(2\pi\hbar)^4 [\delta(0)]^4}. \quad (3.B.7)$$

That is, the constant C becomes zero:

$$C = \frac{1}{(2\pi\hbar)^2 [\delta(0)]^2} \to 0. \quad (3.B.8)$$

Using equations (3.B.3) and (3.B.8), we immediately find with

$$\int_{-\infty}^{\infty} \delta(a - g)\delta(b - g) dg = \delta(a - b)$$

that equation (3.B.6) is fulfilled.

Using (3.B.8) we obtain $\Psi(\mathbf{r}, t) = 0$,[2] and of course

$$\Psi^*(\mathbf{r}, t)\Psi(\mathbf{r}, t) = 0, \quad -\infty \leq \mathbf{r}, t \leq \infty. \quad (3.B.9)$$

Then, equation (3.B.4) makes no sense. (otherwise, the result given by equation (3.B.9) makes no sense in the case of equation (3.B.4).) Due to equation (3.B.9), the detector never registers an event within $-\infty \leq \mathbf{r}, t \leq \infty$ and, therefore, it makes no sense to normalize the function $\Psi^*(\mathbf{r}, t)\Psi(\mathbf{r}, t)$ to unity.[2] Since equation (3.B.9) is a consequence of equation (3.B.4), we are confronted with a contradiction. In other words, the ansatz (3.B.1) cannot be valid within projection theory.

The probability ΔW for an event in the space-time intervals $\Delta \mathbf{r}$ and Δt is given by

$$\Delta W = \Psi^*(\mathbf{r}, t)\Psi(\mathbf{r}, t)\Delta \mathbf{r} \Delta t. \quad (3.B.10)$$

Due to equation (3.B.9) the quantity ΔW becomes uncertain if the intervals $\Delta \mathbf{r}$ and Δt are infinite, that is, ΔW is not defined in this case because we have

$$\Delta W = 0 \cdot \infty. \quad (3.B.11)$$

If the intervals $\Delta \mathbf{r}$ and Δt remain finite, we have $\Delta W = 0$ because $\Psi^*(\mathbf{r}, t)\Psi(\mathbf{r}, t)$ must always be zero in the range $-\infty \leq \mathbf{r}, t \leq \infty$.[3]

Furthermore, the probability in $(\mathbf{p}, E)$-space

$$\Delta W = \Psi^*(\mathbf{p}, E)\Psi(\mathbf{p}, E)\Delta \mathbf{p} \Delta E \quad (3.B.12)$$

of finding certain values $\mathbf{p}$ and E for the momentum and the energy is not properly defined but becomes uncertain in the case of equation (3.B.3). Since the system is permanent in the state $\mathbf{p}_0$ and E_0, i.e., the system does not interact, the intervals $\Delta \mathbf{p}$ and ΔE must be zero,[4] and we get

$$\Delta W = \lim_{\Delta \mathbf{p} \to 0} \lim_{\Delta E \to 0} \Psi^*(\mathbf{p}, E)\Psi(\mathbf{p}, E)\Delta \mathbf{p} \Delta E. \quad (3.B.13)$$

---

[2] See equation (3.B.4).
[3] See equation (3.B.9).
[4] See equation (3.B.12).

Because $\Psi^*(\mathbf{p}, E)\Psi(\mathbf{p}, E)$ becomes infinity[5] we get also in $(\mathbf{p}, E)$-space the property $\Delta W = \infty \cdot 0$, i.e., ΔW also becomes uncertain here and is not defined. Exactly the same result has been found for $(\mathbf{r}, t)$-space.[6] Also this correspondence underlines that the ansatz (3.B.1) makes no sense.

In the case of equation (3.B.2), the integrals

$$\int_{-\infty}^{\infty} \Psi^*(\mathbf{r}, t)\Psi(\mathbf{r}, t)d\mathbf{r}\, dt,$$

and

$$\int_{-\infty}^{\infty} \Psi^*(\mathbf{p}, E)\Psi(\mathbf{p}, E)d\mathbf{p}\, dE,$$

become automatically zero.[7]

However, it is easy to verify that we get the correct mean values $\bar{E}$ and $\bar{\mathbf{p}}$ for the energy E and the momentum $\mathbf{p}$:

$$\bar{\mathbf{p}} = \mathbf{p}_0, \bar{E} = E_0. \quad (3.B.14)$$

Within usual quantum theory an analogous ansatz is used[8] but here the mean values for $\bar{\mathbf{p}}$ for the momentum becomes infinity.[9]

Appendix 3.C

Equation (2.4) connects $\Psi(\mathbf{r}, t)$ with $\Psi(\mathbf{p}, E)$. If we choose for $\Psi(\mathbf{p}, E)$ the relation

$$\Psi(\mathbf{p}, E) = \delta(p_{0x} - p_x)\delta(p_{0y} - p_y)\delta(p_{0z} - p_z)\delta(E - E_0), \quad (3.C.1)$$

we obtain for $\Psi(\mathbf{r}, t)$

$$\Psi(\mathbf{r}, t) = \frac{1}{(2\pi\hbar)^2} \exp\left\{\frac{i}{\hbar}(\mathbf{p}_0 \cdot \mathbf{r} - E_0 t)\right\}. \quad (3.C.2)$$

---

[5] See equation (3.B.3).
[6] See equation (3.B.11).
[7] See Section 3.6.
[8] Equation (3.A.8) is analogous to equation (3.B.1); see Appendix 3.A.
[9] See equation (3.A.14).

This expression for $\Psi(\mathbf{r}, t)$ represents an extension of $\Psi(\mathbf{r})$ which we have treated in connection with usual quantum theory (Appendix 3.A). The extension is due to the existence of the system-specific time t within projection theory. In other words, the use of equation (3.C.1) leads to the transition

$$\Psi(\mathbf{r}) = \frac{1}{(2\pi\hbar)^{3/2}} \exp\left\{\frac{i}{\hbar}(\mathbf{p}_0 \cdot \mathbf{r})\right\} \to \Psi(\mathbf{r}, t)$$

$$= \frac{1}{(2\pi\hbar)^2} \exp\left\{\frac{i}{\hbar}(\mathbf{p}_0 \cdot \mathbf{r} - E_0 t)\right\}. \quad (3.C.3)$$

Thus, by the use of equation (3.C.2) the same problems appear as in usual quantum theory where equation (3.A.8) is used[1]

1. The normalization integrals

$$\int_{-\infty}^{\infty} \Psi^*(\mathbf{r}, t)\Psi(\mathbf{r}, t) d\mathbf{r}\, dt,$$

and

$$\int_{-\infty}^{\infty} \Psi^*(\mathbf{p}, E)\Psi(\mathbf{p}, E) d\mathbf{p}\, dE,$$

become infinity.

$$\int_{-\infty}^{\infty} \Psi^*(\mathbf{r}, t)\Psi(\mathbf{r}, t) d\mathbf{r}\, dt = \frac{1}{(2\pi\hbar)^4} \int_{-\infty}^{\infty} \exp\left\{-\frac{i}{\hbar}\mathbf{p}_0' \cdot \mathbf{r} + E_0' t\right\}$$

$$\times \exp\left\{\frac{i}{\hbar}\mathbf{p}_0 \cdot \mathbf{r} - E_0 t)\right\} d\mathbf{r}\, dt$$

$$= \lim_{\mathbf{p}_0 \to \mathbf{p}_0'} \lim_{E_0 \to E_0'} \delta(\mathbf{p}_0 - \mathbf{p}_0')\delta(E - E_0)$$

$$= [\delta(0)]^4 = \infty, \quad (3.C.4)$$

$$\int_{-\infty}^{\infty} \Psi^*(\mathbf{p}, E)\Psi(\mathbf{p}, E) d\mathbf{p}\, dE = [\delta(0)]^4 = \infty. \quad (3.C.5)$$

The property that the normalization integrals become infinity is not understandable in connection with the probability interpretation of the wave functions that is defined in terms of the absolute probability

---

[1] See Appendix 3.A.

and not with respect to the relative probability. In fact, the probability in $(\mathbf{p}, E)$-space $\Delta W = \Psi^*(\mathbf{p}, E)\Psi(\mathbf{p}, E)\Delta \mathbf{p}\Delta E$ of finding certain values $\mathbf{p}$ and E for the momentum and the energy is not properly defined but becomes uncertain in the case of equation (3.C.1). Since the values for $\mathbf{p}_0$ and E_0 remain constant, i.e., the system does not interact, the intervals $\Delta \mathbf{p}$ and ΔE must be zero and we get $\Delta W = \infty \cdot 0$, i.e., the probability ΔW in $(\mathbf{p}, E)$-space becomes uncertain and is not defined.[2]

Since $(\mathbf{r}, t)$-space and $(\mathbf{p}, E)$-space are completely equivalent, also the expression $\Delta W = \Psi^*(\mathbf{r}, t)\Psi(\mathbf{r}, t)\Delta \mathbf{r}\Delta t$ in $(\mathbf{r}, t)$-space should not be interpreted as probability when equation (3.C.2) is used, and this must hold, for the probabilities must be defined simultaneously[3]

$$\Delta W = \lim_{\Delta \mathbf{p} \to 0} \lim_{\Delta E \to 0} \Psi^*(\mathbf{p}, E)\Psi(\mathbf{p}, E)\Delta \mathbf{p}\Delta E.$$

Also this underlines that equations (3.C.1) and (3.C.2) do not make much sense within projection theory.

Thus, the property expressed by equations (3.C.4) and (3.C.5) should be excluded and, therefore, also the relations (3.C.1) and (3.C.2).

The only quantities in connection with a free non-interacting system are its momentum $\mathbf{p}_0$ and its energy E_0. The repeated measurement of the momentum and energy must lead in the middle to $\mathbf{p}_0$ and E_0 because there is no interaction and, therefore, the momentum and the energy remain constant in the course of time τ, and the wave function must be able to describe that. However, the relations for $\Psi(\mathbf{r}, t)$ with $\Psi(\mathbf{p}, E)$[4] are not able to fulfil this condition: $\bar{\mathbf{p}}$ and $\bar{E}$ become infinity and not $\mathbf{p}_0$ and E_0. We have

$$\bar{\mathbf{p}} = \int_{-\infty}^{\infty} \mathbf{p}\Psi^*(\mathbf{p}, E)\Psi(\mathbf{p} \cdot E) d\mathbf{p}\, dE$$

$$= \mathbf{p}_0 \lim_{\mathbf{p}_0 \to \mathbf{p}_0'} \lim_{E_0 \to E_0'} \delta(\mathbf{p}_0 - \mathbf{p}_0')\delta(E_0 - E_0')$$

$$= \mathbf{p}_0[\delta(0)]^4 = \infty, \qquad (3.C.6)$$

---

[2] See also the discussion in Appendix 3.B.
[3] See equation (3.C.5b).
[4] See equations (3.C.1) and (3.C.2).

and

$$\bar{E} = \int_{-\infty}^{\infty} E\Psi^*(\mathbf{p}, E)\Psi(\mathbf{p} \cdot E)d\mathbf{p}\,dE$$
$$= E_0 \lim_{\mathbf{p}_0 \to \mathbf{p}'_0} \lim_{E_0 \to E'_0} \delta(\mathbf{p}_0 - \mathbf{p}'_0)\delta(E - E_0) = E_0[\delta(0)]^4 = \infty. \tag{3.C.7}$$

What purpose should a wave function have when it is not able to express the physical quantities of a system which are $\bar{\mathbf{p}}$ and $\bar{E}$, or $\mathbf{p}_0$ and E_0, in the case of a free system?

Remark concerning usual quantum theory

Within usual quantum theory we have no choice. The normalization integral becomes infinity, as in the case of equation (3.C.4), and this property comes into play through the superposition ansatz leading to condition (3.A.4) and to expression (3.A.8) for the wave function $\Psi(\mathbf{r})$. However, as we have pointed out in the Appendix 3.A, the superposition ansatz leads to problems because the collapse of the wave function cannot be described by the most basic equation in usual quantum theory [Schrödinger's equation (1.3)], but this collapse has to be considered as "*a consequences of an external a priori metaphysics*"[5] and this is unsatisfactory.

Appendix 3.D
The Stationary Case
3.D.1. Definition

If we assume that the potential is not dependent on time t, we have,

$$V(x, y, z, t) = V(x, y, z), \tag{3.D.1}$$

and we obtain the stationary equations

---
[5] Interested readers might refer to de Witt, B., (1970) Physics today, September issue.

($\mathbf{r}, t$)-space:

$$i\hbar \frac{\partial}{\partial t}\Psi(\mathbf{r},t) = -\frac{\hbar^2}{2m_0}\Delta\Psi(\mathbf{r},t) + V(x,y,z)\Psi(\mathbf{r},t), \quad (3.D.2)$$

($\mathbf{r}, E$)-space:

$$E_S\Psi(\mathbf{r},E_S) = -\frac{\hbar^2}{2m_0}\Delta\Psi(\mathbf{r},E_S) + V(x,y,z)\Psi(\mathbf{r},E_S), \quad (3.D.3)$$

instead of the general cases that are given by equations (2.35) and (2.50). In order to be able to distinguish between the general and the stationary case, we have used in equation (3.D.3) the letter E_S for the energy instead of E (general case). Equation (3.D.3) is identical with the stationary Schrödinger equation of usual quantum theory if $V(x,y,z) = U(x,y,z)$. However, within projection theory the situation is more complex than in conventional quantum theory because we have an equation[1] for ($\mathbf{r}, t$)-space and another for ($\mathbf{r}, E$)-space[2] which has to be compatible. The wave functions $\Psi(\mathbf{r},E_S)$ and $\Psi(\mathbf{r},t)$ are interconnected by a Fourier transform, with respect to the variable $E = E_S$ and t, of the form[3]

$$\Psi(\mathbf{r},t) = \frac{1}{(2\pi\hbar)^{1/2}}\int_{-\infty}^{\infty}\Psi(\mathbf{r},E_S)\exp\left\{-i\frac{E_S}{\hbar}t\right\}dE_S, \quad (3.D.4)$$

which is not known in usual quantum theory and, therefore, we cannot simply take over the results of usual quantum theory. In fact, we will recognize in the following that projection theory leads to a completely different result for the stationary case than that of conventional quantum theory.

3.D.2. Relevant Properties

As we have already remarked, equation (3.D.3) is identical with the stationary Schrödinger equation of usual quantum theory. We know from usual quantum theory that the stationary case leads to a discrete

---

[1] See equation (3.D.2).
[2] See equation (3.D.3).
[3] This is the inverse transformation of equation (2.43).

energy distribution and of course another part which we call continuous spectrum.[4] For example, we have $E_{S1}, E_{S2}, E_{S3}, \ldots$. Thus, one could argue that this behaviour must principally also be valid for projection theory because the basic equations of usual quantum theory and projection theory have the same structure. However, it will turn out in the course of our discussion that systems such as stationary systems are physically not definable on the basis of equations (3.D.3) and (3.D.4). In the following, we want to discuss the stationary case for a certain energy E_S which represents on trial one of the possible solutions $E_{S1}, E_{S2}, E_{S3}, \ldots$.[5]

In the stationary case,[6] there must exist the following relationship between $\Psi(\mathbf{r}, t)$ and $\Psi(\mathbf{r}, E_S)$:

$$\Psi(\mathbf{r}, t) = C\Psi(\mathbf{r}, E_S) \exp\left[-\frac{i}{\hbar} E_S t\right], \qquad (3.D.5)$$

where C is a constant. When we put expression (3.D.5) into equation (3.D.2), we obtain equation (3.D.3). Equation (D.5) contains interesting information. Let us briefly discuss why, first in general form, and after that in connection with equation (3.D.4).

In general, *one* space-time point $(\mathbf{r}, t)$ contains in principle *all* possible values $(\mathbf{p}, E)$, and this property is due to the Fourier transform[7]

$$\Psi(\mathbf{r}, t) = \frac{1}{(2\pi\hbar)^2} \int_{-\infty}^{\infty} \Psi(\mathbf{p}, E) \exp\left\{i\left[\frac{\mathbf{p}}{\hbar} \cdot \mathbf{r} - \frac{E}{\hbar} t\right]\right\} d\mathbf{p}\, dE. \qquad (3.D.6)$$

In connection with equation (3.D.4), we work in $(\mathbf{r}, t)$-space and simultaneously in $(\mathbf{r}, E)$-space. Therefore, the position $\mathbf{r}$ and the momentum $\mathbf{p}$ are not concerned; only the information with respect to the variables t and $E = E_S$ have to be considered. However, not only equation (3.D.4) is valid but also equation (3.D.5) and this is a fundamental point. In contrast to the general case, equation (3.D.5) means that *one* value of t only contains *one* value of the energy, namely

---
[4] The hydrogen atom is a typical example.
[5] This consists of discrete and continuous spectrum.
[6] Within projection theory, defined by equation (3.D.1).
[7] See equation (2.4).

E_S. Moreover, the value E_S determines *all* values t of $\Psi(\mathbf{r},t)$, although other energy values E_S are defined by equation (3.D.3). In other words, the variable E_S is treated as a constant in connection with equation (3.D.5) and, therefore, also in connection with equation (3.D.4) the variable E_S must treated as a constant leading directly to

$$\Psi(\mathbf{r},t) = \frac{1}{(2\pi\hbar)^{1/2}} \Psi(\mathbf{r}, E_S) \exp\left[-\frac{i}{\hbar} E_S t\right] \lim_{\Delta_E \to 0} \int_{E_S}^{E_S + \Delta_{E_S}} dE_S, \quad (3.D.7)$$

where Δ_{E_S} is a small interval in E-space. By comparison with equation (3.D.5), we have

$$C = \frac{1}{(2\pi\hbar)^{1/2}} \lim_{\Delta_E \to 0} \int_{E_S}^{E_S + \Delta_{E_S}} dE_S = 0. \quad (3.D.8)$$

In the case of $\Psi(\mathbf{r}, E_S) < \infty$, we obtain

$$\Psi(\mathbf{r},t) = 0. \quad (3.D.9)$$

On the other hand, with equation (3.D.5) we also have[8]

$$\Psi(\mathbf{r}, E_S) = \frac{1}{(2\pi\hbar)^{1/2}} \int_{-\infty}^{\infty} \Psi(\mathbf{r},t) \exp\left\{i \frac{E_S}{\hbar} t\right\} dt$$

$$= \lim_{E_S' \to E_S} \Psi(\mathbf{r}, E_S') C \frac{1}{(2\pi\hbar)^{1/2}} \lim_{t_E \to \infty} \int_{-t_E}^{t_E} \exp\left\{\frac{i}{\hbar}(E_S - E_S')t\right\} dt, \quad (3.D.10)$$

leading to

$$\Psi(\mathbf{r}, E_S) = \Psi(\mathbf{r}, E_S) C \frac{1}{(2\pi\hbar)^{1/2}} \lim_{E_S' \to E_S} \lim_{t_E \to \infty} 2\hbar \frac{\sin(E_S - E_S')t_E/\hbar}{(E_S - E_S')}. \quad (3.D.11)$$

With

$$\lim_{t_E \to \infty} 2\hbar \frac{\sin(E_S - E_S')t_E/\hbar}{(E_S - E_S')} = 2\pi\delta(E_S/\hbar - E_S'/\hbar), \quad (3.D.12)$$

---

[8] We used the inverse transformation of equation (3.D.4).

we obtain[9]

$$C = \frac{(2\pi\hbar)^{1/2}}{\lim_{E'_S \to E_S} \lim_{t_E \to \infty} 2\hbar \frac{\sin(E_S - E'_S)t_E/\hbar}{(E_S - E'_S)}}$$

$$= \frac{\hbar^{1/2}}{\lim_{E'_S \to E_S} (2\pi)^{1/2} \delta(E_S/\hbar - E'_S/\hbar)}$$

$$= \frac{\hbar^{1/2}}{(2\pi)^{1/2}\delta(0)}, \qquad (3.D.13)$$

and finally $C = 0$ as in the case of equation (3.D.8). With $\Psi(\mathbf{r}, E_s) < \infty$, we get the relation $\Psi(\mathbf{r}, t) = 0$.[10]

3.D.3. The Behaviour of the Wave Functions

3.D.3.1. *Singularities*

If there are certain points $\mathbf{r}$ in $(\mathbf{r}, E_S)$-space where the function $\Psi(\mathbf{r}, E_S)$ becomes infinity, i.e., singularities with $\Psi(\mathbf{r}, E_S) \to \infty$, the function $\Psi(\mathbf{r}, t)$ becomes uncertain, i.e., $\Psi(\mathbf{r}, t) = 0 \cdot \infty$, at those points $\mathbf{r}$ in $(\mathbf{r}, t)$-space where a singularity

$$\Psi(\mathbf{r}, E_S) \to \infty \qquad (3.D.14)$$

appears. This is because the term

$$\lim_{\Delta E \to 0} \int_{E_S}^{E_S + \Delta E_S} dE_S$$

in equation (3.D.7) is zero.

If we denote the position of a singularity by $\mathbf{r}_\infty$ we get: $\Psi(\mathbf{r}_\infty, E_S) \to \infty$ and $\Psi(\mathbf{r}_\infty, t) = 0 \cdot \infty$.

There are the following properties around each singularity. The values for $\Psi(\mathbf{r}, t)$ around a certain singularity $\mathbf{r} = \mathbf{r}_\infty$ are exactly zero

---

[9] See relation (3.53).
[10] See equation (3.D.9).

even when the difference between **r** and $\mathbf{r}_\infty$ is infinitesimal. In other words, the function $\Psi(\mathbf{r},t)$ is zero for all $\mathbf{r} = \mathbf{r}_\infty \pm \varepsilon$ where ε is infinitesimal but different from zero. In this case $\Psi(\mathbf{r}, E_S)$ must be simultaneously expressed by $\Psi(\mathbf{r}, E_S) < \infty$ for all $\mathbf{r} = \mathbf{r}_\infty \pm \varepsilon$ which follows directly from equation (3.D.7).

While $\Psi(\mathbf{r}_\infty \pm \varepsilon, E_S)$ can be different from zero, but must be by definition $\Psi(\mathbf{r}_\infty \pm \varepsilon, E_S) < \infty$ for $\mathbf{r} = \mathbf{r}_\infty \pm \varepsilon \neq \mathbf{r}_\infty$, the function $\Psi(\mathbf{r}_\infty \pm \varepsilon, t)$ is exactly zero for all $\mathbf{r} = \mathbf{r}_\infty \pm \varepsilon \neq \mathbf{r}_\infty$. However, $\Psi(\mathbf{r},t)$ becomes uncertain for $\mathbf{r} = \mathbf{r}_\infty$, i.e., we have $\Psi(\mathbf{r}_\infty, t) = \infty \cdot 0$. In summary, we have the following situation:

$\mathbf{r} = \mathbf{r}_\infty$:

$$\Psi(\mathbf{r}, t) = \infty \cdot 0,$$
$$\Psi(\mathbf{r}, E_S) \to \infty, \qquad (3.D.15)$$

$\mathbf{r} = \mathbf{r}_\infty \pm \varepsilon \neq \mathbf{r}_\infty$:

$$\Psi(\mathbf{r}, t) = 0,$$
$$\Psi(\mathbf{r}, E_S) < \infty. \qquad (3.D.16)$$

Equation (3.D.15) leads to an undefined expression for the basic equation (3.D.3) and such a behaviour has to be excluded because the basic equation (3.D.3) belongs to the fundamentals of projection theory. We may exclude certain solutions but not the fundamental picture, that is, equation (3.D.3). Thus, we have to eliminate the case $\Psi(\mathbf{r}_\infty, E_S) \to \infty$,[11] because it reflects an unphysical situation from the point of view of projection theory and has therefore to be considered as strictly forbidden. In other words, singularities are not allowed. Only the case $\Psi(\mathbf{r}, E_S) < \infty$ has to be admitted for each E_S and all positions **r**. Therefore, the function $\Psi(\mathbf{r},t)$ is zero for all values **r** and t, i.e., $\Psi(\mathbf{r},t) = 0$, and not only between certain singularities.

---

[11] See equations (3.D.14) and (3.D.15).

3.D.3.2. *The probability argument*

We already used the probability argument in connection with the wave function $\Psi(\mathbf{p}_0, E_0)$ for a free non-interacting system.[12] Let us use the same argument also for statements about the behaviour of the stationary function $\Psi(\mathbf{r}, E_S)$.

The wave function $\Psi(\mathbf{r}, E_S)$ cannot exist for $\Psi(\mathbf{r}, E_S) = \infty$.[13] But can the function $\Psi(\mathbf{r}, E_S)$ in the case of the stationary case really exist for $\Psi(\mathbf{r}, E_S) \neq 0$? It cannot for the following simple reason. The probability

$$\Delta W = \Psi^*(\mathbf{r}, E_S)\Psi(\mathbf{r}, E_S)\Delta E_S$$

of finding the stationary system in the state $\mathbf{r}$ and E_S must be zero because we have for the energy interval $\Delta E_S = 0$ and the probability density $\Psi^*(\mathbf{r}, E_S)\Psi(\mathbf{r}, E_S) \neq \infty$.[14] The condition $\Delta E_S = 0$ follows directly from the fact that only the value E_S is defined in connection with the basic equation (3.D.3) but not energy values E with $E \neq E_S$, i.e., we have no uncertainty with respect to the variable E_S in the case of a stationary system and, therefore, we have $\Delta E_S = 0$.

Makes no sense at all to have $\Delta W = 0$ for the probability and, on the other hand, $\Psi^*(\mathbf{r}, E_S)\Psi(\mathbf{r}, E_S) \neq 0 \{\Psi(\mathbf{r}, E_S) \neq 0\}$ for the probability density and the wave function, respectively? Such a behaviour cannot be justified and is rather unphysical. Generally, the property $\Psi^*(\mathbf{r}, E_S)\Psi(\mathbf{r}, E_S) \neq 0$ has to be connected to $\Delta W \neq 0$ but not to $\Delta W = 0$. Therefore, the property $\Delta W = 0$ has to be connected to $\Psi^*(\mathbf{r}, E_S)\Psi(\mathbf{r}, E_S) = 0 \{\Psi(\mathbf{r}, E_S) = 0\}$ and not to the case $\Psi^*(\mathbf{r}, E_S)\Psi(\mathbf{r}, E_S) \neq 0 \{\Psi(\mathbf{r}, E_S) \neq 0\}$. In other words, we should have the following links:

$$\Delta W \neq 0 \leftrightarrow \Psi^*(\mathbf{r}, E_S)\Psi(\mathbf{r}, E_S) \neq 0\{\Psi(\mathbf{r}, E_S) \neq 0\}, \quad (3.D.17)$$

$$\Delta W = 0 \leftrightarrow \Psi^*(\mathbf{r}, E_S)\Psi(\mathbf{r}, E_S) = 0\{\Psi(\mathbf{r}, E_S) = 0\}, \quad (3.D.18)$$

---

[12] See Section 3.5.
[13] See Section 3.D.3.1.
[14] See Section 3.D.3.1.

and nothing else. Since we have $\Delta W = 0$ for a stationary system, also the wave function $\Psi(\mathbf{r}, E_S)$ must be zero in this case:

$$\Psi(\mathbf{r}, E_S) = 0. \qquad (3.D.19)$$

We may also argue as follows. Both spaces, i.e., $(\mathbf{r}, t)$-space and $(\mathbf{r}, E)$-space, are completely equivalent concerning their contents or information. Both spaces only differ in the representation of the results.[15] The result $\Psi(\mathbf{r}, t) = 0$ means *non-existence* in $(\mathbf{r}, t)$-space.[16] Thus, we must also have *non-existence* in $(\mathbf{p}, E)$-space leading directly to $\Psi(\mathbf{r}, E_S) = 0$ which is identical with the result given by equation (3.D.19) that we have found on the basis of probability arguments.

3.D.3.3. More details concerning the potential V(x,y,z)

What does the transition from the t-dependent potential $V(x, y, z, t)$ to the t-independent quantity $V(x, y, z)$ physically mean? What about the quantum effects with respect to $V(x, y, z)$? These questions can only be answered in connection with the basic equations that are formulated for $E = E_S$. Also here we would like to work within $(\mathbf{r}, t)$-space and $(\mathbf{r}, E_S)$-space, respectively. The relevant equations are given by[17]:

$$\left\{ E_S + \frac{\hbar^2}{2m_0} \Delta - U(x, y, z) \right\} \Psi(\mathbf{r}, E_S)$$

$$= \frac{1}{(2\pi\hbar)^{1/2}} \int_{-\infty}^{\infty} \left\{ i\hbar \frac{\partial}{\partial t} + \frac{\hbar^2}{2m_0} \Delta - U(x, y, z) \right\} \Psi(\mathbf{r}, t)$$

$$\times \exp\left[\frac{i}{\hbar} E_S t \right] dt, \qquad (3.D.20)$$

with

$$\left\{ E_S + \frac{\hbar^2}{2m_0} \Delta - U(x, y, z) \right\} \Psi(\mathbf{r}, E_S) = f(\mathbf{r}, E_S), \qquad (3.D.21)$$

---

[15] See Section 2.3.4.
[16] See equations (3.D.9) and (3.D.16).
[17] See Section 2.3.7.3.

and

$$\left\{i\hbar\frac{\partial}{\partial t} + \frac{\hbar^2}{2m_0}\Delta - U(x,y,z)\right\}\Psi(\mathbf{r},t) = f(\mathbf{r},t). \quad (3.D.22)$$

The functions $f(\mathbf{r},t)$ and $f(\mathbf{r},E_S)$ are connected by the Fourier transform[18]

$$f(\mathbf{r},E_S) = \frac{1}{(2\pi\hbar)^{1/2}} \int_{-\infty}^{\infty} f(\mathbf{r},t) \exp\left\{i\frac{E_S}{\hbar}t\right\} dt. \quad (3.D.23)$$

The inverse transformation is

$$f(\mathbf{r},t) = \frac{1}{(2\pi\hbar)^{1/2}} \int_{-\infty}^{\infty} f(\mathbf{r},E_S) \exp\left\{-i\frac{E_S}{\hbar}t\right\} dE_S. \quad (3.D.24)$$

Because the energy E_S is treated as a constant, the last integral can be written as[19]

$$f(\mathbf{r},t) = \frac{1}{(2\pi\hbar)^{1/2}} f(\mathbf{r},E_S) \exp\left[-\frac{i}{\hbar}E_S t\right] \lim_{\Delta E \to 0} \int_{E_S}^{E_S+\Delta E_S} dE_S. \quad (3.D.25)$$

The function $f(\mathbf{r},E_S)$ may have singularities where it becomes infinity:

$$f(\mathbf{r},E_S) \to \infty. \quad (3.D.26)$$

Now we use exactly the same arguments for the estimation of the properties of $f(\mathbf{r},E_S)$ and $f(\mathbf{r},t)$ as we applied in connection with $\Psi(\mathbf{r},t)$ and $\Psi(\mathbf{r},E_S)$.[20] The relationships between $f(\mathbf{r},E_S)$ and $f(\mathbf{r},t)$ are very similar and are deduced as follows.

If we denote the position of a singularity by $\mathbf{r}_\infty$ we get: $f(\mathbf{r}_\infty, E_S) \to \infty$. Because of[21]

$$\lim_{\Delta E \to 0} \int_{E_S}^{E_S+\Delta E_S} dE_S = 0,$$

and the application of $f(\mathbf{r}_\infty, E_S) \to \infty$, we obtain for $f(\mathbf{r}_\infty, t)$ an uncertain expression at position $\mathbf{r}_\infty$: $f(\mathbf{r}_\infty, t) = \infty \cdot 0$. However, the

---

[18] See equation (2.47).
[19] This is in analogy to equation (3.D.7).
[20] See Section 3.D.3.1.
[21] See equation (3.D.25).

values for $f(\mathbf{r}, t)$ around a certain singularity $\mathbf{r} = \mathbf{r}_\infty$ are exactly zero with $f(\mathbf{r}, E_S) < \infty, \mathbf{r} = \mathbf{r}_\infty \pm \varepsilon, E_S$, even when the difference ε between $\mathbf{r}$ and $\mathbf{r}_\infty$ is infinitesimal: $\mathbf{r} = \mathbf{r}_\infty \pm \varepsilon, E_S$.

The function $f(\mathbf{r}_\infty \pm \varepsilon, t)$ is exactly zero for all $\mathbf{r} = \mathbf{r}_\infty \pm \varepsilon \neq \mathbf{r}_\infty$ except for $\mathbf{r} = \mathbf{r}_\infty$ where it becomes uncertain, i.e., $f(\mathbf{r}_\infty, t) = \infty \cdot 0$. On the other hand, the function $f(\mathbf{r}_\infty \pm \varepsilon, E_S)$ can be simultaneously different from zero, but must be by definition $f(\mathbf{r}_\infty \pm \varepsilon, E_S) < \infty$ for $\mathbf{r} = \mathbf{r}_\infty \pm \varepsilon \neq \mathbf{r}_\infty$. In summary, as in the case of $\Psi(\mathbf{r}, t)$ and $\Psi(\mathbf{r}, E_S)$[20] we have the following situation:

$\mathbf{r} = \mathbf{r}_\infty$:

$$f(\mathbf{r}, t) = \infty \cdot 0,$$
$$f(\mathbf{r}, E_S) \to \infty, \qquad (3.D.27)$$

$\mathbf{r} = \mathbf{r}_\infty \pm \varepsilon \neq \mathbf{r}_\infty$:

$$f(\mathbf{r}, t) = 0,$$
$$f(\mathbf{r}, E_S) < \infty. \qquad (3.D.28)$$

Equation (3.D.27) leads to an undefined expression for the basic equation (3.D.22) and such a behaviour has to be excluded because the basic equation (3.D.22) belongs to the fundamentals of projection theory. Thus, the case $f(\mathbf{r}_\infty, E_S) \to \infty$[22] reflects also here an unphysical situation and has therefore to be considered as strictly forbidden. In other words, singularities are not allowed. Only the case $f(\mathbf{r}, E_S) < \infty$[23] has to be admitted for each E_S and all positions r and, therefore, the function $f(\mathbf{r}, t)$ is zero for all values $\mathbf{r}$ and t ($f(\mathbf{r}, t) = 0$) and not only between certain singularities. As in the case of $\Psi(\mathbf{r}, t)$ and $\Psi(\mathbf{r}, E_S)$[24] we can also go a step further by applying the inverse Fourier transformation of (3.D.24) that is given by equation (3.D.23). Since $f(\mathbf{r}, t)$ is zero for all times t and all positions r, the Fourier integral

---
[22] See equations (3.D.26) and (3.D.27).
[23] See equations (3.D.28).
[24] See Section 3.D.3.1.

(3.D.23) must be zero for $E_S \neq 0$ and we obtain

$$f(\mathbf{r}, E_S) = 0, \qquad (3.D.29)$$

instead of the more general expression $f(\mathbf{r}, E_S) < \infty$. However, also in the case of $E_S = 0$ the function $f(\mathbf{r}, E_S)$ must be zero because we have the following situation: Both spaces, i.e. $(\mathbf{r}, t)$-space and $(\mathbf{r}, E_S)$-space, are equivalent and, therefore, the interaction potentials should exactly be the same for both spaces. Since $f(\mathbf{r}, t)$ is zero for all values $\mathbf{r}$ and t, the potential $V(x, y, z)$ takes the form $U(x, y, z)$ in $(\mathbf{r}, t)$-space; see in particular equations (3.D.2) and (3.D.22). Thus, also the potential in $(\mathbf{r}, E_S)$-space in equation (3.D.3) must be given by $V(x, y, z) = U(x, y, z)$. From this and equation (3.D.21) directly follows that the function $f(\mathbf{r}, E_s)$ must also be zero for all values $\mathbf{r}$ and E_s, in particular for $E_s = 0$.

In conclusion, we have $f(\mathbf{r}, t) = 0$ and simultaneously $f(\mathbf{r}, E_S) = 0$, and equations (3.D.21) and (3.D.22) become

$$\left\{i\hbar \frac{\partial}{\partial t} + \frac{\hbar^2}{2m_0}\Delta - U(x, y, z)\right\} \Psi(\mathbf{r}, t) = 0, \qquad (3.D.30)$$

and

$$\left\{E_S + \frac{\hbar^2}{2m_0}\Delta - U(x, y, z)\right\} \Psi(\mathbf{r}, E_S) = 0. \qquad (3.D.31)$$

When we compare these two expressions with the equations (3.D.2) and (3.D.3), we obtain for the stationary potential $V(x, y, z)$ the classical limit $U(x, y, z)$:

$$V(x, y, z) = U(x, y, z). \qquad (3.D.32)$$

In conclusion, the stationary case in projection theory is characterized by the classical potential $U(x, y, z)$, i.e., the quantum-mechanical aspect with respect to the potential $V(x, y, z)$ is eliminated here. In fact, we have seen in Section 2.3.7.3 that the quantum-theoretical aspect comes into play through the system-specific time t and we got $V(x, y, z, t)$, i.e., there is the transition $U(x, y, z) \rightarrow V(x, y, z, t)$ from the classical to the quantum-theoretical aspect but not with respect to $U(x, y, z) \rightarrow V(x, y, z)$.[25]

3.D.3.4. *Mean value for the energy*

The mean value for the energy is expressed by

$$\bar{E} = \int_{-\infty}^{\infty} \Psi^*(\mathbf{r}, t) \left[i\hbar \frac{\partial}{\partial t} \right] \Psi(\mathbf{r}, t) d\mathbf{r} \, dt$$

$$= \int_{-\infty}^{\infty} E\Psi^*(\mathbf{r}, E_S) \Psi(\mathbf{r}, E_S) d\mathbf{r} \, dE. \qquad (3.D.33)$$

With equation (3.D5) we obtain

$$\bar{E} = E_S C^2 B \lim_{E_S' \to E_S} \lim_{t_E \to \infty} \int_{-t_E}^{t_E} \exp\left[\frac{i}{\hbar}(E_S - E_S')t\right] dt, \qquad (3.D.34)$$

with

$$B = \int_{-\infty}^{\infty} \Psi^*(\mathbf{r}, E_S) \Psi(\mathbf{r}, E_S) d\mathbf{r}. \qquad (3.D.35)$$

The parameter C is given by expression (3.D.13). The evaluation of equation (3.D.34) leads to

$$\bar{E} = E_S C^2 B \lim_{E' \to E} \lim_{t_E \to \infty} 2\hbar \frac{\sin(E_S - E_S')t_E/\hbar}{(E_S - E_S')}$$

$$= E_S B \hbar \frac{1}{\lim_{E' \to E} \delta(E_S/\hbar - E_S'/\hbar)}$$

$$= \frac{E_S B \hbar}{\delta(0)}. \qquad (3.D.36)$$

In other words, the mean energy must be zero as B is normalized and not infinite:

$$\bar{E} = 0. \qquad (3.D.37)$$

---

[25] See equation (3.D.32).

Because the energy $E = E_S$ is a constant in the stationary state, the value for the energy must be identical to $\bar{E}$:

$$\bar{E} = E_S, \tag{3.D.38}$$

and, because of $\bar{E} = 0$,[26] E_S must also be zero:

$$E_S = 0. \tag{3.D.39}$$

This means that only solutions for $\Psi(\mathbf{r}, E_S)$ can exist for $E_S = 0$. Then, we get with equation (3.D.19),

$$\Psi(\mathbf{r}, E_S = 0) = 0. \tag{3.D.40}$$

Solutions for $E_S \neq 0$ are forbidden within projection theory. We know that solutions $\Psi(\mathbf{r}, E_S \neq 0)$ are possible within usual quantum theory as, for example, is demonstrated for the hydrogen atom, and the existence of $E_S \neq 0$ has been verified experimentally. We have to analyze this situation.

On the one hand, the stationary case of projection theory cannot lead to solutions $E_S \neq 0$. On the other hand, we just observe states for $E_S \neq 0$ that are within the frame work of usual quantum theory. Solutions for $E_S \neq 0$ are possible here and have been verified experimentally. Projection theory must of course be able to explain this situation.

Within projection theory we have to treat such problems in connection with the system-specific time t, i.e., in terms of $V(x, y, z, t)$ since $V(x, y, z)$ is not suitable to treat such problems within projection theory because only the state $E_S = 0$,[27] can be a solution here. However, in the case of $V(x, y, z, t) \neq V(x, y, z)$ the t-dependence of $V(x, y, z, t)$ must be very smooth and the t-region must have a sufficiently large extension in order to get spectral lines for the energy E that are very small.[28] This indicates that we deal here with long-living systems. In particular, this means that in the case of a spectral line for E with a width of zero, the system must have a life-time of infinity. The fact that we have to use a t-dependent potential $V(x, y, z, t)$ with $V(x, y, z, t) \neq$

---

[26] See equation (3.D.37).
[27] See equation (3.D.39).
[28] In fact, the measurements show that the energy spectral lines are very small.

$V(x,y,z)$ within projection theory means that all systems must have a finite life-time and can never be infinite. This also means that no spectral line for the energy can have a width of exactly zero, but it can be almost zero, i.e., so small that it becomes unobservable. For example, in the case of hydrogen, we may assume that the live time is very large. We may assume that hydrogen has a life-time of the order of the age of the universe. Then, the line width ΔE of a spectral line must be so small that it becomes unobservable. However, the width ΔE may not be zero because of the t-dependence of $V(x,y,z,t)$ which is necessary as we have outlined above. This reflects macroscopic effects at the quantum level. Also within usual quantum theory, macroscopic effects have been discussed in connection with inflation theory.[29]

3.D.3.5. *Normalization condition*

In Section 3.D.2, we found that the wave function $\Psi(\mathbf{r},t)$ becomes zero in the stationary case for all space-time points: $\Psi(\mathbf{r},t) = 0$ with $-\infty \leq \mathbf{r}, t \leq \infty$. Then, the normalization integral

$$\int_{-\infty}^{\infty} \Psi^*(\mathbf{r},t)\Psi(\mathbf{r},t)dt, \qquad (3.D.41)$$

must be zero. This can easily be verified as follows. By the use of equation (3.D.5) we obtain for the normalization integral (3.D.41),

$$\int_{-\infty}^{\infty} \Psi^*(\mathbf{r},t)\Psi(\mathbf{r},t)dt$$

$$= C^2 \lim_{E'_S \to E_S} \Psi^*(\mathbf{r}, E'_S)\Psi(\mathbf{r}, E_S) \lim_{t_E \to \infty} \int_{-t_E}^{t_E} \exp\left[\frac{i}{\hbar}(E_S - E'_S)t\right] dt$$

$$= \Psi^*(\mathbf{r}, E_S)\Psi(\mathbf{r}, E_S) C^2 \lim_{E'_S \to E_S} \lim_{t_E \to \infty} 2\hbar \frac{\sin(E_S - E'_S)t_E/\hbar}{(E_S - E'_S)}.$$

$$(3.D.42)$$

---

[29] Interested readers might refer to Vilenkin, A., (2006) Many worlds in one: Hill and Wang.

With[30]

$$\lim_{t_E \to \infty} 2\hbar \frac{\sin(E_S - E'_S)t_E/\hbar}{(E_S - E'_S)}$$
$$= 2\pi\delta\left(\frac{E_S}{\hbar} - \frac{E'_S}{\hbar}\right),$$

and using the same procedure as we applied in Section 3.D.3.4, we immediately get

$$\int_{-\infty}^{\infty} \Psi^*(\mathbf{r}, t)\Psi(\mathbf{r}, t)dt = \Psi^*(\mathbf{r}, E_S)\Psi(\mathbf{r}, E_S)\frac{\hbar}{\delta(0)}, \qquad (3.\text{D}.43)$$

i.e., we have with $\Psi(\mathbf{r}, E_s) = 0$,[31]

$$\int_{-\infty}^{\infty} \Psi^*(\mathbf{r}, t)\Psi(\mathbf{r}, t)dt = 0. \qquad (3.\text{D}.44)$$

Because of the general relation,

$$\int_{-\infty}^{\infty} \Psi^*(\mathbf{r}, t)\Psi(\mathbf{r}, t)dt = \int_{-\infty}^{\infty} \Psi^*(\mathbf{r}, E_S)\Psi(\mathbf{r}, E_S)dE_S, \qquad (3.\text{D}.45)$$

which must be also valid here, we simultaneously have

$$\int_{-\infty}^{\infty} \Psi^*(\mathbf{r}, E_S)\Psi(\mathbf{r}, E_S)dE_S = 0. \qquad (3.\text{D}.46)$$

3.D.4. Stationary Systems do not Exist

So far we have the following results for the stationary case of projection theory:

1. Only one energy value may exist: $E_S = 0$.[32]

---

[30] See also equation (3.53).
[31] See also equation (3.D.19).
[32] See also equation (3.D.39).

Free, Non-interacting Systems (Particles) 181

2. The wave functions $\Psi(\mathbf{r}, t)$ and $\Psi(\mathbf{r}, E_S)$ become zero.[33] Thus, the probability densities are also zero:

$$\Psi^*(\mathbf{r}, t)\Psi(\mathbf{r}, t) = 0, \qquad (3.D.47)$$

$$\Psi^*(\mathbf{r}, E_S)\Psi(\mathbf{r}, E_S) = 0. \qquad (3.D.48)$$

Consequently, the normalization integrals

$$\int_{-\infty}^{\infty} \Psi^*(\mathbf{r}, t)\Psi(\mathbf{r}, t)dt,$$

and

$$\int_{-\infty}^{\infty} \Psi^*(\mathbf{r}, E_S)\Psi(\mathbf{r}, E_S)dE,$$

are approaching zero as well.[34]

3. The wave function $\Psi(\mathbf{p}, E_S)$ is connected to $\Psi(\mathbf{r}, E_S)$ by the Fourier integral,[35]

$$\Psi(\mathbf{p}, E_S) = \frac{1}{(2\pi\hbar)^{3/2}} \int_{-\infty}^{\infty} \Psi(\mathbf{r}, E_S) \exp\left\{-i\left[\frac{\mathbf{p}}{\hbar} \cdot \mathbf{r}\right]\right\} d\mathbf{r}. \qquad (3.D.49)$$

Because $\Psi(\mathbf{r}, E_S)$ is zero for each value E_S and all positions $\mathbf{r}$, the function $\Psi(\mathbf{p}, E_S)$ and the corresponding probability density $\Psi^*(\mathbf{p}, E_S)\Psi(\mathbf{p}, E_S)$ must be zero for $\mathbf{p} \neq 0$. However, also in the case of $\mathbf{p} = 0$ the function $\Psi(\mathbf{p}, E_S)$ must be zero because we have the following situation: As we have outlined above, the energy takes a value of $E_S = 0$. Thus, in the case of $\mathbf{p} = 0$ we have the property $\mathbf{p} = 0$, $E_S = 0$. We have stated in Section 3.8 that this property, i.e. $\mathbf{p} \neq 0$, $E_S = 0$, reflects non-existence of the system. Therefore, the function $\Psi(\mathbf{p}, E_S)$ must also be zero for $\mathbf{p} = 0$; see in particular Section 3.7. In other words, there can also be no $(\mathbf{p}, E_S)$-state.

4. Summary: There can be no $\mathbf{r}, t$-states and no $\mathbf{p}, E_S$-states. Therefore, from the point of view of projection theory, stationary systems cannot be realized in nature.

---

[33] See also equations (3.D.9) and (3.D.19).
[34] See equations (3.D.44) and (3.D.46).
[35] See Section 2.2.7.3.

3.D.5. Final Remarks

Within projection theory, one can formally define a stationary case, i.e., $V(x,y,z,t) \rightarrow V(x,y,z)$. Here the system is stationary with respect to the system-specific time t. However, our analysis showed that such stationary systems do not exist within projection theory.[36] This result is in contrast to usual quantum theory where the stationary case is definitely defined but with respect to the reference time τ.[37] Within usual quantum theory, the system-specific time t is not defined and, therefore, the Fourier transform (3.D.4) as well as the operator $-i\hbar\partial/\partial E_S$ do not exist within usual quantum theory. In Section 2.3.9, we have also discussed a stationary case within projection theory but with respect to the reference time τ without neglecting the system-specific time t.

The reason why the expression (3.D.5) transforms equation (3.D.2) to equation (3.D.3) is due to the fact that the operator

$$-i\hbar\frac{\partial}{\partial E_S},$$

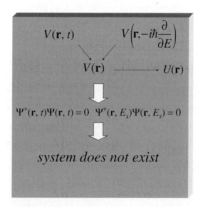

Fig. 3.D.1. From the transition $V(x,y,z,t) \rightarrow V(x,y,z)$, it follows that the potential $V(x,y,z)$ must take the form of the classical function $U(x,y,z)$. The elimination of the t-dependence in connection with $V(x,y,z,t)$ leads to the effect that the probability densities in both spaces, i.e., $(\mathbf{r},t)$-space and $(\mathbf{r},E_S)$-space vanish, that is, we get $\Psi^*(\mathbf{r},t)\Psi(\mathbf{r},t) = 0$ and $\Psi^*(\mathbf{r},E_S)\Psi(\mathbf{r},E_S) = 0$. In other words, there can be no $\mathbf{r},t$-states and no $\mathbf{p},E_S$-states, i.e., a system that is stationary with respect to the system-specific time t cannot exist.

---

[36] A summary is given in Figure 3.D.1.
[37] See Section 2.2.9.

for the time-coordinate t no longer appears within the stationary approximation,[38] i.e., the operator $-i\hbar\partial/\partial E_S$ is eliminated. In $(\mathbf{r},t)$-space we have

$$V(x,y,z,t) \to V(x,y,z). \tag{3.D.50}$$

However, in $(\mathbf{r}, E_S)$-space we get

$$V\left(x,y,z,-i\hbar\frac{\partial}{\partial E_S}\right) \to V(x,y,z) \tag{3.D.51}$$

in the stationary case. The operator $-i\hbar\partial/\partial E_S$ does not appear when we work in $(\mathbf{r}, E_S)$-space. That is, instead of the general case,[39]

$$E_S\Psi(\mathbf{r},E_S) = -\frac{\hbar^2}{2m_0}\Delta\Psi(\mathbf{r},E_S) + V\left(x,y,z,-i\hbar\frac{\partial}{\partial E_S}\right)\Psi(\mathbf{r},E_S), \tag{3.D.52}$$

we obtain equation (3.D.3). Clearly, the ansatz $\Psi(\mathbf{r},t) = C\Psi(\mathbf{r},E_S) \exp[-iE_St/\hbar]$ cannot be applied in the case of equation (3.D.52).

Appendix 3.E
Dependence of Mass on Velocity

Let us consider a classical system, i.e., particle, that is moving with the velocity $\mathbf{v}$ relative to a stationary frame of reference with coordinates x,y,z and time τ, and let us assume that $\mathbf{v}$ is directed to the x-axis: $\mathbf{v} = v_x = v$. The rate of change of momentum along the x-axis is given by the force F_x acting on the particle:[1]

$$F_x = \frac{dp}{d\tau}. \tag{3.E.1}$$

We define the mass m of the particle by

$$\mathbf{p} = m\mathbf{v}. \tag{3.E.2}$$

---
[38] See also equation (3.D.1).
[39] See equation (2.50).
[1] Refer to Newton's second law.

Within Special Theory of Relativity, the mass m varies with the velocity v according to

$$m = \frac{m_0}{\sqrt{1 - v^2/c^2}}. \qquad (3.\text{E}.3)$$

In this Appendix, we want to derive an expression for the dependence of mass on velocity that is more general than that given by equation (3.E.3). In particular, we would like to show that this derivation is not dependent on the postulates of Special Theory of Relativity. The main points of this study have already been published.[2]

We define the kinetic energy E_k of the system as follows. Its differential change is the force F_x times the differential distance moved:

$$dE_k = F_x dx. \qquad (3.\text{E}.4)$$

The derivative of the kinetic energy with respect to time τ is

$$\frac{dE_k}{d\tau} = v\frac{dp}{d\tau} = mv\frac{dv}{d\tau} + v^2\frac{dm}{d\tau}. \qquad (3.\text{E}.5)$$

After some simple manipulations, we obtain, provided $dm/d\tau \neq 0$,

$$\frac{dE_k}{dm} = \frac{m}{2}\frac{dv^2}{dm} + v^2. \qquad (3.\text{E}.6)$$

Furthermore, we assume that the inverse function of $m = m(v)$ exists and the derivative of the inverse function $v^2 = \eta(m)$ is defined over the entire mass range of interest. Then we have $d\eta(m)/dm\, dm/dv^2 = 1$ and we may write equation (3.A.6) as follows:

$$\frac{dE_k}{dm} = \frac{m}{2}\frac{d\eta(m)}{dm} + \eta(m)$$

$$= \frac{m}{2}\frac{1}{dm/dv^2} + v^2. \qquad (3.\text{E}.7)$$

---

[2] Interested readers might refer to Schommers, W. (Ed.), (1989) Quantum theory and pictures of reality: Springer-Verlag.

The integration leads to

$$E_k = \int_{m_0}^{m} \left\{ \frac{m' \, d\eta(m')}{2 \, dm'} + \eta(m') \right\} dm'$$
$$= \gamma(m) - \gamma(m_0), \quad (3.E.8)$$

where $m_0 = m(0)$ is the rest mass and the functions $\gamma(m)$ and $\gamma(m_0)$ are expressed by

$$\gamma(m) = \int_0^m \left\{ \frac{m' \, d\eta(m')}{2 \, dm'} + \eta(m') \right\} dm'$$
$$= E, \quad (3.E.9)$$

and

$$\gamma(m_0) = \int_0^{m_0} \left\{ \frac{m' \, d\eta(m')}{2 \, dm'} + \eta(m') \right\} dm'$$
$$= E_0. \quad (3.E.10)$$

E is the total energy of the system and E_0 is the rest energy. The general expression of the function $\gamma(m)$ is not known. Therefore, let us continue by assuming that a Taylor expansion in powers of the mass m exists for $\gamma(m)$. Provided that the energy E of the particle vanishes if the moving mass m becomes zero, we obtain for the Taylor expansion the expression

$$E = \sum_{k=1}^{\infty} m^k \gamma^{(k)} \frac{1}{k!}, \quad (3.E.11)$$

with

$$\gamma^{(k)} = \left[\frac{d^k \gamma(m)}{dm^k} \right]_{m=0}. \quad (3.E.12)$$

With equation (3.E.8) we get for the coefficients,

$$\gamma^{(1)} = \frac{m \, d\eta(m)}{2 \, dm} + \eta(m) \bigg|_{m=0},$$

$$\gamma^{(2)} = \left[\frac{3 \, d\eta(m)}{2 \, dm} + \frac{m \, d^2\eta(m)}{2 \, dm^2} \right]_{m=0},$$

$$\vdots$$

$$\gamma^{(n)} = \left[\frac{(n+1)}{2}\frac{d^{(n-1)}\eta(m)}{dm^{(n-1)}} + \frac{m}{2}\frac{d^n\eta(m)}{dm^n}\right]_{m=0},$$

$$\vdots \qquad (3.\text{E}.13)$$

Let us approximate the energy by the first two terms of equation (3.E.11):

$$E = m\gamma^{(1)} + \frac{m^2}{2}\gamma^{(2)}. \qquad (3.\text{E}.14)$$

With equation (3.E.14), we obtain for the differential equation (3.E.7)

$$\frac{dm}{dv^2} = \frac{m}{2}\frac{1}{\gamma^{(1)} + m\gamma^{(2)} - v^2}. \qquad (3.\text{E}.15)$$

The integration of equation (3.E.15) over all the masses from m_0 to m and over all velocities from $v = 0$ to v leads to the following equation for the dependence of mass on velocity:

$$m = m(v^2)$$

$$= m_0\left[\frac{1 + 2/3\gamma^{(2)}/\gamma^{(1)}m_0}{1 + 2/3\gamma^{(2)}/\gamma^{(1)}m - v^2/\gamma^{(1)}}\right]^{1/2}. \qquad (3.\text{E}.16)$$

In the case of

$$m \ll 3/2\gamma^{(1)}/\gamma^{(2)}, \qquad (3.\text{E}.17)$$

we get

$$E = m\gamma^{(1)}, \qquad (3.\text{E}.18)$$

and

$$m = m_0\left[\frac{1}{1 - v^2/\gamma^{(1)}}\right]^{1/2}, \qquad (3.\text{E}.19)$$

instead of equations (3.E.14) and (3.E.16). In particular, the rest energy E_{m_0} is given by

$$E_{m_0} = m_0\gamma^{(1)}. \qquad (3.\text{E}.20)$$

Free, Non-interacting Systems (Particles)

Furthermore, for sufficiently small velocities v we obtain with equation (3.E.19) for the energy E,[3]

$$E = m_0 \gamma^{(1)} + \frac{1}{2} m_0 v^2$$

$$= m_0 \gamma^{(1)} + \frac{\mathbf{p}^2}{2 m_0}, \qquad (3.E.21)$$

where equation (3.E.2) has been used.

For the derivation of all these equations, no transformation formulas[4] had to be used, that is, transformations which exist between the space-time coordinates of a resting frame of reference S and those of a moving frame of reference S'. For the derivation of $m_0 c^2$,[5] the Lorentz-transformations are needed which obviously need not be applied within projection theory.[6]

---

[3] See equation (3.E.18).
[4] For example, the Lorentz-transformations.
[5] Refer to the special theory of relativity.
[6] See, in particular, Chapter 4.

4
Interactions

Preliminary Remarks

In Chapter 2, we have introduced the notion "interaction" without needing to discuss details. In this chapter, we would like to give more specific statements about what we call "interaction" and, in particular, its peculiarities in connection with projection theory. In contrast to conventional quantum theory, interactions within projection theory are not restricted to the space variable **r** but it is always an interaction with respect to the space-time variables **r** and t. Let us mention once more that the system-specific time t is not defined in conventional physics but only in a certain kind of external time τ which we have called "reference time" in Chapter 2. τ is the time that we measure with our clocks and reflects the observers time feeling. In projection theory, besides the variables **r** and t, the reference time τ also plays a particular role.[1]

4.1. Interactions within Projection Theory

The interaction takes place in $(\mathbf{p}, E)$-space and, as we will recognize in the forthcoming discussion, we cannot distinguish between the various $\mathbf{p}, E$-fluctuations in connection with a set of individual systems.[2] As we have outlined several times, the picture of usual quantum theory cannot be used in projection theory. Within the framework of projection

---
[1] More details are given in Section 2.3.9.
[2] In usual quantum theory, the real interaction processes take place in space and $(\mathbf{r}, \tau)$-space, respectively. Here it is always assumed that the bodies are really embedded in space-time. Within conventional physics, we principally distinguish between the particles, and the interaction strength is dependent on the distance between the bodies.

theory, the interaction does not take place in $(\mathbf{r}, t)$-space, and there are no real systems embedded in $(\mathbf{r}, t)$-space. In projection theory, reality has to be identified with states in $(\mathbf{p}, E)$-space. The interactions between individual systems exclusively take place in $(\mathbf{p}, E)$-space and, therefore, the interaction mechanisms which are used in conventional physics cannot be applied here. All these peculiarities change the situation fundamentally and have to be analyzed further.

4.2. What does Interaction Mean within Projection Theory?

In contrast to the statements about interactions on the basis of the projection principle, the treatment of interactions in usual quantum theory is based on the classical conception of physics. Everything is embedded in space and the processes which we observe are considered to be real material processes taking place in space. In fact, Born's interpretation of the wave function is based on the assumption that there are material objects embedded in space. According to Born, the probability to find at time τ a real quantum object[3] in the volume element ΔV at a certain position $\mathbf{r}$ is given by $\psi^*(\mathbf{r})\psi(\mathbf{r})\Delta V$.[4] Within projection theory, the situation is quite different: Here measurement means that the properties of $(\mathbf{r}, t)$-space and $(\mathbf{p}, E)$-space are connected and, as we have already outlined in Section 2.2.4.3, this leads to the following picture. The measurement of one of the possible values for $\mathbf{p}$ and for E is done in the space-time intervals $\mathbf{r}, \mathbf{r} + \Delta \mathbf{r}$ and $t, t + \Delta t$ with the probability density of $\Psi^*(\mathbf{r}, t)\Psi(\mathbf{r}, t)$.[5] Again, this statement means that we are working simultaneously on the basis of two spaces, i.e., $(\mathbf{r}, t)$-space and $(\mathbf{p}, E)$-space. Within projection theory, we have to distinguish between basic reality, fictitious reality, and the picture of reality. These are relevant elements and, therefore, let us briefly repeat the main relationships between these basic elements of description.

---

[3] This is assumed to be a point-like particle.
[4] See Section 1.3.
[5] See Section 2.3.8.3.

4.2.1. *Relationships*

The principles of projection theory are different from the conventional conception. That which we call "reality" is here embedded in $(\mathbf{p}, E)$-space. However, in this connection we should not forget that $(\mathbf{p}, E)$-space also cannot be identified with "basic (absolute) reality".[6] The states in $(\mathbf{p}, E)$-space merely reflect a "fictitious reality". From the point of view of projection theory, a human observer has no access to *basic* reality. Reality, i.e., fictitious reality, is embedded in $(\mathbf{p}, E)$-space, and this reality is projected onto space and time, i.e., $(\mathbf{r}, t)$-space, and appears as geometrical positions. In this way we obtain that which we have called in Section 2.3.4.3 "pictures of reality". In this connection, it is important to mention once more that also the measuring instruments (detectors) are positioned in basic reality, and the registration processes take place in basic reality too. However, this situation is exclusively observed by the observer as a picture in $(\mathbf{r}, t)$-space. A measuring instrument and the movements of its pointer has nothing to do with material processes in $(\mathbf{r}, t)$-space but actually take place in basic reality.

Again, basic (absolute) reality cannot be experienced by humans. The starting-point of projection theory are the pictures that we have spontaneously in front of us. These pictures have to be considered as "transformed geometrical structures" and they reflect directly certain situations of basic (absolute) reality. In this connection, the following point is important: *Within the memory of man all essential things are represented within the frame of pictures.* Therefore, we have chosen the pictures, which we have spontaneously in front of us, as the most basic facts, and we have used them as the starting-point for the development of a theoretical framework, i.e., projection theory.

An essential point within projection theory is the feature that (fictitious) reality and the pictures are connected by a Fourier transform. Just this peculiarity makes it necessary to discuss the notion "interaction" in a new light because all interactions exclusively take place in $(\mathbf{p}, E)$-space and not in $(\mathbf{r}, t)$-space, as in conventional physics. In other words, we

---

[6]See Section 2.3.4.3.

are confronted with a completely new situation when we try to analyze interaction processes in projection theory. In the next section, we will deepen that on the basis of some fundamental, qualitative remarks.

4.2.2. Fourier-Effects

A system that is localized in $(\mathbf{r}, t)$-space is delocalized in $(\mathbf{p}, E)$-space. This property follows directly from the Fourier transform. For example, let us consider two systems i and j, which have sharp space-time positions $\mathbf{r}_i, \mathbf{r}_j$ and t_i, t_j in $(\mathbf{r}, t)$-space, and these structures can be characterized by delta-functions

$$\text{system } i:\ \chi_i(\mathbf{r}, t) = \delta(\mathbf{r} - \mathbf{r}_i)\delta(t - t_j), \tag{4.1}$$

$$\text{system } j:\ \chi_j(\mathbf{r}, t) = \delta(\mathbf{r} - \mathbf{r}_j)\delta(t - t_j). \tag{4.2}$$

When we project this information onto $(\mathbf{p}, E)$-space, we obtain

$$\frac{1}{(2\pi\hbar)^2} \int_{-\infty}^{\infty} \delta(\mathbf{r} - \mathbf{r}_k)\delta(t - t_k) \exp\left\{-i\left[\frac{\mathbf{p}}{\hbar} \cdot \mathbf{r} - \frac{E}{\hbar}t\right]\right\} d\mathbf{r}\, dt$$

$$= \chi_k(\mathbf{p}, E), \quad k = i, j, \tag{4.3}$$

with

$$\chi_k(\mathbf{p}, E) = \frac{1}{(2\pi\hbar)^2} \exp\left[-\frac{i}{\hbar}(\mathbf{p} \cdot \mathbf{r}_k - Et_k)\right], \quad k = i, j, \tag{4.4}$$

and, in analogy to that which we have outlined in Section 2.3.4.3, we get for both systems the same probability distribution for the variables $\mathbf{p}$ and E:

$$\chi_k^*(\mathbf{p}, E)\chi_k(\mathbf{p}, E) = \frac{1}{(2\pi\hbar)^4}, \quad k = i, j. \tag{4.5}$$

That is, both systems are uniformly distributed in $(\mathbf{p}, E)$-space. Because of equation (4.5) all $(\mathbf{p}, E)$-points appear with exactly the same probability. Both systems occupy uniformly the whole $(\mathbf{p}, E)$-space, independent of the positions they have in $(\mathbf{r}, t)$-space.[7]

---

[7] In fact, the space-time positions $\mathbf{r}_i, \mathbf{r}_j$ and t_i, t_j do not appear in equation (4.5).

Furthermore, we would like to assume that both systems form a universe, that is, no other systems are within the universe.[8] Since the $\mathbf{p}, E$-values fluctuate for each system in the range $-\infty \leq \mathbf{p}, E \leq \infty$, there must be an interaction, i.e., $\mathbf{p}, E$-fluctuations, between system i and system j. Only in this way, the validity of the conservation laws for momentum and energy can be fulfilled. If both systems would perform $\mathbf{p}, E$-fluctuations independently from the others, these conservation laws would be violated.

Equations (4.1) and (4.2) describe the systems i and j in $(\mathbf{r}, t)$-space and the equivalent description in $(\mathbf{p}, E)$-space is given by the equations (4.3) and (4.4). Both descriptions are equivalent concerning their information contents. There is however a big difference. All interaction processes take place in $(\mathbf{p}, E)$-space and are projected as pictures onto $(\mathbf{r}, t)$-space; the transition from $(\mathbf{p}, E)$-space to $(\mathbf{r}, t)$-space is done without loss of information and is described by a Fourier transform.

Again, the interaction processes between both systems exclusively take place in fictitious reality which is embedded in $(\mathbf{p}, E)$-space, that is, we have $\mathbf{p}, E$-fluctuations between system i, characterized by $\chi_i^*(\mathbf{p}, E)\chi_i(\mathbf{p}, E)$, and system j, which is characterized by the function $\chi_j^*(\mathbf{p}, E)\chi_j(\mathbf{p}, E)$, and we come to the scheme

$$\chi_i^*(\mathbf{p}, E)\chi_i(\mathbf{p}, E) \leftrightarrow \chi_j^*(\mathbf{p}, E)\chi_j(\mathbf{p}, E). \qquad (4.6)$$

In contrast to conventional physics, there can however be no interaction processes within $(\mathbf{r}, t)$-space, that is, we have no interaction processes between system i, characterized in $(\mathbf{r}, t)$-space by $\chi_i^*(\mathbf{r}, t)_i\chi_i^*(\mathbf{r}, t)$, and system j, which is characterized by $\chi_j^*(\mathbf{r}, t)\chi_j(\mathbf{r}, t)$, and we would like to symbolize this situation by the scheme

$$\chi_i^*(\mathbf{r}, t)_i\chi_i^*(\mathbf{r}, t) \leftarrow \vdots \rightarrow \chi_j^*(\mathbf{r}, t)\chi_j(\mathbf{r}, t). \qquad (4.7)$$

---

[8] Instead of an almost empty universe, consisting of only two systems i and j, we may assume that both systems do not interact with any other system in the universe leading to a situation which is equivalent to the case of an almost empty universe.

Equation (4.6) describes the interaction situation in projection theory. Due to the structure of equation (4.5), the systems i and j are completely delocalised in $(\mathbf{p}, E)$-space where the interaction processes take place if the same systems are localized in $(\mathbf{r}, t)$-space which is the case due to equations (4.1) and (4.2). This scenario has always to be kept in mind when we analyze interaction processes in projection theory.

In conclusion, the relative changes in the space-time positions of system i and system j are not due to interaction processes through space-time, i.e., $(\mathbf{r}, t)$-space, but come into play by $\mathbf{p}, E$-fluctuations which take place in $(\mathbf{p}, E)$-space, and the projection of these fluctuations leads to specific structures in $(\mathbf{r}, t)$-space. If the law for the $(\mathbf{p}, E)$-fluctuations are changed, the Fourier transformation leads automatically to the effect that the space-time structures in connection with system i and system j are changed. In contrast to these facts, within conventional physics the material objects are embedded in space and $(\mathbf{r}, t)$-space, respectively, and the interaction between the (localized) objects take place within the same space.

4.3. How Basic is the Notion "Interaction"?

The notion "interaction" should be considered as a basic effect because it was not able in the history of science to explain this notion satisfactorily by the elements of other theoretical pictures. In connection with gravitational and electrical interactions, there exists certain interpretations in conventional physics, but none of them is able to explain where gravity or the electrical forces come from. This is an important point and, therefore, we would like to give some general remarks concerning this topic.

4.3.1. *Classical Force Laws*

As it is well-known, Newton's equations of motion include the forces which act between the bodies. In the case of two bodies we have two masses m_1 and m_2, having the distance r, and the force is proportional to $m_1 m_2 / r^2$ (gravitational law). On the other hand, in the case of two electrical point charges, say q and Q, having also the distance r, the force between the charges is expressed by qQ/r^2 (Coulomb's law). The gravitational law and Coulomb's law of electrostatic force have exactly the same mathematical structure. Therefore, both laws have been interpreted by means of the same physical conceptions. Here two interpretations are of particular relevance: "action-at-a-distance" and the so-called "proximity effect". Both interpretations can be characterized as follows:

1. "Action-at-a-distance"

According to Newton's mechanics it appears to be the case that the forces between N masses act across space instantly. This is because in the gravitational law there is a relationship between the spatially separated positions of the masses and no intermediate position appears in the law. This suggests the view that gravitational forces work at a distance, i.e., the interaction comes about through an "action-at-a-distance" as follows. The gravitational forces reside in body i, but come into effect at the location of the other $(N-1)$ bodies. The space between the bodies in this view is free of gravitation. This "action-at-a-distance" picture can also be applied to electrical interactions. However, it must be emphasized that this effect, i.e., "action-at-a-distance", cannot be proved experimentally, that is, it is a matter of belief, whether such an effect exists in reality or not.

2. Proximity effect

In contrast to the "action-at-a-distance" picture, the so-called "proximity effect" just teaches the opposite. Whereas in the case of "action-at-a-distance" the space between the masses (or charges) is free of gravitation (or electricity), in the case of the "proximity effect" we have space-filling fields: gravitational fields and/or electrical fields. For example, let us consider again the interaction between two charges q and Q. The charge Q is associated with an electrical field $E(r)$ where r is the position of charge q. The charge q "feels" the field $E(r)$ and, therefore, the effect is a "proximity effect" since the force effect on charge **q** is exactly determined by the electrical field at the position at which also the charge q is positioned. However, this space-filling "medium" (field) cannot be proved experimentally and it becomes entirely a matter of belief, whether an electrical field (gravitational field) exists in reality or not.

4.3.2. *Equivalent Conceptions*

The "proximity effect" and the "action-at-a-distance", as qualitatively different as they may be, are nevertheless exactly equivalent. It is not possible to make a choice between these two conceptions, since there is no experimental way to distinguish between them, as they both have the same consequences. In conclusion, the "proximity effect" and the "action-at-a-distance" are equivalent conceptions which do not differ, neither mathematically nor experimentally.

It is of principal interest to note that the "proximity effect" and the "action-at-a-distance" are merely "expressions" or interpretations of the gravitational law to $m_1 m_2/r^2$ and Coulomb's law qQ/r^2. These force laws cannot, however, be derived from these notions. Many people believe that a mechanism, which is composed of many familiar single processes preferably from everyday life, can explain the mathematical structure of the force laws. What mechanism is, for example, responsible for the fact that the forces expressed by $m_1 m_2/r^2$ are inversely proportional to the square of the distance between the masses m_1 and m_2. As already mentioned, the ideas "proximity effect" and "action-at-a-distance" cannot provide the answer to this question, since they interpret the relation $m_1 m_2/r^2$, but are not able to explain the mathematical structure of this force law.

It is therefore not surprising that there are further interpretations in connection with the notion "interaction". In classical mechanics the "principle of least action" is of particular interest. This principle is qualitatively different from Newton's original idea but leads to exactly the same result. Newton's equations of motion describe the trajectory of a body from point to point in space, and they are based on the principle of causality, that is, the body feels a force and yields to it. To this idea there is an alternative. This alternative is the so-called "principle of least action". According to this the path of a body does not develop from place to place on the basis of cause and effect; instead one path is selected from the totality of possible paths, and that is the one whose effect is most minimal. As already stated, although the "principle of least action" and Newton's place-to-place description are principally different from each other, the results obtained using both ideas are exactly equivalent, i.e., we can experimentally not distinguish between both ideas.

4.3.3. Further Remarks

In conclusion, our discussion in connection with the notion "interaction" showed that the situation in conventional physics, where the interaction processes are assumed to take place in space and $(\mathbf{r}, t)$-space, respectively, is not convincing at all. Here, it should be mentioned that Newton himself opposed this idea of "forces acting-at-a-distance". This restraint was soon given up by his successors, partly because of the success of Newtonian mechanics, but also because of the unsuccessful search for a mechanism. In fact, the description of the gravitational law $m_1 m_2/r^2$ on the basis of the familiar mechanisms of everyday life as, for example, by means of impinging particles,[9] was not possible.

In conclusion, specific mechanisms for the description of the force laws to $m_1 m_2/r^2$ and qQ/r^2 could not be found. Therefore, in answering the question "How basic is the notion "interaction?" all the interpretations discussed above do not help further. We have perhaps to take an unusual viewpoint. We may take the viewpoint that the laws $m_1 m_2/r^2$ and qQ/r^2 (or more refined versions) are basic in character when we argue as follows. If it should be possible to model for example the law $m_1 m_2/r^2$ on the basis of principles that are positioned on a deeper level of description than that where the law $m_1 m_2/r^2$ is positioned, then the law $m_1 m_2/r^2$ is at the same time independent of this more basic description, in that it would remain unchanged even if the deeper laws were changed. There is a certain

---

[9] Interested readers might refer to Schommers, W., (1998). The visible and the invisible: World Scientific.

analogy to that which Robert B. Laughlin said about the phenomenon of hydrodynamics:

> *We also know that while a simple and absolute law, such as hydrodynamics, can evolve from the deeper laws underneath, it is at the same time independent of them, in that it would be the same even if the deeper laws are changed.*[10]

4.3.4. Remarks Concerning Quantum Field Theory and the Theory of Strings (Branes)

General theory of relativity deals exclusively with gravitational effects and, therefore, this theory cannot give a general statement on the notion "interaction". On the other hand, also quantum field theory cannot give a general description for that which we call "interaction" because it is not able to describe all phenomena in the universe.[11]

Furthermore, within the theory of strings (branes) the situation with respect to interaction effects is quite different from that in quantum field theory, i.e., elementary particle physics. The possible interactions within string theory is an unavoidable consequence of the string topology. Strings can divide and join themselves etc. A lot of complicated topological processes are possible and constitute interactions between strings. Within string theory, fictitious forces do not have to be introduced, and the existence of new quantum particles does not have to be postulated as is usual in elementary particle physics. In conclusion, interaction processes in the theory of strings (branes) and in connection with elementary particle physics are qualitatively different from each other.

4.3.5. Delocalised Systems in (p, E)-Space

The whole interaction scenario in physics[12] is based on processes between localized systems in space and $(\mathbf{r}, \tau)$-space, respectively. Since

---

[10] Refer to Laughlin, R., (2005) A different universe: Basic Books.
[11] Gravitational effects are excluded.
[12] Such as classical physics, usual quantum theory, quantum field theory, string theory.

the interactions in projection theory take place in $(\mathbf{p}, E)$-space, none of these specific processes can be applied here. Furthermore, in contrast to the conventional conceptions,[13] within projection theory we are confronted with delocalized systems in $(\mathbf{p}, E)$-space, if the same systems are localized in $(\mathbf{r}, t)$-space.[14]

The description of interacting delocalized systems in $(\mathbf{p}, E)$-space is not possible on the basis of interpretations and mechanisms, respectively, we are using in classical physics or usual quantum theory. A mechanism is normally based on our intuition, and man's intuition is tailor-made to localized bodies that are embedded in $(\mathbf{r}, t)$-space and not to delocalized systems that are positioned in $(\mathbf{p}, E)$-space. Our intuition is adjusted to the structures in space which we experience in everyday life, and these space structures are composed of localized bodies that are embedded in space and $(\mathbf{r}, t)$-space, respectively. Our complete intuitive abilities have been developed in connection with situations which we experience in everyday life, and we call a certain process "imaginable" when it can be explained with the means of everyday life. A "mechanism" is normally defined on the basis of an imaginable process. However, interacting delocalized systems which are positioned in $(\mathbf{p}, E)$-space do not belong to the category of imaginable processes. Thus, it makes not much sense to explain these $(\mathbf{p}, E)$-fluctuations by means of such "mechanisms" but we have to accept that they simply "emerge".

4.3.6. *Summary*

In projection theory, interactions exclusively take place in $(\mathbf{p}, E)$-space and the interaction patterns are projected onto $(\mathbf{r}, t)$-space. A pattern that is localized in $(\mathbf{r}, t)$-space is delocalized in $(\mathbf{p}, E)$-space, and reflects a property of the Fourier transformation. In conventional physics, the interaction between localized systems are considered. In contrast to this view, projection theory deals with the interaction of delocalized systems in $(\mathbf{p}, E)$-space. Mechanisms that are used in

---

[13] Localized systems are used here.
[14] See Section 4.2.2.

connection with conventional physics for the explanation of interaction effects cannot be applied to the typical situations dictated by the projection principle.

We will continue our discussion with the following topics: 1. Distance-dependent interactions; 2. Distance-independent interactions; 3. Can systems be elementary in character?; 4. Self-creating processes; 5. Arbitrary motions through space-time and Mach's principle; and 6. "Hierarchy of the parts in a part". Let us start with point 1: Distance-dependent interactions within the framework of projection theory.

4.4. Description of Interactions within Projection Theory: Principal Remarks

In the description of interaction processes one normally starts from non-interacting, free systems (particles) and the interaction between these systems is introduced by switching on a suitable interaction potential. In the forthcoming sections, we will talk about space-time limiting interactions, distance-dependent interactions and form interactions which are distance-independent.

All these types of interactions can be assumed to be independent from each other, and each of them may be applied without the existence of the others. However, for the existence of a certain system, interaction processes are absolutely needed, i.e., one of the three interaction types listed above must be switched on; a system cannot be existent without interaction process. This is a necessary condition. We have outlined in Chapter 3 that static elementary systems (building blocks) with a definite, not fluctuating momentum p_0 and a definite, not fluctuating energy E_0 cannot exist. Within projection theory, these interactions are *quantum processes* and are described by p, E-fluctuations. In other words, quantum processes are a primary factor for the existence of the world.

4.4.1. Space-Time Limiting Interactions

Let us consider N systems having the momenta $p_1, p_2, \ldots, p_N$ and the energies $E_1, E_2, \ldots, E_N$. The pictures of these N systems are supposed to be projected onto $(\mathbf{r}, t)$-space with an extension of $(\pm \mathbf{r}_a, \pm t_a)$ which corresponds to an interaction in $(\mathbf{p}, E)$-space.[15] However, we would like to suppose that there is no mutual interaction between the N systems at the beginning. At time τ, the geometrical positions in $(\mathbf{r}, t)$-space are $\mathbf{r}_1, \mathbf{r}_2, \ldots, \mathbf{r}_N, t_1, t_2, \ldots, t_N$.

Due to the space-time limiting global interaction, which is completely determined by the parameters $(\pm \mathbf{r}_a, \pm t_a)$, the N systems are equally confined in the same space-time which is limited for them by $(\pm \mathbf{r}_a, \pm t_a)$. Then, we have

$$\Psi(\mathbf{r}_1, t_1) = C_1 \exp\left\{\frac{i}{\hbar}(\mathbf{p}_1 \cdot \mathbf{r}_1 - E_1 t_1)\right\},$$
$$-\mathbf{r}_a, -t_a \leq \mathbf{r}_1, t_1 \leq \mathbf{r}_a, t_a,$$

$$\Psi(\mathbf{r}_2, t_2) = C_2 \exp\left\{\frac{i}{\hbar}(\mathbf{p}_2 \cdot \mathbf{r}_2 - E_2 t_2)\right\},$$
$$-\mathbf{r}_a, -t_a \leq \mathbf{r}_2, t_2 \leq \mathbf{r}_a, t_a, \qquad (4.8)$$

$$\vdots$$

$$\Psi(\mathbf{r}_N, t_N) = C_N \exp\left\{\frac{i}{\hbar}(\mathbf{p}_N \cdot \mathbf{r}_N - E_N t_N)\right\},$$
$$-\mathbf{r}_a, -t_a \leq \mathbf{r}_N, t_N \leq \mathbf{r}_a, t_a.$$

In other words, the systems are not able to look behind $-\mathbf{r}_a, -t_a$ and $\mathbf{r}_a, t_a$, i.e., the space-time positions $\mathbf{r}_k, t_k < -\mathbf{r}_a, -t_a$ and $\mathbf{r}_k, t_k > \mathbf{r}_a, t_a$ ($k = 1, \ldots, N$) are not allowed. Again, all systems experience the same $(\pm \mathbf{r}_a, \pm t_a)$-interaction leading to the effect that all N systems[16] are embedded in a space-time which is equally limited by $(\pm \mathbf{r}_a, \pm t_a)$ for all these systems. However, there are no material walls at $(\pm \mathbf{r}_a, \pm t_a)$ but, due to the projection principle, the space-limiting effect comes into existence through an interaction process ($\mathbf{p}, E$-fluctuations) in $(\mathbf{p}, E)$-space which is projected onto $(\mathbf{r}, t)$-space leading to the geometrical structures limited in space-time.

---

[15] See also Chapter 3.
[16] More precisely, their geometrical structure.

Since the parameters $(\pm \mathbf{r}_a, \pm t_a)$ are exactly the same for all N systems, the constants $C_1, C_2, \ldots, C_N$ also must be the same:

$$C = C_1 = C_2 = \cdots = C_N. \tag{4.9}$$

In the case of $(\mathbf{r}_a = \pm\infty, t_a = \pm\infty)$, all N systems are completely free and cannot exist as we have outlined in Chapter 3, and instead of equation (4.9) we obtain

$$C = C_1 = C_2 = \cdots = C_N = 0. \tag{4.10}$$

The existence of the space-time limitations $(\pm \mathbf{r}_a, \pm t_a)$ lead to interactions, i.e., we have $\mathbf{p}, E$-fluctuations, in $(\mathbf{p}, E)$-space. These fluctuations are expressed by the wave functions $\Psi(\mathbf{p}, E)_k$ with $k = 1, 2, \ldots, N$ which can easily be determined by a Fourier transform of the functions $\Psi(\mathbf{r}_1, t_1), \ldots, \Psi(\mathbf{r}_N, t_N)$[17]:

$$\Psi(\mathbf{p}, E)_1 = C \frac{(2\hbar)^2}{\pi^2} \frac{\sin((p_{1x} - p_x)x_a/\hbar)}{(p_{1x} - p_x)} \frac{\sin((p_{1y} - p_y)y_a/\hbar)}{(p_{1y} - p_y)}$$
$$\times \frac{\sin((p_{1z} - p_z)z_a/\hbar)}{(p_{1z} - p_z)} \frac{\sin((E - E_1)t_a/\hbar)}{(E - E_1)}, \tag{4.11}$$

$$\Psi(\mathbf{p}, E)_2 = C \frac{(2\hbar)^2}{\pi^2} \frac{\sin((p_{2x} - p_x)x_a/\hbar)}{(p_{2x} - p_x)} \frac{\sin((p_{2y} - p_y)y_a/\hbar)}{(p_{2y} - p_y)}$$
$$\times \frac{\sin((p_{2z} - p_z)z_a/\hbar)}{(p_{2z} - p_z)} \frac{\sin((E - E_2)t_a/\hbar)}{(E - E_2)}, \tag{4.12}$$

$\vdots$

$$\Psi(\mathbf{p}, E)_N = C \frac{(2\hbar)^2}{\pi^2} \frac{\sin((p_{Nx} - p_x)x_a/\hbar)}{(p_{Nx} - p_x)} \frac{\sin((p_{Ny} - p_y)y_a/\hbar)}{(p_{Ny} - p_y)}$$
$$\times \frac{\sin((p_{Nz} - p_z)z_a/\hbar)}{(p_{Nz} - p_z)} \frac{\sin((E - E_N)t_a/\hbar)}{(E - E_N)}, \tag{4.13}$$

---

[17] See equation (4.8).

with

$$\mathbf{r}_a = (x_a, y_a, z_a),$$
$$\mathbf{p}_k = (p_{kx}, p_{ky}, p_{kz}), \quad k = 1, \ldots, N,$$
$$\mathbf{p} = (p_x, p_y, p_z).$$

In other words, we have $\mathbf{p}, E$-fluctuations around the constant values $\mathbf{p}_1, \mathbf{p}_2, \ldots, \mathbf{p}_N$ and $E_1, E_2, \ldots, E_N$. We may otherwise state that just these $\mathbf{p}, E$-fluctuations create the space-time limiting effect which we have expressed by $(\pm \mathbf{r}_a, \pm t_a)$. These $\mathbf{p}, E$-fluctuations, which define the interaction within projection theory, can only come into play through an interaction of all N systems with an external system.[18] The most direct way to express these interactions, i.e., $\mathbf{p}, E$-fluctuations, is the use of the probability densities $\Psi^*(\mathbf{p}, E)_k \Psi(\mathbf{p}, E)_k, k = 1, \ldots, N$.

4.4.2. Mutual (Distance-Dependent) Interactions

Now let us switch on, at time τ, an interaction between the N systems which are equally confined within $(\pm \mathbf{r}_a, \pm t_a)$. What does that mean? This in particular means that there exist two independent characteristic parts for times smaller than τ : 1. The N systems without mutual interaction, and 2. the interaction without the N systems, which is responsible for the mutual interaction. At time τ, both parts are put together and we get *one* system consisting of N interacting subsystems.

However, this construction seems to be artificial. Who puts the N systems and the interaction together? In other words, who switches the interaction on? In order to avoid such questions, it is more realistic to assume that the N subsystems, which are under mutual interaction, simply exist (emerge) in this form, i.e., in the form of *one* system consisting of N interacting subsystems, and we should assume that this system is not divisible into two parts, i.e., the N systems without mutual interaction and with interaction.

---

[18] This can be identical with the environment or with a part of it.

Then, the N subsystem with the momenta $\mathbf{p}_1, \mathbf{p}_2, \ldots, \mathbf{p}_N$ and the energies $E_1, E_2, \ldots, E_N$ exchange momentum and energy in the course of time, that is, we have $\mathbf{p}, E$-fluctuations (interactions) between the N subsystems, and the momenta $\mathbf{p}_1, \mathbf{p}_2, \ldots, \mathbf{p}_N$ and the energies $E_1, E_2, \ldots, E_N$ vary in the course of time τ with respect to the conservation laws for momentum and energy. This physical situation is described in $(\mathbf{p}, E)$-space by the wave function $\Psi(\mathbf{p}_1, \mathbf{p}_2, \ldots, \mathbf{p}_N, E_1, E_2, \ldots, E_N)$. Then, the corresponding information in $(\mathbf{r}, t)$-space is expressed by the Fourier transform

$$\Psi(\mathbf{r}_1; \mathbf{r}_2, \ldots, \mathbf{r}_N, t_1, t_2, \ldots, t_N)$$
$$= \frac{1}{(2\pi\hbar)^{2N}} \int_{-\infty}^{\infty} \Psi(\mathbf{p}_1, \mathbf{p}_2, \ldots, \mathbf{p}_N, E_1, E_2, \ldots, E_N)$$
$$\times \exp\left\{\frac{i}{\hbar}[\mathbf{p}_1 \cdot \mathbf{r}_1 + \mathbf{p}_2 \cdot \mathbf{r}_2 + \cdots + \mathbf{p}_N \cdot \mathbf{r}_N\right.$$
$$\left. - (E_1 t_1 + E_2 t_2 + \cdots + E_N t_N)]\right\}$$
$$\times d\mathbf{p}_1 d\mathbf{p}_2 \cdots d\mathbf{p}_N dE_1 dE_2 \cdots dE_N. \qquad (4.14)$$

Then, it is easy to verify that the inverse transformation is expressed by

$$\Psi(\mathbf{p}_1, \mathbf{p}_2, \ldots, \mathbf{p}_N, E_1, E_2, \ldots, E_N)$$
$$= \frac{1}{(2\pi\hbar)^{2N}} \int_{-\infty}^{\infty} \Psi(\mathbf{r}_1; \mathbf{r}_2, \ldots, \mathbf{r}_N, t_1, t_2, \ldots, t_N)$$
$$\times \exp\left\{-\frac{i}{\hbar}[\mathbf{p}_1 \cdot \mathbf{r}_1 + \mathbf{p}_2 \cdot \mathbf{r}_2 + \cdots + \mathbf{p}_N \cdot \mathbf{r}_N\right.$$
$$\left. - (E_1 t_1 + E_2 t_2 + \cdots + E_N t_N)]\right\}$$
$$\times d\mathbf{r}_1 d\mathbf{r}_2 \cdots d\mathbf{r}_N dt_1 dt_2 \cdots dt_N. \qquad (4.15)$$

Within projection theory, interaction means that there is an exchange of momentum and energy between the N subsystems, that is, we have certain fluctuations

$$\Delta \mathbf{p}_1, \Delta \mathbf{p}_2, \ldots, \Delta \mathbf{p}_N, \Delta E_1, \Delta E_2, \ldots, \Delta E_N,$$

in connection with the quantities $\mathbf{p}_1, \mathbf{p}_2, \ldots, \mathbf{p}_N, E_1, E_2, \ldots, E_N$. If there are fluctuations around the values

$$\mathbf{p}'_1, \mathbf{p}'_2, \ldots, \mathbf{p}'_N, E'_1, E'_2, \ldots, E'_N$$

we have with

$$\mathbf{p}_1, \mathbf{p}_2, \ldots, \mathbf{p}_N, E_1, E_2, \ldots, E_N$$

the relations

$$\begin{aligned}
\mathbf{p}_1 &= \mathbf{p}'_1 + \Delta \mathbf{p}_1, \\
\mathbf{p}_2 &= \mathbf{p}'_2 + \Delta \mathbf{p}_2, \\
&\vdots \\
\mathbf{p}_N &= \mathbf{p}'_N + \Delta \mathbf{p}_N,
\end{aligned} \tag{4.16}$$

$$\begin{aligned}
E_1 &= E'_1 + \Delta E_1, \\
E_2 &= E'_2 + \Delta E_2, \\
&\vdots \\
E_N &= E'_N + \Delta E_N,
\end{aligned} \tag{4.17}$$

where $\mathbf{p}_1, \mathbf{p}_2, \ldots, \mathbf{p}_N, E_1, E_2, \ldots, E_N$ are again the values at time τ. Using equations (4.16) and (4.17), equation (4.14) takes the form

$$\begin{aligned}
&\Psi(\mathbf{r}_1, \mathbf{r}_2, \ldots, \mathbf{r}_N, t_1, t_2, \ldots, t_N) \\
&= \frac{1}{(2\pi\hbar)^{2N}} \int_{-\infty}^{\infty} \Psi(\mathbf{p}'_1 + \Delta \mathbf{p}_1, \mathbf{p}'_2 + \Delta \mathbf{p}_2, \ldots, \mathbf{p}'_N \\
&\quad + \Delta \mathbf{p}_N, E'_1 + \Delta E_1, E'_2 + \Delta E_2, \ldots, E'_N + \Delta E_N) \\
&\quad \times \exp\left\{\frac{i}{\hbar}[\mathbf{p}'_1 \cdot \mathbf{r}_1 + \mathbf{p}'_2 \cdot \mathbf{r}_2 + \cdots + \mathbf{p}'_N \cdot \mathbf{r}_N \right. \\
&\quad \left. - (E'_1 t_1 + E'_2 t_2 + \cdots + E'_N t_N)]\right\} \\
&\quad \times \exp\left\{\frac{i}{\hbar}[\Delta \mathbf{p}_1 \cdot \mathbf{r}_1 + \Delta \mathbf{p}_2 \cdot \mathbf{r}_2 + \cdots + \Delta \mathbf{p}_N \cdot \mathbf{r}_N \right.
\end{aligned}$$

$$-(\Delta E_1 t_1 + \Delta E_2 t_2 + \cdots + \Delta E_N t_N)]\Big\}$$

$$\times d\Delta \mathbf{p}_1 d\Delta \mathbf{p}_2 \cdots d\Delta \mathbf{p}_N d\Delta E_1 d\Delta E_2 \cdots d\Delta E_N. \quad (4.18)$$

Let us analyze this equation in more detail, but first we want to give a general remark concerning a specific property of the Fourier transform expressed by equation (4.18).

General remark

The following relations are valid:

$$\Psi(\mathbf{a},b) = \frac{1}{(2\pi\hbar)^2} \int_{-\infty}^{\infty} \Psi(\mathbf{g},f) \exp\left\{\frac{i}{\hbar}[\mathbf{g}\cdot\mathbf{a} - fb]\right\} d\mathbf{g}\, df, \quad (4.19)$$

$$\Psi(\mathbf{a},b) = \frac{1}{(2\pi\hbar)^2} \int_{-\infty}^{\infty} \Psi(\mathbf{g}+\Delta\mathbf{g}, f+\Delta f) \exp\left\{\frac{i}{\hbar}[\Delta\mathbf{g}\cdot\mathbf{a} - \Delta f b]\right\}$$

$$\times \exp\left\{\frac{i}{\hbar}[\mathbf{g}\cdot\mathbf{a} - fb]\right\} d\mathbf{g}\, df. \quad (4.20)$$

In the case of equation (4.20), the function $\Psi(\mathbf{g},f)$ is shifted by $\Delta \mathbf{g}$ and Δf.[19] Both functions

$$\Psi(\mathbf{g},f), \Psi(\mathbf{g}+\Delta\mathbf{g}, f+\Delta f)\exp\left\{\frac{i}{\hbar}[\Delta\mathbf{g}\cdot\mathbf{a} - \Delta f b]\right\}, \quad (4.21)$$

produce exactly the same function $\Psi(\mathbf{a},b)$ and, therefore, they are equivalent[20]:

$$\Psi(\mathbf{g},f) \leftrightarrow \Psi(\mathbf{g}+\Delta\mathbf{g}, f+\Delta f)\exp\left\{\frac{1}{\hbar}[\Delta\mathbf{g}\cdot\mathbf{a} - \Delta f b]\right\}. \quad (4.22)$$

In particular, we have

$$\Psi^*(\mathbf{g},f)\Psi(\mathbf{g},f) \leftrightarrow \Psi^*(\mathbf{g}+\Delta\mathbf{g}, f+\Delta f)\Psi(\mathbf{g}+\Delta\mathbf{g}, f+\Delta f). \quad (4.23)$$

That is, the form of the probability density remains also constant; it is only shifted by $\Delta \mathbf{g}$ and Δf.

---

[19] Its form remains constant.
[20] See Section 2.3.7.3.

On the basis of these general results[21] we get in connection with equation (4.18) the following relationships:

$$\Psi(\Delta\mathbf{p}_1, \Delta\mathbf{p}_2, \ldots, \Delta\mathbf{p}_N, \Delta E_1, \Delta E_2, \ldots, \Delta E_N)$$
$$\leftrightarrow \Psi(\mathbf{p}'_1 + \Delta\mathbf{p}_1, \mathbf{p}'_2 + \Delta\mathbf{p}_2, \ldots, \mathbf{p}'_N + \Delta\mathbf{p}_2, E'_1$$
$$+ \Delta E_1, E'_2 + \Delta E_2, \ldots, E'_N + \Delta E_N)$$
$$\times \exp\left\{\frac{i}{\hbar}[\mathbf{p}'_1 \cdot \mathbf{r}_1 + \mathbf{p}'_2 \cdot \mathbf{r}_2 + \cdots + \mathbf{p}'_N \cdot \mathbf{r}_N\right.$$
$$\left. - (E'_1 t_1 + E'_2 t_2 + \cdots + E'_N t_N)]\right\}, \qquad (4.24)$$

and

$$\Psi^*(\Delta\mathbf{p}_1, \Delta\mathbf{p}_2, \ldots, \Delta\mathbf{p}_N, \Delta E_1, \Delta E_2, \ldots, \Delta E_N)$$
$$\times \Psi(\Delta\mathbf{p}_1, \Delta\mathbf{p}_2, \ldots, \Delta\mathbf{p}_N, \Delta E_1, \Delta E_2, \ldots, \Delta E_N)$$
$$\leftrightarrow \Psi^*(\mathbf{p}'_1 + \Delta\mathbf{p}_1, \mathbf{p}'_2 + \Delta\mathbf{p}_2, \ldots, \mathbf{p}'_N + \Delta\mathbf{p}_2, E'_1$$
$$+ \Delta E_1, E'_2 + \Delta E_2, \ldots, E'_N + \Delta E_N)$$
$$\times \Psi(\mathbf{p}'_1 + \Delta\mathbf{p}_1, \mathbf{p}'_2 + \Delta\mathbf{p}_2, \ldots, \mathbf{p}'_N + \Delta\mathbf{p}_2, E'_1$$
$$+ \Delta E_1, E'_2 + \Delta E_2, \ldots, E'_N + \Delta E_N). \qquad (4.25)$$

In particular, we have instead of equation (4.14)

$$\Psi(\mathbf{r}_1, \mathbf{r}_2, \ldots, \mathbf{r}_N, t_1, t_2, \ldots, t_N)$$
$$= \frac{1}{(2\pi\hbar)^{2N}} \int_{-\infty}^{\infty} \Psi(\Delta\mathbf{p}_1, \Delta\mathbf{p}_2, \ldots, \Delta\mathbf{p}_N, \Delta E_1, \Delta E_2, \ldots, \Delta E_N)$$
$$\times \exp\left\{\frac{i}{\hbar}[\Delta\mathbf{p}_1 \cdot \mathbf{r}_1 + \Delta\mathbf{p}_2 \cdot \mathbf{r}_2 + \cdots + \Delta\mathbf{p}_N \cdot \mathbf{r}_N\right.$$
$$\left. - (\Delta E_1 t_1 + \Delta E_2 t_2 + \cdots + \Delta E_N t_N)]\right\}$$
$$\times d\Delta\mathbf{p}_1 d\Delta\mathbf{p}_2 \cdots d\Delta\mathbf{p}_N d\Delta E_1 d\Delta E_2 \cdots d\Delta E_N, \qquad (4.26)$$

---

[21] See equations (4.22) and (4.23).

and we obtain for the inverse transformation

$$\Psi(\Delta \mathbf{p}_1, \Delta \mathbf{p}_2, \ldots, \Delta \mathbf{p}_N, \Delta E_1, \Delta E_2, \ldots, \Delta E_N)$$
$$= \frac{1}{(2\pi\hbar)^{2N}} \int_{-\infty}^{\infty} \Psi(\mathbf{r}_1; \mathbf{r}_2, \ldots, \mathbf{r}_N, t_1, t_2, \ldots, t_N)$$
$$\times \exp\left\{-\frac{i}{\hbar}[\Delta \mathbf{p}_1 \cdot \mathbf{r}_1 + \Delta \mathbf{p}_2 \cdot \mathbf{r}_2 + \cdots + \Delta \mathbf{p}_N \cdot \mathbf{r}_N\right.$$
$$\left. - (\Delta E_1 t_1 + \Delta E_2 t_2 + \cdots + \Delta E_N t_N)]\right\}$$
$$\times d\mathbf{r}_1 d\mathbf{r}_2 \cdots d\mathbf{r}_N dt_1 dt_2 \cdots dt_N. \tag{4.27}$$

The transition from equation (4.14) to equation (4.26), that is, the possibility for the use of the variables $\Delta \mathbf{p}_1, \Delta \mathbf{p}_2, \ldots, \Delta \mathbf{p}_N$, $\Delta E_1, \Delta E_2, \ldots, \Delta E_N$ instead of $\mathbf{p}_1, \mathbf{p}_2, \ldots, \mathbf{p}_N, E_1, E_2, \ldots, E_N$ is of particular relevance. Let us briefly discuss this point.

The equivalence of the wave functions

$$\Psi(\mathbf{p}_1, \mathbf{p}_2, \ldots, \mathbf{p}_N, E_1, E_2, \ldots, E_N)$$

and

$$\Psi(\Delta \mathbf{p}_1, \Delta \mathbf{p}_2, \ldots, \Delta \mathbf{p}_N, \Delta E_1, \Delta E_2, \ldots, \Delta E_N)$$

means that the wave function does not depend on the absolute values of the momenta $\mathbf{p}_1, \mathbf{p}_2, \ldots, \mathbf{p}_N$ and the energies $E_1, E_2, \ldots, E_N$ but only on the fluctuations $\Delta \mathbf{p}_1, \Delta \mathbf{p}_2, \ldots, \Delta \mathbf{p}_N, \Delta E_1, \Delta E_2, \ldots, \Delta E_N$. In other words, the wave function exclusively describes interaction processes and nothing else.

Let us briefly summarize the situation: Within projection theory the interaction is characterized by $\mathbf{p}, E$-fluctuations, that is, by an exchange of momentum and energy between individual systems. The $\mathbf{p}, E$-fluctuations are fixed by the magnitudes of the $\mathbf{p}, E$-values and their relative rates. Just these characteristics are described by the probability density $\Psi^*(\mathbf{p}, E)\Psi(\mathbf{p}, E)$. In other words, the function $\Psi^*(\mathbf{p}, E)\Psi(\mathbf{p}, E)$ is relevant and is the basic quantity. Since the shape of $\Psi^*(\mathbf{p}, E)\Psi(\mathbf{p}, E)$

is not dependent on the absolute values of the momenta and energies, also the interaction processes are not dependent on these absolute values. This property is expressed by equation (4.25).

4.4.3. Specific Treatment in Connection with the Exchange of Momentum and Energy

Within projection theory, interaction exclusively takes place in $(\mathbf{p}, E)$-space, and interaction here means that there is an exchange of momentum and energy between the N subsystem, that is, we have certain fluctuations $\Delta \mathbf{p}_1, \Delta \mathbf{p}_2, \ldots, \Delta \mathbf{p}_N, \Delta E_1, \Delta E_2, \ldots, \Delta E_N$ in connection with the quantities $\mathbf{p}_1, \mathbf{p}_2, \ldots, \mathbf{p}_N, E_1, E_2, \ldots, E_N$.[22] This is a very general statement and needs further considerations.

As in conventional physics, also within projection theory, each system interacts with all the others, i.e., we have $(N-1)$ interaction processes in connection with a certain subsystem. How can we formulate this fact theoretically? Let us first explain the situation in conventional physics.

Conventional physics

In contrast to projection theory, within conventional physics the interaction takes place in $(\mathbf{r}, \tau)$-space. Let us also here consider a system which consists of N subsystems having the positions $r_m, m = 1, \ldots, N$. Then, the potential V for a certain subsystem, say k, is normally expressed by

$$V(r_k) = \sum_{\substack{i=1 \\ i \neq k}}^{N} V(r_i - r_k). \quad (4.28)$$

In other words, the kth subsystem interacts with all the other subsystems and we get $(N-1)$ interaction terms. Exactly the same arguments have to be applied in $(\mathbf{p}, E)$-space, i.e., within projection theory.

---

[22] See Section 4.4.2.

Projection theory

As in conventional physics, each system interacts with all the others, i.e., we have $(N-1)$ interaction processes in connection with a certain subsystem which is positioned in $(\mathbf{p}, E)$-space. Then, we must have $N(N-1)$ interaction processes with respect to all N subsystems. In other words, the quantities $\Delta \mathbf{p}_1, \Delta \mathbf{p}_2, \ldots, \Delta \mathbf{p}_N, \Delta E_1, \Delta E_2, \ldots, \Delta E_N$ must still be splitted. A certain momentum–energy fluctuation, say $\Delta \mathbf{p}_k, \Delta E_k$, has to be splitted into $(N-1)$ terms. Such a momentum–energy fluctuation $\Delta \mathbf{p}_k, \Delta E_k$ corresponds to the potential $V(r_k)$,[23] and we have the following equivalence:

$$V(r_k) \leftrightarrow \Delta \mathbf{p}_k, \Delta E_k. \tag{4.29}$$

Then, in analogy to equation (4.28) we get for all terms $\Delta \mathbf{p}_k, \Delta E_k$, $k = 1, \ldots, N$,

$$\begin{aligned}
\Delta \mathbf{p}_1 &= \Delta \mathbf{p}_{12} + \Delta \mathbf{p}_{13} + \cdots + \Delta \mathbf{p}_{1N}, \\
\Delta \mathbf{p}_2 &= \Delta \mathbf{p}_{21} + \Delta \mathbf{p}_{23} + \cdots + \Delta \mathbf{p}_{2N}, \\
\Delta \mathbf{p}_3 &= \Delta \mathbf{p}_{31} + \Delta \mathbf{p}_{32} + \cdots + \Delta \mathbf{p}_{3N}, \\
&\vdots \\
\Delta \mathbf{p}_{(N-1)} &= \Delta \mathbf{p}_{(N-1)1} + \Delta \mathbf{p}_{(N-1)2} + \cdots + \Delta \mathbf{p}_{(N-1)N}, \\
\Delta \mathbf{p}_N &= \Delta \mathbf{p}_{N1} + \Delta \mathbf{p}_{N2} + \cdots + \Delta \mathbf{p}_{N(N-1)},
\end{aligned} \tag{4.30}$$

and

$$\begin{aligned}
\Delta E_1 &= \Delta E_{12} + \Delta E_{13} + \cdots + \Delta E_{1N}, \\
\Delta E_2 &= \Delta E_{21} + \Delta E_{23} + \cdots + \Delta E_{2N}, \\
\Delta E_3 &= \Delta E_{31} + \Delta E_{32} + \cdots + \Delta E_{3N}, \\
&\vdots \\
\Delta E_{(N-1)} &= \Delta E_{(N-1)1} + \Delta E_{(N-1)2} + \cdots + \Delta E_{(N-1)N}, \\
\Delta E_N &= \Delta E_{N1} + \Delta E_{N2} + \cdots + \Delta E_{N(N-1)},
\end{aligned} \tag{4.31}$$

---

[23] See equation (4.28).

where the terms $\Delta \mathbf{p}_{km}, \Delta E_{km}, k, m = 1, \ldots, N, k \neq m$ describe the $\mathbf{p}, E$-fluctuations between subsystem k and subsystem m. The terms $\Delta \mathbf{p}_{km}, \Delta E_{km}, k, m = 1, \ldots, N, k \neq m$, must be chosen so that the conservation laws for the momentum and energy are fulfilled. This is the case if

$$\begin{aligned}
\Delta \mathbf{p}_{12} &= -\Delta \mathbf{p}_{21}, \\
\Delta \mathbf{p}_{23} &= -\Delta \mathbf{p}_{32}, \\
&\vdots \\
\Delta \mathbf{p}_{(N-1)N} &= -\Delta \mathbf{p}_{N(N-1)},
\end{aligned} \tag{4.32}$$

and

$$\begin{aligned}
\Delta E_{12} &= -\Delta E_{21}, \\
\Delta E_{23} &= -\Delta E_{32}, \\
&\vdots \\
\Delta E_{(N-1)N} &= -\Delta E_{N(N-1)}.
\end{aligned} \tag{4.33}$$

Then, we obtain instead of equation (4.26)

$$\begin{aligned}
\Psi(\mathbf{r}_1, \mathbf{r}_2, &\ldots, \mathbf{r}_N, t_1, t_2, \ldots, t_N) \\
&= \Psi(\mathbf{r}_{12}, \mathbf{r}_{13}, \ldots, \mathbf{r}_{(N-1)N}, t_{12}, t_{13}, \ldots, t_{(N-1)N}) \\
&= \frac{1}{(2\pi\hbar)^{2N}} \int_{-\infty}^{\infty} \Psi(\Delta \mathbf{p}_1, \Delta \mathbf{p}_2, \ldots, \Delta \mathbf{p}_N, \Delta E_1, \Delta E_2, \ldots, \Delta E_N) \\
&\quad \times \exp\left\{\frac{i}{\hbar}[\Delta \mathbf{p}_{12} \cdot \mathbf{r}_{12} + \Delta \mathbf{p}_{13} \cdot \mathbf{r}_{13} + \cdots + \Delta \mathbf{p}_{(N-1)N} \cdot \mathbf{r}_{(N-1)N} \right. \\
&\quad \left. - (\Delta E_{12} t_{12} + \Delta E_{13} t_{13} + \cdots + \Delta E_{(N-1)N} t_{(N-1)N})]\right\} \\
&\quad \times d\Delta \mathbf{p}_1 d\Delta \mathbf{p}_2 \cdots d\Delta \mathbf{p}_N d\Delta E_1 d\Delta E_2 \cdots d\Delta E_N,
\end{aligned} \tag{4.34}$$

with

$$\begin{aligned}
\mathbf{r}_{12} &= \mathbf{r}_1 - \mathbf{r}_2, \\
\mathbf{r}_{13} &= \mathbf{r}_1 - \mathbf{r}_3, \\
&\vdots \\
\mathbf{r}_{(N-1)N} &= \mathbf{r}_{N-1} - \mathbf{r}_N,
\end{aligned} \tag{4.35}$$

and

$$t_{12} = t_1 - t_2,$$
$$t_{13} = t_1 - t_3,$$
$$\vdots \qquad (4.36)$$
$$t_{(N-1)N} = t_{N-1} - t_N.$$

Using equations (4.30) and (4.31) a new set of variables are introduced. $(\mathbf{r}, t)$-space: Instead of the $2N$ variables,

$$\mathbf{r}_1, \mathbf{r}_2, \ldots, \mathbf{r}_N, t_1, t_2, \ldots, t_N,$$

we have the $N(N-1)$ variables

$$\mathbf{r}_{12}, \mathbf{r}_{13}, \ldots, \mathbf{r}_{(N-1)N}, t_{12}, t_{13}, \ldots, t_{(N-1)N}.$$

$(\mathbf{p}, E)$-space: Instead of the $2N$ variables

$$\Delta \mathbf{p}_1, \Delta \mathbf{p}_2, \ldots, \Delta \mathbf{p}_N, \Delta E_1, \Delta E_2, \ldots, \Delta E_N,$$

we have the $N(N-1)$ variables

$$\Delta \mathbf{p}_{12}, \Delta \mathbf{p}_{13}, \ldots, \Delta \mathbf{p}_{(N-1)N}, \Delta E_{12}, \Delta E_{13}, \ldots, \Delta E_{(N-1)N}.$$

In order to obtain proper Fourier transforms, the following replacements have to be done in connection with equation (4.34):

$$\frac{1}{(2\pi\hbar)^{2N}} \to \frac{1}{(2\pi\hbar)^{N(N-1)}}, \qquad (4.37)$$

$$\Psi(\Delta \mathbf{p}_1, \Delta \mathbf{p}_2, \ldots, \Delta \mathbf{p}_N, \Delta E_1, \Delta E_2, \ldots, \Delta E_N)$$
$$\to \Psi(\Delta \mathbf{p}_{12}, \Delta \mathbf{p}_{13}, \ldots, \Delta \mathbf{p}_{(N-1)N}, \Delta E_{12},$$
$$\Delta E_{13}, \ldots, \Delta E_{(N-1)N}), \qquad (4.38)$$

$$d\Delta \mathbf{p}_1 d\Delta \mathbf{p}_2 \cdots d\Delta \mathbf{p}_N d\Delta E_1 d\Delta E_2 \cdots d\Delta E_N$$
$$\to d\Delta \mathbf{p}_{12} d\Delta \mathbf{p}_{13} \cdots d\Delta \mathbf{p}_{(N-1)N} d\Delta E_{12}$$
$$\times d\Delta E_{13} \cdots d\Delta E_{(N-1)N}. \qquad (4.39)$$

Using these replacements, equation (4.34) takes the form

$$\Psi(\mathbf{r}_1, \mathbf{r}_2, \ldots, \mathbf{r}_N, t_1, t_2, \ldots, t_N)$$
$$= \Psi(\mathbf{r}_{12}, \mathbf{r}_{13}, \ldots, \mathbf{r}_{(N-1)N}, t_{12}, t_{13}, \ldots, t_{(N-1)N})$$

$$= \frac{1}{(2\pi\hbar)^{N(N-1)}} \int_{-\infty}^{\infty} \Psi(\Delta p_{12}, \Delta p_{13}, \ldots, \Delta p_{(N-1)N},$$

$$\Delta E_{12}, \Delta E_{13}, \ldots, \Delta E_{(N-1)N}) \exp\left\{\frac{i}{\hbar}[\Delta p_{12} \cdot r_{12}\right.$$

$$+ \Delta p_{13} \cdot r_{13} + \cdots + \Delta p_{(N-1)N} \cdot r_{(N-1)N} - (\Delta E_{12} t_{12}$$

$$+ \Delta E_{13} t_{13} + \cdots + \Delta E_{(N-1)N} t_{(N-1)N})]\Big\}$$

$$\times d\Delta p_{12} d\Delta p_{13} \cdots d\Delta p_{(N-1)N} d\Delta E_{12}$$

$$\times d\Delta E_{13} \cdots d\Delta E_{(N-1)N}. \tag{4.40}$$

Exactly the same arguments can be applied to the inverse transformation (4.27). Using again the equations (4.32) and (4.33) we obtain instead of equation (4.27)

$$\Psi(\Delta p_1, \Delta p_2, \ldots, \Delta p_N, \Delta E_1, \Delta E_2, \ldots, \Delta E_N)$$

$$= \Psi(\Delta p_{12}, \Delta p_{13}, \ldots, \Delta p_{(N-1)N}, \Delta E_{12}, \Delta E_{13}, \ldots, \Delta E_{(N-1)N})$$

$$= \frac{1}{(2\pi\hbar)^{2N}} \int_{-\infty}^{\infty} \Psi(r_1; r_2, \ldots, r_N, t_1, t_2, \ldots, t_N)$$

$$\times \exp\left\{-\frac{i}{\hbar}[\Delta p_{12} \cdot r_{12} + \Delta p_{13} \cdot r_{13} + \cdots + \Delta p_{(N-1)N}\cdot\right.$$

$$\times r_{(N-1)N} - (\Delta E_{12} t_{12} + \Delta E_{13} t_{13}$$

$$+ \cdots + \Delta E_{(N-1)N} t_{(N-1)N})]\Big\} dr_1 dr_2 \cdots dr_N dt_1 dt_2 \cdots dt_N. \tag{4.41}$$

Also in order to get a proper inverse transformation, the following replacements have to be done in connection with equation (4.41):

$$\frac{1}{(2\pi\hbar)^{2N}} \to \frac{1}{(2\pi\hbar)^{N(N-1)}}, \tag{4.42}$$

$$\Psi(r_1; r_2, \ldots, r_N, t_1, t_2, \ldots, t_N) \to \Psi(r_{12}, r_{13}, \ldots, r_{(N-1)N},$$

$$t_{12}, t_{13}, \ldots, t_{(N-1)N}), \tag{4.43}$$

$$dr_1 dr_2 \cdots dr_N dt_1 dt_2 \cdots dt_N \to dr_{12} dr_{13} \cdots dr_{(N-1)N}$$

$$\times dt_{12} dt_{13} \cdots dt_{(N-1)N}. \tag{4.44}$$

Then, the inverse transformation to equation (4.40) is

$$\Psi(\Delta \mathbf{p}_1, \Delta \mathbf{p}_2, \ldots, \Delta \mathbf{p}_N, \Delta E_1, \Delta E_2, \ldots, \Delta E_N)$$
$$= \Psi(\Delta \mathbf{p}_{12}, \Delta \mathbf{p}_{13}, \ldots, \Delta \mathbf{p}_{(N-1)N}, \Delta E_{12}, \Delta E_{13}, \ldots, \Delta E_{(N-1)N})$$
$$= \frac{1}{(2\pi\hbar)^{N(N-1)}} \int_{-\infty}^{\infty} \Psi(\mathbf{r}_{12}, \mathbf{r}_{13}, \ldots, \mathbf{r}_{(N-1)N},$$
$$t_{12}, t_{13}, \ldots, t_{(N-1)N}) \exp\left\{-\frac{i}{\hbar}[\Delta \mathbf{p}_{12} \cdot \mathbf{r}_{12} + \Delta \mathbf{p}_{13} \cdot \mathbf{r}_{13}\right.$$
$$+ \cdots + \Delta \mathbf{p}_{(N-1)N} \cdot \mathbf{r}_{(N-1)N} - (\Delta E_{12} t_{12} + \Delta E_{13} t_{13}$$
$$\left. + \cdots + \Delta E_{(N-1)N} t_{(N-1)N})]\right\}$$
$$\times d\mathbf{r}_{12} d\mathbf{r}_{13} \cdots d\mathbf{r}_{(N-1)N} dt_{12} dt_{13} \cdots dt_{(N-1)N}. \qquad (4.45)$$

4.4.4. The p, E-Concert

The next step within the frame of analysis is to make specific assumptions for the quantities $\Delta \mathbf{p}_{12}, \Delta \mathbf{p}_{13}, \ldots, \Delta \mathbf{p}_{(N-1)N}, \Delta E_{12}, \Delta E_{13}, \ldots, \Delta E_{(N-1)N}$. The simplest model is to use exactly the same p, E-fluctuations between all pairs of the ensemble. That is, we have the following conditions at each time τ:

$$\begin{aligned} \Delta \mathbf{p}_{12} &= \Delta \mathbf{p}, \\ \Delta \mathbf{p}_{23} &= \Delta \mathbf{p}, \\ &\vdots \\ \Delta \mathbf{p}_{(N-1)N} &= \Delta \mathbf{p}, \end{aligned} \qquad (4.46)$$

and

$$\begin{aligned} \Delta E_{12} &= \Delta E, \\ \Delta E_{23} &= \Delta E, \\ &\vdots \\ \Delta E_{(N-1)N} &= \Delta E. \end{aligned} \qquad (4.47)$$

Interactions

We will still recognize that these conditions mean that the type of interactions between subsystem k and subsystem m is exactly the same as the interaction type for subsystem k and subsystem n with $k \neq n \neq m$ and $k, m, n = 1, \ldots, N$. In the case of (4.46) and (4.47), the $\mathbf{p}, E$-fluctuations between all pairs of the ensemble are exactly the same at time τ, and this property is valid for all times τ. In other words, there is a perfect "concert" in connection with the system consisting of N subsystems.

With equations (4.46) and (4.47) we obtain for equation (4.40)

$$\Psi(\mathbf{r}_1, \mathbf{r}_2, \ldots, \mathbf{r}_N, t_1, t_2, \ldots, t_N)$$
$$= \Psi(\mathbf{r}_{12}, \mathbf{r}_{13}, \ldots, \mathbf{r}_{(N-1)N}, t_{12}, t_{13}, \ldots, t_{(N-1)N})$$
$$= \frac{1}{(2\pi\hbar)^{N(N-1)}} \int_{-\infty}^{\infty} \Psi(\Delta\mathbf{p}, \Delta\mathbf{p}, \ldots, \Delta\mathbf{p}, \Delta E, \Delta E, \ldots, \Delta E)$$
$$\times \exp\left\{\frac{i}{\hbar}[\Delta\mathbf{p} \cdot (\mathbf{r}_{12} + \mathbf{r}_{13} + \cdots + \mathbf{r}_{(N-1)N})\right.$$
$$\left. - \Delta E(t_{12} + t_{13} + \cdots + t_{(N-1)N})]\right\}$$
$$\times d\Delta\mathbf{p}\, d\Delta\mathbf{p} \cdots d\Delta\mathbf{p}\, d\Delta E\, d\Delta E \cdots d\Delta E. \tag{4.48}$$

If we perform $(N-1)N/2 - 1$ integrations with respect to the variable $\Delta\mathbf{p}$ and also $(N-1)N/2 - 1$ integrations with respect to the variable ΔE. Furthermore, if we replace the variables $\Delta\mathbf{p}, \Delta E$ by $\mathbf{p}, E$ we immediately get the following relationship:

$$\Psi(\mathbf{r}_1, \mathbf{r}_2, \ldots, \mathbf{r}_N, t_1, t_2, \ldots, t_N)$$
$$= \Psi(\mathbf{r}_{12}, \mathbf{r}_{13}, \ldots, \mathbf{r}_{(N-1)N}, t_{12}, t_{13}, \ldots, t_{(N-1)N})$$
$$= \frac{1}{(2\pi\hbar)^{N(N-1)}} \Omega_\mathbf{p}^{(N-1)N/2-1} \Omega_E^{(N-1)N/2-1} \int_{-\infty}^{\infty} \Psi(\mathbf{p}, E)$$
$$\times \exp\left\{\frac{i}{\hbar}[\mathbf{p} \cdot (\mathbf{r}_{12} + \mathbf{r}_{13} + \cdots + \mathbf{r}_{(N-1)N})\right.$$
$$\left. - E(t_{12} + t_{13} + \cdots + t_{(N-1)N})]\right\} d\mathbf{p}\, dE, \tag{4.49}$$

where $\Omega_\mathbf{p}$ is the volume of **p**-space and Ω_E is the volume of E-space.

When we apply equations (4.46) and (4.47) to equation (4.45), we obtain

$$\Psi(\Delta\mathbf{p}_{12}, \Delta\mathbf{p}_{13}, \ldots, \Delta\mathbf{p}_{(N-1)N}, \Delta E_{12}, \Delta E_{13}, \ldots, \Delta E_{(N-1)N})$$

$$= \Psi(\mathbf{p}, E) = \frac{1}{(2\pi\hbar)^{N(N-1)}}$$

$$\times \int_{-\infty}^{\infty} \Psi(\mathbf{r}_{12}, \mathbf{r}_{13}, \ldots, \mathbf{r}_{(N-1)N}, t_{12}, t_{13}, \ldots, t_{(N-1)N})$$

$$\times \exp\left\{-\frac{i}{\hbar}[\mathbf{p} \cdot (\mathbf{r}_{12} + \mathbf{r}_{13} + \cdots + \mathbf{r}_{(N-1)N})\right.$$

$$\left. - E(t_{12} + t_{13} + \cdots + t_{(N-1)N})]\right\}$$

$$\times d\mathbf{r}_{12} d\mathbf{r}_{13} \cdots d\mathbf{r}_{(N-1)N} dt_{12} dt_{13} \cdots dt_{(N-1)N}. \quad (4.50)$$

However, equation (4.49) is not compatible with equation (4.50), and this is because we cannot deduce equation (4.49) on the basis of equation (4.50) by applying the inverse Fourier transform. It is straightforward to recognize that equation (4.50) is correct but equation (4.49) can in general not be valid. Thus, equation (4.49) should not be applied.

On the one hand, the function $\Psi(\mathbf{p}, E)$ contains the entire information about the system, and this follows directly from equation (4.50). On the other hand, the function $\Psi(\mathbf{p}, E)$ and $\Psi^*(\mathbf{p}, E)\Psi(\mathbf{p}, E)$, respectively, must equally describe the interaction processes, i.e., **p**, E-fluctuations, between any pair of the ensemble, and this is because the function $\Psi(\mathbf{p}, E)$ can no longer distinguish between the various subsystems as in the case of

$$\Psi(\Delta\mathbf{p}_{12}, \Delta\mathbf{p}_{13}, \ldots, \Delta\mathbf{p}_{(N-1)N}, \Delta E_{12}, \Delta E_{13}, \ldots, \Delta E_{(N-1)N}).$$

In conclusion, all individual interaction processes that take place at time τ in connection with N subsystems of the ensemble are equally determined by $\Psi(\mathbf{p}, E)$ and $\Psi^*(\mathbf{p}, E)\Psi(\mathbf{p}, E)$, respectively.

4.4.5. Individual Processes

Under the conditions formulated in Section 4.4.4, the general wave function $\Psi(\mathbf{p}_1, \mathbf{p}_2, \ldots, \mathbf{p}_N, E_1, E_2, \ldots, E_N)$[24] is completely expressed by $\Psi(\mathbf{p}, E)$. Each subsystem interacts with $(N-1)$ subsystems, i.e., we have $(N-1)$ $\mathbf{p}, E$-processes and each of them is described by $\Psi(\mathbf{p}, E)$ and $\Psi^*(\mathbf{p}, E)\Psi(\mathbf{p}, E)$, respectively.

We have $N(N-1)$ $\mathbf{p}, E$-processes with respect to the entire system, which are all independent of each other, and each process is described by the function $\Psi(\mathbf{p}, E)$ and $\Psi^*(\mathbf{p}, E)\Psi(\mathbf{p}, E)$, respectively. In the course of time τ, the N subsystems equally run through all possible $\mathbf{p}, E$-states with a certain probability which is described by the probability density $\Psi^*(\mathbf{p}, E)\Psi(\mathbf{p}, E)$. Then, we have $N(N-1)$ individual wave functions

$$\Psi_{12}(\mathbf{p}, E), \Psi_{13}(\mathbf{p}, E), \ldots, \Psi_{(N-1)N}(\mathbf{p}, E), \qquad (4.51)$$

with

$$\begin{aligned}\Psi_{12}(\mathbf{p}, E) &= \Psi(\mathbf{p}, E), \\ \Psi_{13}(\mathbf{p}, E) &= \Psi(\mathbf{p}, E), \\ &\vdots \\ \Psi_{(N-1)N}(\mathbf{p}, E) &= \Psi(\mathbf{p}, E).\end{aligned} \qquad (4.52)$$

In connection with $\mathbf{p}, E$-fluctuations (interactions), the probability densities

$$\Psi_{ij}^*(\mathbf{p}, E)\Psi_{ij}(\mathbf{p}, E), \quad i, j = 1, \ldots, N, i \neq j,$$

are relevant because the $\mathbf{p}, E$-fluctuations are directly described by $\Psi_{ij}^*(\mathbf{p}, E)\Psi_{ij}(\mathbf{p}, E)$.

Each of the $N(N-1)$ functions $\Psi_{ij}^*(\mathbf{p}, E)\Psi_{ij}(\mathbf{p}, E)$ is of course normalized to unity:

$$\int_{-\infty}^{\infty} \Psi_{ij}^*(\mathbf{p}, E)\Psi_{ij}(\mathbf{p}, E) d\mathbf{p}\, dE = \int_{-\infty}^{\infty} \Psi^*(\mathbf{p}, E)\Psi(\mathbf{p}, E) d\mathbf{p}\, dE = 1, \qquad (4.53)$$

for $i, j = 1, \ldots, N, i \neq j$.

---

[24] See equation (4.14).

The function $\Psi_{ij}^*(\mathbf{p}, E)\Psi_{ij}(\mathbf{p}, E)$ belongs to subsystem i which interacts by $\mathbf{p}, E$-fluctuations with subsystem j. On the other hand, $\Psi_{ji}^*(\mathbf{p}, E)\Psi_{ji}(\mathbf{p}, E)$ belongs to subsystem j and interacts with subsystem i. However, both fluctuations represent one process because they are strongly correlated. If subsystem i *emits* the quantities $\Delta \mathbf{p}_{ij}, \Delta E_{ij}$, subsystem j *absorbs* exactly these quantities and we get $\Delta \mathbf{p}_{ij} = -\Delta \mathbf{p}_{ji}, \Delta E_{ij} = -\Delta E_{ji}$, and we must always have

$$\Psi_{ij}^*(\mathbf{p}, E)\Psi_{ij}(\mathbf{p}, E) = \Psi_{ij}^*(-\mathbf{p}, -E)\Psi_{ji}(-\mathbf{p}, -E), \qquad (4.54)$$

at each time τ. It is easy to recognize that condition (4.54) must hold. Otherwise, we get problems with the conservation laws for momentum and energy and, furthermore, the distribution function $\Psi_{ij}^*(\mathbf{p}, E)\Psi_{ij}(\mathbf{p}, E)$ for subsystem 1 would be different from the function $\Psi_{ji}^*(\mathbf{p}, E)\Psi_{ji}(\mathbf{p}, E)$ for subsystem 2. Equation (4.54) means that the values $\mathbf{p}_a, E_a$ for subsystem 1 at time τ must be correlated to the values $\mathbf{p}_b, E_b$ for subsystem 1 at the same time τ and we must have

$$\mathbf{p}_b = -\mathbf{p}_a, E_b = -E_a. \qquad (4.55)$$

This has to be the case for each time τ, i.e., we must have

$$\tau_a : \Psi_{ij}^*(\mathbf{p}_a, E_a)\Psi_{ij}(\mathbf{p}_a, E_a) = \Psi_{ji}^*(-\mathbf{p}_a, -E_a)\Psi_{ji}(-\mathbf{p}_a, -E_a),$$
$$\tau_b : \Psi_{ij}^*(\mathbf{p}_b, E_b)\Psi_{ij}(\mathbf{p}_b, E_b) = \Psi_{ji}^*(-\mathbf{p}_b, -E_b)\Psi_{ji}(-\mathbf{p}_b, -E_b),$$
$$\vdots$$
$$\tau_n : \Psi_{ij}^*(\mathbf{p}_n, E_n)\Psi_{ij}(\mathbf{p}_n, E_n) = \Psi_{ji}^*(-\mathbf{p}_n, -E_n)\Psi_{ji}(-\mathbf{p}_n, -E_n),$$
$$\vdots \qquad (4.56)$$

If condition (4.56) is not fulfilled, the distribution functions $\Psi_{ij}^*(\mathbf{p}, E)\Psi_{ij}(\mathbf{p}, E)$ and $\Psi_{ji}^*(\mathbf{p}, E)\Psi_{ji}(\mathbf{p}, E)$ would be different from each other because both subsystems would run through the $\mathbf{p}, E$-states with different probability densities. This would inevitably lead to internal contradictions. If, for example, subsystem 1 fulfils the condition[25]

$$\Psi^*_{ij}(\mathbf{p}, E)\Psi_{ij}(\mathbf{p}, E) = \Psi^*(\mathbf{p}, E)\Psi(\mathbf{p}, E), \qquad (4.57)$$

but subsystem 2 runs through $\mathbf{p}, E$-states in the course of time τ producing a probability density $\Psi^*_{ji}(\mathbf{p}, E)\Psi_{ji}(\mathbf{p}, E)$ different from $\Psi^*_{ij}(\mathbf{p}, E)\Psi_{ij}(\mathbf{p}, E)$. We obtain for subsystem 2 the inequality $\Psi^*_{ji}(\mathbf{p}, E)\Psi_{ji}(\mathbf{p}, E) \neq \Psi^*(\mathbf{p}, E)\Psi(\mathbf{p}, E)$, and this is in contradiction to equation (4.52). The same effect would appear if the functions $\Psi^*_{ij}(\mathbf{p}, E)\Psi_{ij}(\mathbf{p}, E)$ and $\Psi^*_{ji}(\mathbf{p}, E)\Psi_{ji}(\mathbf{p}, E)$ would not be symmetrical with respect to the internal $\mathbf{p}, E$-fluctuations.

4.4.6. Analogy to Conventional Physics

The term $V(r_i - r_k)$ of equation (4.28) describes *one* interaction process. $V(r_i - r_k)$ corresponds to two distribution functions: $\Psi^*_{ik}(\mathbf{p}, E)\Psi_{ik}(\mathbf{p}, E)$ and $\Psi^*_{ki}(\mathbf{p}, E)\Psi_{ki}(\mathbf{p}, E)$, i.e., both functions describe *one* process, as in the case of $V(r_i - r_k)$. As we have already pointed out in Section 4.4.5, the function $\Psi^*_{ik}(\mathbf{p}, E)\Psi_{ik}(\mathbf{p}, E)$ belongs to subsystem i, which interacts by $\mathbf{p}, E$-fluctuations with subsystem k. On the other hand, $\Psi^*_{ki}(\mathbf{p}, E)\Psi_{ki}(\mathbf{p}, E)$ belongs to subsystem k and interacts with subsystem i. However, both fluctuations represent *one* process because they are strongly correlated. If subsystem i *emits* the quantities $\Delta \mathbf{p}_{ik} = \mathbf{p}$, $\Delta E_{ik} = E$, subsystem k *absorbs* exactly these quantities and we have $\Delta \mathbf{p}_{ki} = -\mathbf{p}$, $\Delta E_{ij} = -\Delta E_{ki} = -E$. In conclusion, the function $V(r_i - r_k)$ of conventional physics corresponds to *two* probability densities $\Psi^*_{ik}(\mathbf{p}, E)\Psi_{ik}(\mathbf{p}, E)$ and $\Psi^*_{kji}(\mathbf{p}, E)\Psi_{ki}(\mathbf{p}, E)$, and this fact can be symbolically expressed by

$$V(r_i - r_k) \leftrightarrow \Psi^*_{ik}(\mathbf{p}, E)\Psi_{ik}(\mathbf{p}, E), \Psi^*_{ki}(\mathbf{p}, E)\Psi_{ki}(\mathbf{p}, E). \qquad (4.58)$$

Equation (4.58) reflects the law for individual processes between two subsystems. For the entire process[26] the analogy to conventional

---

[25] See equation (4.52).
[26] All the N subsystems are involved.

physics must be based on the total potential energy

$$V = \sum_{\substack{i,k=1 \\ i<k}}^{N} V(r_i - r_k), \qquad (4.59)$$

and we get

$$V \leftrightarrow \begin{cases} \Psi_{12}^*(\mathbf{p}, E)\Psi_{12}(\mathbf{p}, E), \Psi_{21}^*(\mathbf{p}, E)\Psi_{21}(\mathbf{p}, E), \\ \Psi_{13}^*(\mathbf{p}, E)\Psi_{13}(\mathbf{p}, E), \Psi_{31}^*(\mathbf{p}, E)\Psi_{31}(\mathbf{p}, E), \\ \vdots \\ \Psi_{(N-1)N}^*(\mathbf{p}, E)\Psi_{(N-1)N}(\mathbf{p}, E), \\ \Psi_{N(N-1)}^*(\mathbf{p}, E)\Psi_{N(N-1)}(\mathbf{p}, E). \end{cases} \qquad (4.60)$$

As in the case of conventional physics, where V consists of $(N-1)N/2$ terms,[27] also within projection theory we have $(N-1)N/2$ pairs of probability distributions.[28]

4.4.7. Total Momentum and Total Energy

Each system interacts at a time with all the other $(N-1)$ subsystems of the ensemble, and all these $(N-1)$ interactions, i.e., $\mathbf{p}, E$-fluctuations are equally described by the same values $\pm \mathbf{p}, \pm E$. Let us denote the net fluctuations for system k at time τ by $\mathbf{p}_{net,k}, E_{ne,kt}$. Since $\mathbf{p}_{net,k}$ and $E_{ne,kt}$ consist of $(N-1)$ terms, we obtain

$$\mathbf{p}_{net,k}, E_{ne,kt} = \pm \mathbf{p} + \cdots, \pm E + \cdots. \qquad (4.61)$$

The values $\mathbf{p}_{net,k}, E_{ne,kt}$ are identical with $\Delta \mathbf{p}_k; \Delta E_{k1}$[29] and we get for the total momenta $\mathbf{p}_k$ and the total energies E_k, with $k = 1, \ldots, N$, at

---

[27] See equation (4.59).
[28] See equation (4.60).
[29] $\mathbf{p}_{net,k} \equiv \Delta \mathbf{p}_k, E_{net,k} \equiv \Delta E_{k1}$; see equations (4.16) and (4.17).

time τ,[30]

$$\mathbf{p}_k = \mathbf{p}'_k + \mathbf{p}_{net,k}, \quad E_k = E'_k + E_{ne,kt}, \quad k = 1, \ldots, N. \quad (4.62)$$

Because of the validity of the conservation laws for the momentum and the energy, the following conditions have to be fulfilled:

$$\sum_{k=1}^{N} \mathbf{p}_{net,k} = 0, \quad (4.63)$$

$$\sum_{k=1}^{N} E_{net,k} = 0. \quad (4.64)$$

Thus, the total momentum and the total energy of the system consisting of N subsystems remain constant in the course of time τ, and let us denote these constants by $const_\mathbf{p}$ and $const_E$. Then, in addition to equations (4.63) and (4.64) we have the relations:

$$\sum_{k=1}^{N} \mathbf{p}_k = const_\mathbf{p}, \quad (4.65)$$

$$\sum_{k=1}^{N} E_k = const_E. \quad (4.66)$$

Equations (4.63), (4.64), (4.65) and (4.66) are the conditions for the interactions, i.e., $\mathbf{p}, E$-fluctuations, in $(\mathbf{p}, E)$-space.

4.5. Pair Distributions

The function $\Psi(\mathbf{p}, E)$ has only two variables, $\mathbf{p}$ and E. In addition to the wave function $\Psi(\mathbf{r}_1, \mathbf{r}_2, \ldots, \mathbf{r}_N, t_1, t_2, \ldots, t_N)$ with $2N$ variables[31] we may formally define a wave function in $(\mathbf{r}, t)$-space with two variables,

---
[30] See equations (4.16) and (4.17).
[31] See equation (4.14).

say $\mathbf{r}_a, t_a$, by means of the often used Fourier transform

$$\frac{1}{(2\pi\hbar)^2} \int_{-\infty}^{\infty} \Psi(\mathbf{p}, E) \exp\left\{\frac{i}{\hbar}[\mathbf{p}\cdot\mathbf{r}_a - E\, t_a]\right\} d\mathbf{p}\, dE = \Psi(\mathbf{r}_a, t_a). \tag{4.66}$$

What do the variables $\mathbf{r}_a, t_a$ mean in connection with our system consisting of N subsystems? Since the function $\Psi(\mathbf{p}, E)$ and the variables $\mathbf{p}, E$ only make sense when we consider at least two subsystems,[32] the quantities $\mathbf{r}_a, t_a$ may not simply reflect an isolated space-time point but should contain the information about two space-time points because two subsystems must be involved when we generate $\Psi(\mathbf{r}_a, t_a)$ from $\Psi(\mathbf{p}, E)$.[33] Let us briefly repeat the main facts.

Suppose that a certain subsystem, say k, is at time τ_1 in the state $\mathbf{p}_1, E_1$. This state is realized with a definite probability, and the probability density $\Psi^*(\mathbf{p}_1, E_1)\Psi(\mathbf{p}_1, E_1)$ is relevant. If the same subsystem is at time $\tau_2 > \tau_1$ in the state $\mathbf{p}_2, E_2$, we have for this state the probability density $\Psi^*(\mathbf{p}_2, E_2)\Psi(\mathbf{p}_2, E_2)$. The transition from $\mathbf{p}_1, E_1$ to $\mathbf{p}_2, E_2$ can only take place in connection with another subsystem, say l. This is required by the conservation laws for the momentum and the energy. That is, we require a $\mathbf{p}, E$-fluctuation $\Delta\mathbf{p} = \mathbf{p}_1 - \mathbf{p}_2, \Delta E = E_1 - E_2$ between subsystem k and subsystem l. In other words, the function $\Psi(\mathbf{p}, E)$ describes the $\mathbf{p}, E$-fluctuations between two subsystems and cannot be understood on the basis of only one subsystem. Therefore, also the Fourier transform $\Psi(\mathbf{r}_a, t_a)$[33] of $\Psi(\mathbf{p}, E)$ can only be explained in terms of the two subsystems k and l. Let us briefly analyze the situation in somewhat more detail.

4.5.1. *Information About the Interaction*

While the discussion in connection with $\Psi(\mathbf{p}, E)$ was an analysis in $(\mathbf{p}, E)$-space, the function $\Psi(\mathbf{r}_a, t_a)$ needs to be analyzed in $(\mathbf{r}, t)$-space. The subsystems k and l are characterized in $(\mathbf{r}, t)$-space by certain space-time positions, say $\mathbf{r}_k, \mathbf{r}_l, t_k, t_l$, i.e., the variables $\mathbf{r}_a, t_a$

---

[32] See Section 4.4.
[33] See equation (4.66).

must contain the information $\mathbf{r}_k, \mathbf{r}_l, t_k, t_l$:

$$(\mathbf{r}_a, t_a) = (\mathbf{r}_k, \mathbf{r}_l, t_k, t_l). \tag{4.67}$$

If the properties are dependent on the space-time distance between the subsystems k and l, we have

$$\begin{aligned}\mathbf{r}_a &= \mathbf{r}_k - \mathbf{r}_l = \mathbf{r}_{kl},\\ t_a &= t_k - t_l = t_{kl}\end{aligned} \tag{4.68}$$

Then, instead of equation (4.66) we get

$$\frac{1}{(2\pi\hbar)^2}\int_{-\infty}^{\infty} \Psi(\mathbf{p}, E)\exp\left\{\frac{i}{\hbar}[\mathbf{p}\cdot\mathbf{r}_{kl} - E\,t_{kl}]\right\} d\mathbf{p}\,dE = \Psi(\mathbf{r}_{kl}, t_{kl}). \tag{4.69}$$

This equation is valid for all $k, l = 1, \ldots, N, k \neq l$. Because each $\Psi(\mathbf{r}_{kl}, t_{kl})$, $k, l = 1, \ldots, N, k \neq l$, is described by the same function $\Psi(\mathbf{p}, E)$ we must have exactly the same form for all wave functions $\Psi(\mathbf{r}_{kl}, t_{kl})$ and probability densities $\Psi^*(\mathbf{r}_{kl}, t_{kl})\Psi(\mathbf{r}_{kl}, t_{kl})$. In conclusion, all pairs $k, l = 1, \ldots, N, k \neq l$ of the ensemble contain exactly the same information about the interaction processes, i.e., $\mathbf{p}$, E-fluctuations. Therefore, the general wave function

$$\Psi(\mathbf{r}_{12}, \mathbf{r}_{13}, \ldots, \mathbf{r}_{(N-1)N}, t_{12}, t_{13}, \ldots, t_{(N-1)N}),$$

and the general probability density

$$\Psi^*(\mathbf{r}_{12}, \mathbf{r}_{13}, \ldots, \mathbf{r}_{(N-1)N}, t_{12}, t_{13}, \ldots, t_{(N-1)N})$$
$$\times \Psi(\mathbf{r}_{12}, \mathbf{r}_{13}, \ldots, \mathbf{r}_{(N-1)N}, t_{12}, t_{13}, \ldots, t_{(N-1)N}),$$

introduced above,[34] may not contain more information about the interaction processes than those described by one pair k, l of the system consisting of N subsystems. In fact, as in the case of equation (4.69), the function $\Psi(\mathbf{r}_{12}, \mathbf{r}_{13}, \ldots, \mathbf{r}_{(N-1)N}, t_{12}, t_{13}, \ldots, t_{(N-1)N})$ is also connected to $\Psi(\mathbf{p}, E)$ by a Fourier transform.

---

[34] See equation (4.40).

4.5.2. Collective Effects in Connection with $\Psi(\mathbf{p}, E)$

Again, all pairs k, l of the ensemble are equally described by the function $\Psi(\mathbf{p}, E)$. However, the $\mathbf{p}, E$-fluctuations with respect to all pairs are strictly correlated and, therefore, the ensemble forms a collective $\mathbf{p}, E$-system. This behaviour follows directly from the conditions (4.46) and (4.47) that we have introduced in Section 4.4.4. At time τ, all pairs $k, l = 1, \ldots, N, k \neq l$ are exactly in the same $(\mathbf{p}, E)$-state, and all pairs are coupled. Both the collective effect and the shape (form) of $\Psi(\mathbf{p}, E)$ do not emerge independently from each other. Both effects should come into existence by only *one* basic process. Therefore, each configuration k, l, described by $\Psi(\mathbf{r}_{kl}, t_{kl})$, contains the information of the whole system consisting of N subsystems. This behaviour within $(\mathbf{p}, E)$-space can also be recognized by an analysis in $(\mathbf{r}, t)$-space, and we will do that below. There is however one point that should be mentioned before.

Dependency of the number of subsystems

In addition to the collective fluctuation effect in connection with the wave function $\Psi(\mathbf{p}, E)$, that we have just discussed, another effect can be superimposed. In principle, the form of $\Psi(\mathbf{p}, E)$ can be dependent on the number of subsystems. In other words, the wave function $\Psi(\mathbf{p}, E)$ for a system consisting of N_a subsystems can be different from that for a system consisting of N_b subsystems with $N_a \neq N_b$, and we would have for the corresponding functions $\Psi(\mathbf{p}, E)_{N_a}$ and $\Psi(\mathbf{p}, E)_{N_b}$ the relation

$$\Psi(\mathbf{p}, E)_{N_a} \neq \Psi(\mathbf{p}, E)_{N_b}. \qquad (4.70)$$

For example, in solid state physics a lot of systems show such a behaviour. Clearly, also those effects that lead to equation (4.70) belong to the class of collective properties, and each pair k, l of the ensemble, described by $\Psi(\mathbf{r}_{kl}, t_{kl})$, must contain this kind of information because the function $\Psi(\mathbf{r}_{kl}, t_{kl})$ is also related to $\Psi(\mathbf{p}, E)_{N_a}$ or $\Psi(\mathbf{p}, E)_{N_b}$ via equation (4.69).

Interactions with past and future events

The next step in our analysis is to develop the basic equations further in order to be able to study the behaviour of the system with N subsystems in $(\mathbf{r}, t)$-space. So far, we have mainly discussed the $\mathbf{p}, E$-states ($\mathbf{p}, E$-fluctuations) in connection with $\Psi(\mathbf{p}, E)$ and the probability density $\Psi^*(\mathbf{p}, E)\Psi(\mathbf{p}, E)$. What can we say about the space-time structure of such systems? In the following, we will give more details. Not only the space-structure at time τ is meant here but in particular, effects with respect to the system-specific time t, that is, the interactions of the subsystems with past and future events. Again, the formalism of usual physics only allows us to investigate the space-structure at time τ because the system-specific time t is not defined here. Within projection theory, the formalism is extended and the space-time structure at time τ is concerned.

4.5.3. Analysis in $(\mathbf{r}, t)$-Space

The function $\Psi(\mathbf{p}, E)$, i.e., $(\mathbf{p}, E)$-space information, is not only expressed by equation (4.50) on the basis of $\Psi(\mathbf{r}_{12}, \mathbf{r}_{13}, \ldots, \mathbf{r}_{(N-1)N}, t_{12}, t_{13}, \ldots, t_{(N-1)N})$, but also by the inverse transformation of equation (4.69) on the basis of $\Psi(\mathbf{r}_{kl}, t_{kl})$:

$$\Psi(\mathbf{p}, E) = \frac{1}{(2\pi\hbar)^2} \int_{-\infty}^{\infty} \Psi(\mathbf{r}_{kl}, t_{kl}) \exp\left\{-\frac{i}{\hbar}[\mathbf{p} \cdot \mathbf{r}_{kl} - E\, t_{kl}]\right\} d\mathbf{r}_{kl}\, dt_{kl}. \tag{4.71}$$

Then, we may argue as follows. We may treat the space-time positions $\mathbf{r}_k, \mathbf{r}_l, t_k, t_l$, with $\mathbf{r}_{kl} = \mathbf{r}_k - \mathbf{r}_l$ and $t_{kl} = t_k - t_l$, used in equation (4.69) as constants $\mathbf{r}_{k*}, \mathbf{r}_{l\bullet}, t_{k\bullet}, t_{l*}$, i.e., we may replace in equation (4.69) the variables $\mathbf{r}_k, \mathbf{r}_l, t_k, t_l$ by the constants $\mathbf{r}_{k*}, \mathbf{r}_{l\bullet}, t_{k\bullet}, t_{l*}$. Since no pair k, l of the ensemble is privileged, we would like to continue our analysis for $k = 1, l = 2$, i.e., we have

$$\begin{aligned}\mathbf{r}_k &\to \mathbf{r}_{1\bullet}, \mathbf{r}_l \to \mathbf{r}_{2\bullet}, \\ t_k &\to t_{1\bullet}, t_l \to t_{2\bullet},\end{aligned} \tag{4.72}$$

with

$$\mathbf{r}_{1\bullet 2\bullet} = \mathbf{r}_{1\bullet} - \mathbf{r}_{2\bullet},$$
$$t_{1\bullet 2\bullet} = t_{1\bullet} - t_{2\bullet}. \qquad (4.73)$$

Then, instead of equation (4.69) we obtain

$$\frac{1}{(2\pi\hbar)^2} \int_{-\infty}^{\infty} \Psi(\mathbf{p}, E) \exp\left\{\frac{i}{\hbar}[\mathbf{p} \cdot \mathbf{r}_{1\bullet 2\bullet} - Et_{1\bullet 2\bullet}]\right\} d\mathbf{p}\, dE$$
$$= \Psi(\mathbf{r}_{1\bullet 2\bullet}, t_{1\bullet 2\bullet}). \qquad (4.74)$$

Because the space-time positions $\mathbf{r}_{k*}, \mathbf{r}_{l\bullet}, t_{k\bullet}, t_{l\bullet}$ are constants in equation (4.74), instead of the function $\Psi(\mathbf{r}_{12}, \mathbf{r}_{13}, \mathbf{r}_{23}, \ldots, \mathbf{r}_{(N-1)N}, t_{12}, t_{13}, t_{23}, \ldots, t_{(N-1)N})$, we have

$$\Psi(\mathbf{r}_{1\bullet 2\bullet}, \mathbf{r}_{1\bullet 3}, \mathbf{r}_{2\bullet 3}, \ldots, \mathbf{r}_{(N-1)N}, t_{1\cdot 2}, t_{1\cdot 3}, t_{2\cdot 3}, \ldots - t_{(N-1)N}, \quad (4.75)$$

that is, we have the formal transition

$$\Psi(\mathbf{r}_{12}, \mathbf{r}_{13}, \mathbf{r}_{23}, \ldots, \mathbf{r}_{(N-1)N}, t_{12}, t_{13}, t_{23}, \ldots, t_{(N-1)N})$$
$$\to \Psi(\mathbf{r}_{1\bullet 2\bullet}, \mathbf{r}_{1\bullet 3}, \mathbf{r}_{2\bullet 3}, \ldots, \mathbf{r}_{(N-1)N}, t_{1\bullet 2\bullet},$$
$$t_{1\bullet 3}, t_{2\bullet 3}, \ldots, t_{(N-1)N}). \qquad (4.76)$$

The quantities $\mathbf{r}_{k\bullet m}, t_{k\bullet m}$ are defined by $\mathbf{r}_{k\bullet m} = \mathbf{r}_{k\bullet} - \mathbf{r}_m$ and $t_{k\bullet m\bullet} = t_{k\bullet} - t_{m\bullet}$ where $\mathbf{r}_{k\bullet}, t_{k\bullet}$ are constant values and the space-time positions $\mathbf{r}_m, t_m$ are variables.

We have stated above that all pairs $k, l = 1, \ldots, N, k \neq l$ of the ensemble contain exactly the same information about the interaction processes. Therefore, the wave function

$$\Psi(\mathbf{r}_{1\bullet 2\bullet}, \mathbf{r}_{1\bullet 3}, \mathbf{r}_{2\bullet 3}, \ldots, \mathbf{r}_{(N-1)N}, t_{1\bullet 2\bullet}, t_{1\bullet 3}, t_{2\bullet 3}, \ldots, t_{(N-1)N}$$

also cannot contain more information about the interaction processes than those described by one pair k, l of the system consisting of N subsystems. This also means that the information contents about the interaction processes which is expressed by

$$\Psi(\mathbf{r}_{12}, \mathbf{r}_{13}, \mathbf{r}_{23}, \ldots, \mathbf{r}_{(N-1)N}, t_{12}, t_{13}, t_{23}, \ldots, t_{(N-1)N}),$$

and

$$\Psi(\mathbf{r}_{1\bullet 2\bullet}, \mathbf{r}_{1\bullet 3}, \mathbf{r}_{2\bullet 3}, \ldots, \mathbf{r}_{(N-1)N}, t_{1\bullet 2\bullet}, t_{1\bullet 3}, t_{2\bullet 3}, \ldots, t_{(N-1)N}),$$

are exactly the same. However, each of these functions create a wave function $\Psi(\mathbf{p}, E)$ by a Fourier transform that are different from each other. $\Psi(\mathbf{p}, E)$ for the general case is given by equation (4.50). $\Psi(\mathbf{p}, E)$ for the case of constant space-time positions as, for example, $\mathbf{r}_{1\bullet 2\bullet}$ and $t_{1\bullet 2\bullet}$, must take the form

$$\Psi(\mathbf{p}, E) = \frac{1}{(2\pi\hbar)^{(N-1)N-2}} \int_{-\infty}^{\infty} \Psi(\mathbf{r}_{1\bullet 2\bullet}, \mathbf{r}_{1\bullet 3}, \mathbf{r}_{2\bullet 3}, \mathbf{r}_{34}, \ldots, \mathbf{r}_{(N-1)N},$$

$$t_{1\bullet 2\bullet}, t_{1\bullet 3}, t_{2\bullet 3}, t_{34}, \ldots, t_{(N-1)N})$$

$$\times \exp\left\{-\frac{i}{\hbar}[\mathbf{p} \cdot (\mathbf{r}_{1\bullet 2\bullet} + \mathbf{r}_{1\bullet 3} + \mathbf{r}_{2\bullet 3} + \mathbf{r}_{34} + \cdots + \mathbf{r}_{(N-1)N})\right.$$

$$\left. - E(t_{1\bullet 2\bullet} + t_{1\bullet 3} + t_{2\bullet 3} + t_{34} + \cdots + t_{(N-1)N})]\right\}$$

$$\times d\mathbf{r}_{1\bullet 3} d\mathbf{r}_{2\bullet 3} d\mathbf{r}_{34} \cdots d\mathbf{r}_{(N-1)N} dt_{1\bullet 2\bullet}$$

$$\times dt_{1\bullet 3} dt_{2\bullet 3} dt_{34} \cdots dt_{(N-1)N}. \tag{4.77}$$

$\Psi(\mathbf{p}, E)$ of equation (4.77) is dependent on the constant space-time positions $\mathbf{r}_{1\bullet 2\bullet}$ and $t_{1\bullet 2\bullet}$: $\Psi(\mathbf{p}, E) = \Psi(\mathbf{p}, E)_{r_{1\bullet 2\bullet}, t_{1\bullet 2\bullet}}$. However, the dependence of $\Psi(\mathbf{p}, E) = \Psi(\mathbf{p}, E)_{r_{1\bullet 2\bullet}, t_{1\bullet 2\bullet}}$ on $\mathbf{p}$ and E must be the same as in the case of the general function $\Psi(\mathbf{p}, E)$ defined by equation (4.50). This is a condition because the interaction, i.e., the $\mathbf{p}, E$-fluctuations, between the N subsystems is not changed by the consideration of various configurations with and without constant space-time positions. In other words, $\Psi(\mathbf{p}, E)_{r_{1\bullet 2\bullet}, t_{1\bullet 2\bullet}}$ must be directly proportional to the general function $\Psi(\mathbf{p}, E)$: $\Psi(\mathbf{p}, E)_{r_{1\bullet 2\bullet}, t_{1\bullet 2\bullet}} = A\Psi(\mathbf{p}, E)$, where the constant A must be dependent on $\mathbf{r}_{1\bullet 2\bullet}$ and $t_{1\bullet 2\bullet}$, that is, we have $A = A(\mathbf{r}_{1\bullet 2\bullet}, t_{1\bullet 2\bullet})$. Since the connection between $\Psi(\mathbf{p}, E)_{r_{1\bullet 2\bullet}, t_{1\bullet 2\bullet}}$ and $\Psi(\mathbf{p}, E)$ is trivial, in the following we will formally not distinguish between $\Psi(\mathbf{p}, E)_{r_{1\bullet 2\bullet}, t_{1\bullet 2\bullet}}$ and $\Psi(\mathbf{p}, E)$ or other configurations different from $\mathbf{r}_{1\bullet 2\bullet}$ and $t_{1\bullet 2\bullet}$, because the constant $A = A(\mathbf{r}_{1\bullet 2\bullet}, t_{1\bullet 2\bullet})$ does not lead to physical effects and has merely to be considered in the normalization process of the wave functions. In conclusion, instead of $\Psi(\mathbf{p}, E)_{\mathbf{r}_{1\bullet 2\bullet}, t_{1\bullet 2\bullet}}$ or similar functions we will use in the following the marking $\Psi(\mathbf{p}, E)$. With equations (4.74) and (4.77) we immediately obtain for $\Psi(\mathbf{p}, E)_{\mathbf{r}_{1\bullet 2\bullet}, t_{1\bullet 2\bullet}}$ the following relation:

$$\Psi(\mathbf{r}_{1\bullet 2\bullet}, t_{1\bullet 2\bullet}) = \frac{1}{(2\pi\hbar)^{(N-1)N}} \int_{-\infty}^{\infty} \Psi(\mathbf{r}_{1\bullet 2\bullet}, \mathbf{r}_{1\bullet 3}, \mathbf{r}_{2\bullet 3},$$
$$\mathbf{r}_{34}, \ldots, \mathbf{r}_{(N-1)N}, t_{1\bullet 2\bullet}, t_{1\bullet 3}, t_{2\bullet 3}, t_{34}, \ldots, t_{(N-1)N})$$
$$\times \exp\left\{-\frac{i}{\hbar}[\mathbf{p} \cdot (\mathbf{r}_{1\bullet 2\bullet} + \mathbf{r}_{1\bullet 3} + \mathbf{r}_{2\bullet 3} + \mathbf{r}_{34}\right.$$
$$+ \cdots + \mathbf{r}_{(N-1)N}) - E(t_{1\bullet 2\bullet} + t_{1\bullet 3} + t_{2\bullet 3} + t_{34}$$
$$\left.+ \cdots + t_{(N-1)N})]\right\} \exp\left\{\frac{i}{\hbar}[\mathbf{p} \cdot \mathbf{r}_{1\bullet 2\bullet} - Et_{1\bullet 2\bullet}]\right\}$$
$$\times d\mathbf{p}\, dE\, d\mathbf{r}_{1\bullet 3} d\mathbf{r}_{2\bullet 3}\, d\mathbf{r}_{34} \cdots d\mathbf{r}_{(N-1)N} dt_{1\bullet 2\bullet}$$
$$\times dt_{1\bullet 3} dt_{2\bullet 3} dt_{34} \cdots dt_{(N-1)N}. \qquad (4.78)$$

What can we do with equation (4.78)? There are two unknown functions in equation (4.78): $\Psi(\mathbf{r}_{1\bullet 2\bullet}, t_{1\bullet 2\bullet})$ and $\Psi(\mathbf{r}_{1\bullet 2\bullet}, \mathbf{r}_{1\bullet 3}, \mathbf{r}_{2\bullet 3}, \ldots, \mathbf{r}_{(N-1)N}, t_{1\bullet 2\bullet}, t_{1\bullet 3}, t_{2\bullet 3}, \ldots, t_{(N-1)N})$. Therefore, equation (4.78) alone cannot be used for the determination of the space-time structure of the system consisting of N subsystems. A possible way is to use relation equation (4.74) beside equation (4.78). Furthermore, since none of the configurations $\mathbf{r}_{k\bullet} - \mathbf{r}_{l\bullet}, t_{k\bullet} - t_{l\bullet}, k, l = 1, \ldots, N$, $k \neq l$ is privileged, we get the following equivalent expressions which combine equation (4.74) with equation (4.78).

Configuration $\mathbf{r}_{1\bullet 2\bullet}, t_{1\bullet 2\bullet}$:

$$\frac{1}{(2\pi\hbar)^2} \int_{-\infty}^{\infty} \Psi(\mathbf{p}, E) \exp\left\{\frac{i}{\hbar}[\mathbf{p} \cdot \mathbf{r}_{1\bullet 2\bullet} - Et_{1\bullet 2\bullet}]\right\} d\mathbf{p}\, dE$$
$$= \Psi(\mathbf{r}_{1\bullet 2\bullet}, t_{1\bullet 2\bullet}) = \frac{1}{(2\pi\hbar)^{(N-1)N}}$$

$$\times \int_{-\infty}^{\infty} \Psi(\mathbf{r}_{1\bullet 2\bullet}, \mathbf{r}_{1\bullet 3}, \mathbf{r}_{2\bullet 3}, \mathbf{r}_{34}, \ldots, \mathbf{r}_{(N-1)N},$$

$$t_{1\bullet 2\bullet}, t_{1\bullet 3}, t_{2\bullet 3}, t_{34}, \ldots, t_{(N-1)N})$$

$$\times \exp\left\{-\frac{i}{\hbar}[\mathbf{p} \cdot (\mathbf{r}_{1\bullet 2\bullet} + \mathbf{r}_{1\bullet 3} + \mathbf{r}_{2\bullet 3} + \mathbf{r}_{34} + \cdots + \mathbf{r}_{(N-1)N})\right.$$

$$\left. - E(t_{1\bullet 2\bullet} + t_{1\bullet 3} + t_{2\bullet 3} + t_{34} + \cdots + t_{(N-1)N})]\right\}$$

$$\times \exp\left\{\frac{i}{\hbar}[\mathbf{p} \cdot \mathbf{r}_{1\bullet 2\bullet} - Et_{1\bullet 2\bullet}]\right\} d\mathbf{p}\, dE\, d\mathbf{r}_{1\bullet 3}\, d\mathbf{r}_{2\bullet 3}$$

$$\times d\mathbf{r}_{34} \cdots d\mathbf{r}_{(N-1)N}\, dt_{1\bullet 2\bullet}\, dt_{1\bullet 3}\, dt_{2\bullet 3}\, dt_{34} \cdots dt_{(N-1)N}. \quad (4.79)$$

Configuration $\mathbf{r}_{1\bullet 3\bullet}, t_{1\bullet 3\bullet}$:

$$\frac{1}{(2\pi\hbar)^2} \int_{-\infty}^{\infty} \Psi(\mathbf{p}, E) \exp\left\{\frac{i}{\hbar}[\mathbf{p} \cdot \mathbf{r}_{1\bullet 3\bullet} - Et_{1\bullet 3\bullet}]\right\} d\mathbf{p}\, dE$$

$$= \Psi(\mathbf{r}_{1\bullet 3\bullet}, t_{1\bullet 3\bullet}) = \frac{1}{(2\pi\hbar)^{(N-1)N}}$$

$$\times \int_{-\infty}^{\infty} \Psi(\mathbf{r}_{1\bullet 2}, \mathbf{r}_{1\bullet 3\bullet}, \mathbf{r}_{23\bullet}, \ldots, d\mathbf{r}_{(N-1)N},$$

$$t_{1\bullet 2}, t_{1\bullet 3\bullet}, t_{23\bullet}, \ldots, t_{(N-1)N})$$

$$\times \exp\left\{-\frac{i}{\hbar}[\mathbf{p} \cdot (\mathbf{r}_{1\bullet 2} + \mathbf{r}_{1\bullet 3\bullet} + \cdots + \mathbf{r}_{(N-1)N})\right.$$

$$\left. - E(t_{1\bullet 2} + t_{1\bullet 3\bullet} + \cdots + t_{(N-1)N})]\right\}$$

$$\times \exp\left\{\frac{i}{\hbar}[\mathbf{p} \cdot \mathbf{r}_{1\bullet 3\bullet} - Et_{1\bullet 3\bullet}]\right\} d\mathbf{p}\, dE\, d\mathbf{r}_{1\bullet 2}\, d\mathbf{r}_{23\bullet}$$

$$\times d\mathbf{r}_{24} \cdots d\mathbf{r}_{(N-1)N}\, dt_{1\bullet 2}\, dt_{23\bullet}\, dt_{24} \cdots dt_{(N-1)N}. \quad (4.80)$$

$\vdots$

Configuration: $\mathbf{r}_{(N-1)\bullet N\bullet}, t_{(N-1)\bullet N\bullet}$:

$$\frac{1}{(2\pi\hbar)^2} \int_{-\infty}^{\infty} \Psi(\mathbf{p}, E) \exp\left\{\frac{i}{\hbar}[\mathbf{p} \cdot (\mathbf{r}_{(N-1)\bullet N\bullet} - Et_{(N-1)\bullet N\bullet})]\right\} d\mathbf{p}\, dE$$

$$= \Psi(\mathbf{r}_{(N-1)\bullet N\bullet}, t_{(N-1)\bullet N\bullet}) = \frac{1}{(2\pi\hbar)^{N(N-1)}}$$

$$\times \int_{-\infty}^{\infty} \Psi(\mathbf{r}_{12}, \mathbf{r}_{13}, \ldots, \mathbf{r}_{(N-2)(N-1)\bullet}, \mathbf{r}_{(N-2)(N)\bullet},$$
$$t_{12}, t_{13}, \ldots, t_{(N-2)(N-1)\bullet}, t_{(N-2)N\bullet})$$
$$\times \exp\left\{-\frac{i}{\hbar}[\mathbf{p} \cdot (\mathbf{r}_{12} + \mathbf{r}_{13} + \cdots + \mathbf{r}_{(N-2)(N-1)\bullet} + \mathbf{r}_{(N-2)N\bullet})\right.$$
$$\left. - E(t_{12} + (t_{13} + \cdots + t_{(N-2)(N-1)\bullet} + t_{(N-2)N\bullet}))]\right\}$$
$$\times \exp\left\{\frac{i}{\hbar}[\mathbf{p} \cdot \mathbf{r}_{(N-1)\bullet N\bullet} - E\, t_{(N-1)\bullet N\bullet}]\right\}$$
$$\times d\mathbf{p}\, dE\, d\mathbf{r}_{12}\, d\mathbf{r}_{13} \cdots d\mathbf{r}_{(N-2)(N-1)\bullet}\, d\mathbf{r}_{(N-2)N\bullet}\, dt_{12}$$
$$\times dt_{13} \cdots dt_{(N-2)(N-1)\bullet}\, dt_{(N-2)N\bullet}. \tag{4.81}$$

Using one of these configurations, as, for example, configuration $\mathbf{r}_{1\bullet 2\bullet}, t_{1\bullet 2\bullet}$ we can determine $\Psi(\mathbf{r}_{1\bullet 2\bullet}, t_{1\bullet 2\bullet})$ when the function $\Psi(\mathbf{p}, E)$ is given. Then, the space-time structure of the system with N subsystems is expressed by

$$\Psi(\mathbf{r}_{1\bullet 2\bullet}, \mathbf{r}_{1\bullet 3}, \mathbf{r}_{2\bullet 3}, \mathbf{r}_{34}, \ldots, \mathbf{r}_{(N-1)N},$$
$$t_{1\bullet 2\bullet}, t_{1\bullet 3}, t_{2\bullet 3}, t_{34}, \ldots, t_{(N-1)N},$$

and this function has to be varied so that the function $\Psi(\mathbf{r}_{1\bullet 2\bullet}, t_{1\bullet 2\bullet})$ is reproduced by equation (4.79). We will discuss more details below.

4.5.4. Example for N = 4

In this section, we would like to give an example for the application of the procedure introduced in Section 4.5.3. It is the goal to find the wave function

$$\Psi(\mathbf{r}_{1\bullet 2\bullet}, \mathbf{r}_{1\bullet 3\bullet}, \ldots, \mathbf{r}_{(N-1)\bullet N\bullet}, t_{1\bullet 2\bullet}, t_{1\bullet 3\bullet}, \ldots, t_{(N-1)\bullet N\bullet}), \tag{4.82}$$

for certain configurations with constant space-time positions $\mathbf{r}_{k\bullet}, t_{k\bullet}$ and distances $\mathbf{r}_{k\bullet m\bullet} = \mathbf{r}_{k\bullet} - \mathbf{r}_{m\bullet}$, $t_{k\bullet m\bullet} = t_{k\bullet} - t_{m\bullet}$, respectively. Let us start with equation (4.50) for $N = 4$,

$$\Psi(\mathbf{p}, E) = \frac{1}{(2\pi\hbar)^{12}}$$

$$\times \int_{-\infty}^{\infty} \Psi(\mathbf{r}_{12}, \mathbf{r}_{13}, \mathbf{r}_{14}, \mathbf{r}_{23}, \mathbf{r}_{24}, \mathbf{r}_{34}, t_{12}, t_{13}, t_{14}, t_{23}, t_{24}, t_{34})$$

$$\times \exp\left\{\frac{i}{\hbar}[\mathbf{p} \cdot (\mathbf{r}_{12} + \mathbf{r}_{13} + \mathbf{r}_{14} + \mathbf{r}_{23} + \mathbf{r}_{24} + \mathbf{r}_{34})\right.$$

$$\left. - E(t_{12} + t_{13} + t_{14} + t_{23} + t_{24} + t_{34})]\right\}$$

$$\times d\mathbf{r}_{12} d\mathbf{r}_{13} d\mathbf{r}_{14} d\mathbf{r}_{23} d\mathbf{r}_{24} d\mathbf{r}_{34} dt_{12} dt_{13} dt_{14} dt_{23} dt_{24} dt_{34}.$$
(4.83)

Specific configurations

The quantities $\mathbf{r}_{1\bullet}, \mathbf{r}_{2\bullet}, t_{1\bullet}, t_{2\bullet}$ and $\mathbf{r}_{1\bullet 2\bullet}, t_{1\bullet 2\bullet}$ are constants. Then, we obtain with equation (4.79) the following equation:

$$\frac{1}{(2\pi\hbar)^2} \int_{-\infty}^{\infty} \Psi(\mathbf{p}, E) \exp\left\{\frac{i}{\hbar}[\mathbf{p} \cdot \mathbf{r}_{1\bullet 2\bullet} - E t_{1\bullet 2\bullet}]\right\} d\mathbf{p}\, dE$$

$$= \Psi(\mathbf{r}_{1\bullet 2\bullet}, t_{1\bullet 2\bullet}) = \frac{1}{(2\pi\hbar)^{(N-1)N}} \times \int_{-\infty}^{\infty} \Psi(\mathbf{r}_{1\bullet 2\bullet}, \mathbf{r}_{1\bullet 3}, \mathbf{r}_{1\bullet 4},$$

$$\mathbf{r}_{2\bullet 3}, \mathbf{r}_{2\bullet 4}, \mathbf{r}_{34}, t_{1\bullet 2\bullet}, t_{1\bullet 3}, t_{1\bullet 4}, t_{2\bullet 3}, t_{2\bullet 4}, t_{34})$$

$$\times \exp\left\{-\frac{i}{\hbar}[\mathbf{p} \cdot (\mathbf{r}_{1\bullet 2\bullet} + \mathbf{r}_{1\bullet 3} + \mathbf{r}_{1\bullet 4} + \mathbf{r}_{2\bullet 3} + \mathbf{r}_{2\bullet 4} + \mathbf{r}_{34})\right.$$

$$\left. - E(t_{1\bullet 2\bullet} + t_{1\bullet 3} + t_{1\bullet 4} + t_{2\bullet 3} + t_{2\bullet 4} + t_{34})]\right\}$$

$$\times \exp\left\{\frac{i}{\hbar}[\mathbf{p} \cdot \mathbf{r}_{1\bullet 2\bullet} - E t_{1\bullet 2\bullet}]\right\} d\mathbf{p} dE d\mathbf{r}_{1\bullet 3} d\mathbf{r}_{1\bullet 4} d\mathbf{r}_{2\bullet 3} d\mathbf{r}_{2\bullet 4}$$

$$\times d\mathbf{r}_{34} dt_{1\bullet 3} dt_{1\bullet 4} dt_{2\bullet 3} dt_{2\bullet 4} dt_{34}.$$
(4.84)

In addition to $\mathbf{r}_{1\bullet}, \mathbf{r}_{2\bullet}, t_{1\bullet}, t_{2\bullet}$, also the variables $\mathbf{r}_3, t_3$ can be treated as constants. That is, we have $\mathbf{r}_{3\bullet}, t_{3\bullet}$ instead of $\mathbf{r}_3, t_3$. Then, not only the distances $\mathbf{r}_{1\bullet 2\bullet}, t_{1\bullet 2\bullet}$ have to be considered as constants but $\mathbf{r}_{1\bullet 3\bullet}, t_{1\bullet 3\bullet}$ and $\mathbf{r}_{2\bullet 3\bullet}, t_{2\bullet 3\bullet}$ as well. For this configuration we have to construct an equation in analogy to the equation (4.77), and we obtain the following relationship:

$$\frac{1}{(2\pi\hbar)^2} \int_{-\infty}^{\infty} \Psi(\mathbf{p}, E) \exp\left\{\frac{i}{\hbar}[\mathbf{p} \cdot \mathbf{r}_{1\bullet 3\bullet} - Et_{1\bullet 3\bullet}]\right\} d\mathbf{p}\, dE$$

$$= \Psi(\mathbf{r}_{1\bullet 3\bullet}, t_{1\bullet 3\bullet}) = \frac{1}{(2\pi\hbar)^{(N-1)N-4}} \times \int_{-\infty}^{\infty} \Psi(\mathbf{r}_{1\bullet 2\bullet}, \mathbf{r}_{1\bullet 3\bullet}, \mathbf{r}_{1\bullet 4},$$

$$\mathbf{r}_{2\bullet 3\bullet}, \mathbf{r}_{2\bullet 4}, \mathbf{r}_{3\bullet 4}, t_{1\bullet 2\bullet}, t_{1\bullet 3\bullet}, t_{1\bullet 4}, t_{2\bullet 3\bullet}, t_{2\bullet 4}, t_{3\bullet 4})$$

$$\times \exp\left\{-\frac{i}{\hbar}[\mathbf{p} \cdot (\mathbf{r}_{1\bullet 2\bullet} + \mathbf{r}_{1\bullet 3\bullet} + \mathbf{r}_{1\bullet 4} + \mathbf{r}_{2\bullet 3\bullet} + \mathbf{r}_{2\bullet 4\bullet} + \mathbf{r}_{3\bullet 4})\right.$$

$$\left. -E(t_{1\bullet 2\bullet} + t_{1\bullet 3\bullet} + t_{1\bullet 4} + t_{2\bullet 3\bullet} + t_{2\bullet 4} + t_{3\bullet 4})]\right\}$$

$$\times \exp\left\{\frac{i}{\hbar}[\mathbf{p} \cdot \mathbf{r}_{1\bullet 3\bullet} - Et_{1\bullet 3\bullet}]\right\} d\mathbf{p}\, dE\, d\mathbf{r}_{1\bullet 4}$$

$$\times d\mathbf{r}_{2\bullet 4}\, d\mathbf{r}_{3\bullet 4}\, dt_{1\bullet 4}\, dt_{2\bullet 4}\, dt_{3\bullet 4}. \tag{4.85}$$

Then, the wave function for a system consisting of $N = 4$ subsystems is expressed at reference time τ by

$$\Psi(\mathbf{r}_{1\bullet 2\bullet}, \mathbf{r}_{1\bullet 3\bullet}, \mathbf{r}_{1\bullet 4\bullet}, \mathbf{r}_{2\bullet 3\bullet}, \mathbf{r}_{2\bullet 4\bullet}, \mathbf{r}_{3\bullet 4\bullet},$$
$$t_{1\bullet 2\bullet}, t_{1\bullet 3\bullet}, t_{1\bullet 4\bullet}, t_{2\bullet 3\bullet}, t_{2\bullet 4\bullet}, t_{3\bullet 4\bullet}), \tag{4.86}$$

where the values $\mathbf{r}_{1\bullet}, \mathbf{r}_{2\bullet}, t_{1\bullet}, t_{2\bullet}, \mathbf{r}_{3\bullet}, t_{3\bullet}$ are fixed, and the space-time positions $\mathbf{r}_{4\bullet}, t_{4\bullet}$ can be chosen arbitrarily. The quantity given by equation (4.86) is not the absolute wave function of the system and this is because we have not considered the constant A for this configuration; we have introduced the factor A in connection with equation (4.77). Clearly, the constant A varies from configuration to configuration.

The probability density for this case ($N = 4$) is directly given by

$$\Psi^{\bullet}(\mathbf{r}_{1\bullet 2\bullet}, \mathbf{r}_{1\bullet 3\bullet}, \mathbf{r}_{1\bullet 4\bullet}, \mathbf{r}_{2\bullet 3\bullet}, \mathbf{r}_{2\bullet 4\bullet}, \mathbf{r}_{3\bullet 4\bullet}, t_{1\bullet 2\bullet}, t_{1\bullet 3\bullet}, t_{1\bullet 4\bullet},$$
$$t_{2\bullet 3\bullet}, t_{2\bullet 4\bullet}, t_{3\bullet 4\bullet})$$
$$\times \Psi(\mathbf{r}_{1\bullet 2\bullet}, \mathbf{r}_{1\bullet 3\bullet}, \mathbf{r}_{1\bullet 4\bullet}, \mathbf{r}_{2\bullet 3\bullet}, \mathbf{r}_{2\bullet 4\bullet}, \mathbf{r}_{3\bullet 4\bullet},$$
$$t_{1\bullet 2\bullet}, t_{1\bullet 3\bullet}, t_{1\bullet 4\bullet}, t_{2\bullet 3\bullet}, t_{2\bullet 4\bullet}, t_{3\bullet 4\bullet}). \tag{4.87}$$

Clearly, formulas (4.83)–(4.87) can be easily extended for $N > 4$.

4.5.5. Further Discussions

What does the relationship (4.79) mean? It means that the wave function $\Psi(\mathbf{r}_{1\bullet 2\bullet}, t_{1\bullet 2\bullet})$ and the probability density $\Psi^*(\mathbf{r}_{1\bullet 2\bullet}, t_{1\bullet 2\bullet}) \Psi(\mathbf{r}_{1\bullet 2\bullet}, t_{1\bullet 2\bullet})$, respectively, for a certain configuration $\mathbf{r}_{1\bullet 2\bullet}, t_{1\bullet 2\bullet}$ must be compatible with the space-time structure of the other $(N-2)$ subsystems around the configuration $\mathbf{r}_{1\bullet 2\bullet}, t_{1\bullet 2\bullet}$ or any other configuration $\mathbf{r}_{k\bullet m\bullet}, t_{k\bullet m\bullet}$ within the ensemble. That is, the wave function $\Psi(\mathbf{r}_{1\bullet 2\bullet}, t_{1\bullet 2\bullet})$ and the probability density $\Psi^*(\mathbf{r}_{1\bullet 2\bullet}, t_{1\bullet 2\bullet})\Psi(\mathbf{r}_{1\bullet 2\bullet}, t_{1\bullet 2\bullet})$, respectively, is dependent on the space-time structure of the other $(N-2)$ subsystems. In particular, there is not only an interaction process, i.e., $\mathbf{p}, E$-fluctuations in $(\mathbf{p}, E)$-space, between subsystem 1 and subsystem 2 but both subsystems also interact simultaneously with all the other $(N-2)$ subsystems which are described by exactly the same function $\Psi(\mathbf{p}, E)$. As we have already pointed out in Section 4.4.4, the whole interaction scenario between the N subsystems is described in $(\mathbf{p}, E)$-space by strongly coupled pairs $k, m = 1, N, k \neq m$. At time τ all pairs are in exactly the same $\mathbf{p}, E$-state,[35] and these states are simultaneously changed by $\mathbf{p}, E$-fluctuations. For example, if all pairs are at time τ_1 in the $\mathbf{p}_1, E_1$-state, all these pairs are in the $\mathbf{p}_2, E_2$-state at time τ_2 after certain $\mathbf{p}, E$-fluctuations.

This interaction scenario also follows directly from the fact that $\Psi(\mathbf{p}, E)$[36] contains the information about the whole system and, therefore, also the configuration $\mathbf{r}_{1\bullet 2\bullet}, t_{1\bullet 2\bullet}$ must contain the information about the whole system. The $\mathbf{p}, E$-fluctuations between the subsystems 1 and 2 are described by the function $\Psi(\mathbf{p}, E)$ and the probability density $\Psi^*(\mathbf{p}, E)\Psi(\mathbf{p}, E)$, respectively. However, all the other configurations $\mathbf{r}_{k\bullet m\bullet}, t_{k\bullet m\bullet}$ are described simultaneously by exactly the same wave function $\Psi(\mathbf{p}, E)$,[37] Therefore, all configurations $\mathbf{r}_{k\bullet m\bullet}, t_{k\bullet m\bullet}, k, m = 1, N, k \neq m$, contain the information about the whole system.

---

[35] See Section 4.4.5.
[36] See equation (4.50).
[37] See, in particular, Section 4.5.4.

Equation (4.79) must be read as follows. For the description of the space-time structure of the system with N subsystems, we have considered two subsystems that can be chosen arbitrarily. The configuration $\mathbf{r}_{k \bullet m \bullet}, t_{k \bullet m \bullet}$ of these two subsystems, as, for example, subsystem 1 and subsystem 2, is formulated by equation (4.79) and, if the function $\Psi(\mathbf{p}, E)$ is given, the configuration $\mathbf{r}_{1 \bullet 2 \bullet}, t_{1 \bullet 2 \bullet}$ is directly expressed by equation (4.79):

$$\Psi(\mathbf{r}_{1 \bullet 2 \bullet}, t_{1 \bullet 2 \bullet}) = \frac{1}{(2\pi\hbar)^2} \int_{-\infty}^{\infty} \Psi(\mathbf{p}, E)$$

$$\times \exp\left\{\frac{i}{\hbar}[\mathbf{p} \cdot \mathbf{r}_{1 \bullet 2 \bullet} - E t_{1 \bullet 2 \bullet}]\right\} d\mathbf{p}\, dE. \quad (4.88)$$

In this case the function $\Psi(\mathbf{p}, E)$ is given by $\Psi(\mathbf{p}, E)_{\mathbf{r}_{1 \bullet 2 \bullet}, t_{1 \bullet 2 \bullet}}$ with $\Psi(\mathbf{p}, E)_{\mathbf{r}_{1 \bullet 2 \bullet}, t_{1 \bullet 2 \bullet}} = A\Psi(\mathbf{p}, E)$; see in particular Section 4.5.3. As we have outlined above, not only the complex $\mathbf{r}_{1 \bullet 2 \bullet}, t_{1 \bullet 2 \bullet}$ is described by $\Psi(\mathbf{p}, E)$ but also the other configurations $\mathbf{r}_{k \bullet m \bullet}, t_{k \bullet m \bullet}, k, m = 1, N$, $k \neq m$ are equally described by $\Psi(\mathbf{p}, E)$ and, therefore, the wave functions $\Psi(\mathbf{r}_{k \bullet m \bullet}, t_{k \bullet m \bullet})$ of all configurations are exactly the same, only differing in the space-time distances, i.e., none of the configurations $\mathbf{r}_{k \bullet m \bullet}, t_{k \bullet m \bullet}, k, m = 1, N, k \neq m$ is privileged and, therefore, we may choose any pair of subsystems. In fact, the $\mathbf{p}, E$-fluctuations between all subsystems are exactly the same at each time τ.[38] Therefore, we may use the specific configuration $\mathbf{r}_{1 \bullet 2 \bullet}, t_{1 \bullet 2 \bullet}$ within our general consideration. We get the wave function $\Psi(\mathbf{r}_{1 \bullet 2 \bullet}, t_{1 \bullet 2 \bullet})$ from equation (4.88) if the function $\Psi(\mathbf{p}, E)$ is given.

4.6. Basic Equations
4.6.1. *The Main Features*

Due to the transition from $\Psi(\Delta \mathbf{p}_1, \Delta \mathbf{p}_2, \ldots, \Delta \mathbf{p}_N, \Delta E_1, \Delta E_2, \ldots, \Delta E_N)$[39] to $\Psi(\mathbf{p}, E)$ we were able to divide the ensemble into small

---
[38] See Section 4.5.4.
[39] See equation (4.50).

complexes and each of them consists of only two subsystems. We have demonstrated this point in connection with the configurations[40]

$$\mathbf{r}_{1\bullet 2\bullet}, t_{1\bullet 2\bullet}, \mathbf{r}_{1\bullet 3\bullet}, t_{1\bullet 3\bullet}, \ldots, \mathbf{r}_{(N-1)\bullet N\bullet}, t_{(N-1)\bullet N\bullet}.$$

We obtained for each complex exactly the same basic equation,[41] i.e., we have to consider all configurations (pairs) as equivalent only differing in the space-time distances. There are $N(N-1)$ equivalent configurations with the wave functions $\Psi(\mathbf{r}_{k\bullet m\bullet}, t_{k\bullet m\bullet})$, $k, m = 1, N$, $k \neq m$, for a system with N subsystems.

Analysis in $(\mathbf{p}, E)$-space

Since we have chosen for all interaction processes between all N subsystems at time τ exactly the same momentum transfers and also exactly the same energy transfers,[42] the wave function with the variables $\mathbf{p}_1, \mathbf{p}_2, \ldots, \mathbf{p}_N, E_1, E_2, \ldots, E_N$ reduces to a wave function with only two variables, namely $\mathbf{p}$ and E, that is, we have a transition of the form

$$\Psi(\mathbf{p}_1, \mathbf{p}_2, \ldots, \mathbf{p}_N, E_1, E_2, \ldots, E_N) \to \Psi(\mathbf{p}, E). \qquad (4.89)$$

Again, $\Psi(\mathbf{p}, E)$ with the property $\Psi^*(\mathbf{p}, E)\Psi(\mathbf{p}, E) = \Psi^*(-\mathbf{p}, -E)\Psi(-\mathbf{p}, -E)$ means the following.[43] If subsystem i of two subsystems i and j of the ensemble with N systems has momentum $\mathbf{p}$ and the energy E, subsystem j has simultaneously the momentum $-\mathbf{p}$ and the energy $-E$. This is the case for all pairs of the ensemble. All pairs are at each time τ in exactly the same $\mathbf{p}, E$-state. There are strong correlations with respect to the $\mathbf{p}, E$-fluctuations among the subsystems which come into existence through the conservation laws of momentum and energy. This is the situation in $(\mathbf{p}, E)$-space. All these points have been discussed in the last few sections.

---

[40] See Section 4.5.4.
[41] See equation (4.79).
[42] See equation (4.46).
[43] See Section 4.4.5.

Analysis in $(\mathbf{r}, t)$-space

Each of the wave functions $\Psi(\mathbf{r}_{k\bullet m\bullet}, t_{k\bullet m\bullet})$, $k, m = 1, N$, $k \neq m$ contains the information about the entire system, consisting of N subsystems, in condensed form. Any $\Psi(\mathbf{r}_{k\bullet m\bullet}, t_{k\bullet m\bullet})$ is expressed by $\Psi(\mathbf{r}_{12}, \mathbf{r}_{13}, \mathbf{r}_{23}, \ldots, \mathbf{r}_{k\bullet m\bullet}, \ldots, \mathbf{r}_{(N-1)N}, t_{12}, t_{13}, t_{23}, \ldots, t_{k\bullet m\bullet}, \ldots, t_{(N-1)N})$ as well as by $\Psi(\mathbf{p}, E) = \Psi(\mathbf{p}, E)_{\mathbf{r}_{k\bullet m\bullet}, t_{k\bullet m\bullet}}$. This can be recognized through equation (4.79) in the case of configuration $\mathbf{r}_{1\bullet 2\bullet}, t_{1\bullet 2\bullet}$. This property can be formulated symbolically by

$$\Psi(\mathbf{p}, E) = \Psi(\mathbf{p}, E)_{\mathbf{r}_{k\bullet m\bullet}, t_{k\bullet m\bullet}} \rightarrow \Psi(\mathbf{r}_{k\bullet m\bullet}, t_{k\bullet m\bullet})$$
$$\leftarrow \Psi(\mathbf{r}_{12}, \mathbf{r}_{13}, \mathbf{r}_{23}, \ldots, \mathbf{r}_{k\bullet m\bullet}, \ldots, \mathbf{r}_{(N-1)N},$$
$$t_{12}, t_{13}, t_{23}, \ldots, t_{k\bullet m\bullet}, \ldots, t_{(N-1)N}). \tag{4.90}$$

In order to be able to say something about the interaction potential within a complex, consisting of the two subsystems k and m, we need, as we learned in Chapter 2, a suitable basic equation which has the form of equation (4.88) and is expressed for the configuration $\mathbf{r}_{k\bullet m\bullet}, t_{k\bullet m\bullet}$ by

$$\Psi(\mathbf{r}_{k\bullet m\bullet}, t_{1\bullet 2\bullet}) = \frac{1}{(2\pi\hbar)^2} \int_{-\infty}^{\infty} \Psi(\mathbf{p}, E)$$
$$\times \exp\left\{\frac{i}{\hbar}[\mathbf{p}\cdot\mathbf{r}_{k\bullet m\bullet} - Et_{k\bullet m\bullet}]\right\} d\mathbf{p}\, dE,$$
$$\text{for } k, m = 1, \ldots, N, k \neq m. \tag{4.91}$$

with $\Psi(\mathbf{p}, E) = \Psi(\mathbf{p}, E)_{\mathbf{r}_{k\bullet m\bullet}, t_{k\bullet m\bullet}}$. In the general case no space-time position $\mathbf{r}_{km}, t_{km}$ is fixed and quantities like $\mathbf{r}_{k\bullet m\bullet}, t_{k\bullet m\bullet}$ do not appear. In this case $\Psi(\mathbf{p}, E)$ must take the general form given by equation (4.50) and is not given by $\Psi(\mathbf{p}, E)_{\mathbf{r}_{k\bullet m}, t_{k\bullet m}}$. Then, we may express the general case by

$$\Psi(\mathbf{r}_{km}, t_{km}) = \frac{1}{(2\pi\hbar)^2} \int_{-\infty}^{\infty} \Psi(\mathbf{p}, E) \exp\left\{\frac{i}{\hbar}[\mathbf{p}\cdot\mathbf{r}_{km} - Et_{km}]\right\} d\mathbf{p}\, dE,$$
$$\text{for } k, m = 1, \ldots, N, k \neq m. \tag{4.92}$$

However, in the case of equation (4.92) the function $\Psi(\mathbf{p}, E)_{\mathbf{r}_{1\bullet 2\bullet}, t_{1\bullet 2\bullet}}$ is not applicable but only the general function $\Psi(\mathbf{p}, E)$, defined by equation (4.50). It is relatively easy to verify that. In accordance with the principles introduced in Chapter 2, equation (4.92) must necessarily lead to the definition of a pair potential

$$\upsilon(x_{km}, y_{km}, z_{km}, t_{km}), \quad k, m = 1, \ldots, N, k \neq m, \quad (4.93)$$

with $\mathbf{r}_{km} = (x_{km}, y_{km}, z_{km})$. The pair potentials $\upsilon(x_{km}, y_{km}, z_{km}, t_{km})$ must have the same form for all complexes k and m, and this is because for all complexes, the same function $\Psi(\mathbf{p}, E)$ is effective, i.e., the $\mathbf{p}, E$-fluctuations are exactly the same for all pairs (complexes) of the system consisting of N subsystems. Since the $\mathbf{p}, E$-fluctuations reflect interaction processes, we must therefore have the same pair potential $\upsilon(x_{km}, y_{km}, z_{km}, t_{km})$ for each complex consisting of subsystem k and subsystem m. In summary, the pair potentials for all complexes have exactly the same form. They only differ in the space-time distances $x_{km}, y_{km}, z_{km}, t_{km}$. This property has to be considered in the construction of the pair potentials $\upsilon(x_{km}, y_{km}, z_{km}, t_{km})$ with $k, m = 1, \ldots, N, k \neq m$. In Section 4.6.3 we will come back to this point.

In conclusion, instead of the general potential $V(x_1, y_1, z_1, \ldots, x_N, y_N, z_N, t_1, \ldots, t_N)$, we get $\upsilon(x_{km}, y_{km}, z_{km}, t_{km})$ and we have the transition

$$V(x_1, y_1, z_1, \ldots, x_N, y_N, z_N, t_1, \ldots, t_N) \to \upsilon(x_{km}, y_{km}, z_{km}, t_{km}), \quad (4.94)$$

in the case of the specific interaction processes in $(\mathbf{p}, E)$-space.[44] However, in the description of $\upsilon(x_{km}, y_{km}, z_{km}, t_{km})$, we may not assume that the two subsystems k and l are detached from the remaining $(N-2)$ subsystems but it is a pair potential for the two subsystems which are embedded in the environment of the other $(N-2)$ subsystems. Thus, also this environment must be reflected in $\upsilon(x_{km}, y_{km}, z_{km}, t_{km})$, and when we formally express this dependence on the environment by the quantity α_{env}, we necessarily have

$$\upsilon(x_{km}, y_{km}, z_{km}, t_{km}) = \upsilon(x_{km}, y_{km}, z_{km}, t_{km}, \alpha_{env}). \quad (4.95)$$

---

[44] See transition (4.89).

The next step in our analysis is to deepen these connections and to define the pair potential $v(x_{km}, y_{km}, z_{km}, t_{km}, \alpha_{env})$ on the basis of equation (4.92). However, first we have to clarify the role of the potential energy within projection theory.

Determination of the total potential energy

As we have already pointed out, both subsystems k and m, described by the wave function $\Psi(\mathbf{r}_{km}, t_{km})$, not only interact with each other but simultaneously with all the other $(N-2)$ subsystems. However, in the case of $\Psi(\mathbf{r}_{km}, t_{km})$ only the processes between the subsystems k and m are considered, which leads to the pair potential effect. Again, for all the other pairs $i \neq k$ and $j \neq m$ with $\Psi(\mathbf{r}_{ij}, t_{ij})$ must have exactly the same form for the pair potential. In other words, no pair of the ensemble is preferred. This has been demonstrated in Section 4.5.3 in connection with the configurations $\mathbf{r}_{1\bullet 2\bullet}, t_{1\bullet 2\bullet}$, $\mathbf{r}_{1\bullet 3\bullet}, t_{1\bullet 3\bullet}$ and $\mathbf{r}_{(N-1)\bullet N\bullet}, t_{(N-1)\bullet N\bullet}$. Thus, we may state that the contribution of each pair i, j of the ensemble to the total potential energy V is $v(x_{km}, y_{km}, z_{km}, t_{km}, \alpha_{env})$ and we get

$$V = \frac{1}{2} \sum_{i,j=1}^{N} v(x_{ij}, y_{ij}, z_{ij}, t_{ij}, \alpha_{env}), \qquad (4.96)$$

with

$$v(x_{km}, y_{km}, z_{km}, t_{km}, \alpha_{env}) = v(x_{mk}, y_{mk}, z_{mk}, t_{km}, \alpha_{env}). \qquad (4.97)$$

That is, V is simply given by the superposition of pair potentials $v(x_{km}, y_{km}, z_{km}, t_{km}, \alpha_{env})$ which is also a typical feature of conventional physics. However, as we have mentioned several times, the system-specific time t is not defined in conventional physics but only the external time τ which is used within projection theory as reference time.

Remark

The $(\mathbf{r}, t)$-space merely contains geometrical structures and there can be no energies definable in $(\mathbf{r}, t)$-space as, for example, the potential energy V. The projection principle requires that any form of energy belongs to $(\mathbf{p}, E)$-space. Nevertheless, we may express V and $\upsilon(x_{km}, y_{km}, z_{km}, t_{km}, \alpha_{env})$, respectively, by the variables of $(\mathbf{r}, t)$-space. That however does not mean that there are energies embedded in $(\mathbf{r}, t)$-space as is the case of conventional physics. We calculate $\Psi(\mathbf{r}, t)$, and the probability density $\Psi^*(\mathbf{r}, t)\Psi(\mathbf{r}, t)$ can be represented as a geometrical structure in $(\mathbf{r}, t)$-space. This geometrical structure has to be identified with that which we have in everyday life in front of us. The equation for the determination of $\Psi(\mathbf{r}, t)$ is, for example,[45]

$$i\hbar \frac{\partial}{\partial t} \Psi(\mathbf{r}, t) = -\frac{\hbar^2}{2m} \Delta \Psi(\mathbf{r}, t) + V(x, y, z, t)\Psi(\mathbf{r}, t).$$

In any case, it is not depictable in $(\mathbf{r}, t)$-space.[46]

4.6.2. *Some Additional Statements*

The pair potential $\upsilon(x_{km}, y_{km}, z_{km}, t_{km}, \alpha_{env})$ describes the interaction in $(\mathbf{r}, t)$-space. $\upsilon(x_{km}, y_{km}, z_{km}, t_{km}, \alpha_{env})$ can be determined on the basis of fundamental equations.[47] However, we know from Chapter 2 that an equivalent description in $(\mathbf{p}, E)$-space must exist. In $(\mathbf{p}, E)$-space, the potential υ becomes an operator[47]

$$\upsilon\left[i\hbar\frac{\partial}{\partial p_x}, i\hbar\frac{\partial}{\partial p_y}, i\hbar\frac{\partial}{\partial p_z}, -i\hbar\partial/\partial E, \alpha_{env}\right].$$

So far we have discussed the interaction for the system consisting of N subsystems in terms of $\mathbf{p}, E$-fluctuations on the basis of $\Psi(\mathbf{p}, E)$. In the following, we will formulate the basic equations for the determination of the functions $\Psi(\mathbf{r}_{km}, t_{km})$ and $\Psi(\mathbf{p}, E)$ on the basis of $\upsilon(x_{km}, y_{km}, z_{km}, t_{km}, \alpha_{env})$ and $\upsilon[i\hbar\partial/\partial p_x, i\hbar\partial/\partial p_y, i\hbar\partial/\partial p_z, -i\hbar\partial/\partial E, \alpha_{env}]$, just in connection with the system consisting of N subsystems. But let us first repeat the main features with respect to $\mathbf{p}, E$-fluctuations.

---

[45] See Chapter 2, equation (2.35).
[46] Interested readers might refer to Schommers, W., Cosmic secrets: World Scientific, in preparation.
[47] See Section 2.3.7.

A brief summary concerning p, E-fluctuations

The system consisting of N subsystems is described in $(\mathbf{p}, E)$-space by the function $\Psi(\mathbf{p}, E)$, and all subsystems equally experience $\Psi(\mathbf{p}, E)$. The $\mathbf{p}, E$-fluctuations for each of the N subsystems are exactly the same.[48] Then, we obtain the following picture. The wave function[49]

$$\Psi(\mathbf{p}_1, \mathbf{p}_2, \ldots, \mathbf{p}_N, E_1, E_2, \ldots, E_N),$$

with the variables $\mathbf{p}_1, \mathbf{p}_2, \ldots, \mathbf{p}_N, E_1, E_2, \ldots, E_N$ can be expressed in terms of the fluctuations $\Delta\mathbf{p}_1, \Delta\mathbf{p}_2, \ldots, \Delta\mathbf{p}_N, \Delta E_1, \Delta E_2, \ldots, \Delta E_N$ and we get[50,51]

$$\Psi(\Delta\mathbf{p}_1, \Delta\mathbf{p}_2, \ldots, \Delta\mathbf{p}_N, \Delta E_1, \Delta E_2, \ldots, \Delta E_N).$$

The variables $\Delta\mathbf{p}_i, \Delta E_i, 1 = 1, \ldots, N$ have to be considered as net fluctuations, and this is because each subsystem interacts with all the other $(N-1)$ subsystems leading to the effect that the quantities $\Delta\mathbf{p}_k, \Delta E_k, k = 1, \ldots, N$ are composed of $(N-1)$ fluctuation terms that we have expressed in Section 4.4.3 by $\Delta\mathbf{p}_{ij}, \Delta E_{ij}, i, j = 1, \ldots, N, i \neq j$. Furthermore, we assumed that all these fluctuations are equally given by exactly the same quantities $\Delta\mathbf{p} \equiv \mathbf{p}$ and $\Delta E \equiv E$ and we obtained the following schemes:

$$\begin{aligned}\Delta\mathbf{p}_1 &\to \Delta\mathbf{p}_{12}, \Delta\mathbf{p}_{13}, \ldots, \Delta\mathbf{p}_{1N} \leftarrow \pm\mathbf{p}, \\ \Delta\mathbf{p}_2 &\to \Delta\mathbf{p}_{21}, \Delta\mathbf{p}_{23}, \ldots, \Delta\mathbf{p}_{2N} \leftarrow \pm\mathbf{p}, \\ &\vdots \\ \Delta\mathbf{p}_N &\to \Delta\mathbf{p}_{N1}, \Delta\mathbf{p}_{N2}, \ldots, \Delta\mathbf{p}_{N(N-1)} \leftarrow \pm\mathbf{p},\end{aligned} \quad (4.98)$$

and

$$\begin{aligned}\Delta E_1 &\to \Delta E_{12}, \Delta E_{13}, \ldots, \Delta E_{1N} \leftarrow \pm E, \\ \Delta E_2 &\to \Delta E_{21}, \Delta E_{23}, \ldots, \Delta E_{2N} \leftarrow \pm E, \\ &\vdots \\ \Delta E_N &\to \Delta E_{N1}, \Delta E_{N2}, \ldots, \Delta E_{N(N-1)} \leftarrow \pm E.\end{aligned} \quad (4.99)$$

---

[48] See Section 4.5.
[49] See equation (4.14).
[50] See Section 4.4.2.
[51] See equation (4.27).

A more detailed discussion concerning $\mathbf{p}, E$-fluctuations in connection with the wave function $\Psi(\mathbf{p}, E)$ is given in the sections above.

Further remarks

Instead of the $\mathbf{p}, E$-fluctuations in $(\mathbf{p}, E)$-space, the interaction in $(\mathbf{r}, t)$-space is described by the pair potentials

$$v(x_{ij}, y_{ij}, z_{ij}, t_{ij}, \alpha_{env}), i, j = 1, \ldots, N, i \neq j.$$

Each pair potential $v(x_{ij}, y_{ij}, z_{ij}, t_{ij}, \alpha_{env})$ corresponds to two probability densities. In the case of $v(x_{ij}, y_{ij}, z_{ij}, t_{ij}, \alpha_{env})$, we have the two probability distributions $\Psi_{ij}^*(\mathbf{p}, E)\Psi_{ij}(\mathbf{p}, E)$ and $\Psi_{ji}^*(\mathbf{p}, E)\Psi_{ji}(\mathbf{p}, E)$, one for subsystem i and another for subsystem j. The $\mathbf{p}, E$-fluctuation between subsystem i and subsystem j changes the states of both subsystems simultaneously. If system i is in the state $\mathbf{p}', E'$ at time τ' and if it is in the state $\mathbf{p}'', E''$ at time τ'', subsystem j must be in the state $-\mathbf{p}', -E'$ at time τ' and in the state $-\mathbf{p}'', -E''$ at time τ''. Therefore, we have[52]

$$\Psi_{ij}^*(\mathbf{p}', E')\Psi_{ij}(\mathbf{p}', E') = \Psi_{ji}^*(-\mathbf{p}', -E')\Psi_{ji}(-\mathbf{p}', -E'), \quad (4.100)$$

and

$$\Psi_{ij}^*(\mathbf{p}'', E'')\Psi_{ij}(\mathbf{p}'', E'') = \Psi_{ji}^*(-\mathbf{p}'', -E'')\Psi_{ji}(-\mathbf{p}'', -E''). \quad (4.101)$$

Because of equation (4.54) we also have

$$\Psi^*(\mathbf{p}, E)\Psi(\mathbf{p}, E) = \Psi^*(-\mathbf{p}, -E)\Psi(-\mathbf{p}, -E). \quad (4.102)$$

On the basis of equation (4.102) the wave function $\Psi(\mathbf{p}, E)$ can be expressed in general by

$$\Psi(\mathbf{p}, E) = \Psi_{ph}(\mathbf{p}, E) \exp[i\varphi(\mathbf{p}, E)], \quad (4.103)$$

with

$$\Psi_{ph}(\mathbf{p}, E) = \Psi_{ph}(-\mathbf{p}, -E), \quad (4.104)$$

where the phase factor $\exp[i\varphi(\mathbf{p}, E)]$ is determined by the function $\varphi(\mathbf{p}, E)$.

---

[52] See equation (4.54).

In summary, in the case of N subsystems we have $(N-1)$ interactions for all subsystems. For subsystem k we have in $(\mathbf{r}, t)$-space the following connections:

$$v(x_{k1}, y_{k1}, z_{k1}, t_{k1}, \alpha_{env}), v(x_{k2}, y_{k2}, z_{k2}, t_{k2}, \alpha_{env}), \ldots,$$

$$v(x_{kN}, y_{kN}, z_{kN}, t_{kN}, \alpha_{env}), \quad \text{for } k \neq 1, 2, \ldots, N. \quad (4.105)$$

In $(\mathbf{p}, E)$-space we have, instead of the $(N-1)$ pair interactions in $(\mathbf{r}, t)$-space, $(N-1)$ distribution functions

$$\Psi_{k1}^*(\mathbf{p}, E)\Psi_{k1}(\mathbf{p}, E), \Psi_{k2}^*(\mathbf{p}, E)\Psi_{k2}(\mathbf{p}, E), \ldots, \Psi_{kN}^*(\mathbf{p}, E)\Psi_{kN}(\mathbf{p}, E),$$

$$k \neq 1, 2, \ldots, N, \quad (4.106)$$

which exactly corresponds to the situation in $(\mathbf{r}, t)$-space.[53]

System with two subsystems

In the case of two subsystems ($N = 2$), we have the wave functions $\Psi_{12}(\mathbf{p}, E) = \Psi(\mathbf{p}, E)$ and $\Psi_{21}(\mathbf{p}, E) = \Psi(\mathbf{p}, E)$ with $\Psi(\mathbf{p}, E) = \Psi(-\mathbf{p}, -E)$. If subsystem 1 takes at time τ the values $\mathbf{p}$ and E, subsystem 2 must take the values $-\mathbf{p}$ and $-E$ and, therefore, we obtain $\Psi_{12}(\mathbf{p}, E) = \Psi_{21}(-\mathbf{p}, -E)$ and we have for both subsystems one distribution function which are identical. It is $\Psi_{12}^*(\mathbf{p}, E)\Psi_{12}(\mathbf{p}, E)$ for subsystem 1 and it is $\Psi_{21}^*(\mathbf{p}, E)\Psi_{21}(\mathbf{p}, E)$ for subsystem 2. Each of these distribution functions are normalized to unity.[54]

Instead of the two distribution functions in $(\mathbf{p}, E)$-space, we have one pair potential in $(\mathbf{r}, t)$-space which is expressed by $v(x_{12}, y_{12}, z_{12}, t_{12}) = v(x_{21}, y_{21}, z_{21}, t_{21})$. The parameter α_{env} is not defined in the case of $N = 2$ because the two systems do not interact with other systems. We may formally assign to each distribution function the quantity $v(x_{12}, y_{12}, z_{12}, t_{12})/2$ and $v(x_{21}, y_{21}, z_{21}, t_{21})/2$,

---

[53] See equation (4.105).
[54] See equation (4.53).

respectively. Then, we come to a symmetrical correspondence:

$$\Psi_{12}^{*}(\mathbf{p}, E)\Psi_{12}(\mathbf{p}, E) \leftrightarrow v(x_{12}, y_{12}, z_{12}, t_{12})/2, \qquad (4.107)$$

$$\Psi_{21}^{*}(\mathbf{p}, E)\Psi_{21}(\mathbf{p}, E) \leftrightarrow v(x_{21}, y_{21}, z_{21}, t_{21})/2. \qquad (4.108)$$

We have already discussed this point in Section 4.4.6.

4.6.3. Classical Formulation

On the basis of the transformation formula (4.92) which connects the functions $\Psi(\mathbf{p}, E)$ and $\Psi(\mathbf{r}_{km}, t_{km})$ we can deduce potential functions for each complex. As we have already outlined in Section 4.6.1, these potential functions must have the same form for each specific complex and, furthermore, these potential functions have to be *pair potentials* because each complex consists of only two subsystems for the reason discussed above. The connection between $\Psi(\mathbf{r}_{km}, t_{km})$ and $\Psi(\mathbf{p}, E)$, respectively, and this pair potential can be formulated on the basis of a suitable classical expression for the variables $\mathbf{p}, E$ and $\mathbf{r}_{km}$.[55] Again, the time t, in this case t_{km}, is a strict quantum-theoretical variable and is therefore not defined within the classical formulation.[56]

How can we find for the specific problem discussed here a suitable classical expression that connects the variables $\mathbf{p}, E$ and $\mathbf{r}_{km}$? Let us briefly discuss this question on the basis of a complex consisting of two subsystems that we have discussed in Section 4.5 and let us choose, as a typical example, configuration $\mathbf{r}_{k\bullet m\bullet}, t_{k\bullet m\bullet}$ consisting of subsystem k and subsystem m. Clearly, the result will hold for each other's complex.

Relationships

First, let us give some general remarks. In Section 2.3.7.1 we have introduced the classical equation

$$E = \frac{\mathbf{p}^2}{2m_0} + v_{cl}(x, y, z), \qquad (4.109)$$

---

[55] See also Section 2.3.7.1.
[56] See Chapter 2.

with $v_{cl}(x,y,z) = U(x,y,z)$ where the function $v_{cl}(x,y,z)$ plays the role of a classical pair potential. Let us repeat here the main facts concerning the notion "classical description" within projection theory.

Equation (4.109) is a classical expression and connects the variables $E, \mathbf{p}$ and $\mathbf{r}$, and these variables take well-defined values at time τ:

$$\tau : E, \mathbf{p}, \mathbf{r}. \qquad (4.110)$$

The system-specific time t is not defined here. It is important to mention once more that within conventional physics the time τ plays the role of an external parameter and has nothing to do with the process under investigation. Within projection theory the time τ is used as reference time.

Equation (4.109) holds for all possible constellations of $E, \mathbf{p}, \mathbf{r}$ at time τ, and a certain constellation is determined by certain initial conditions. Then, the system under investigation develops in accordance with the equations of motions.[57] Here we apply the basic classical equation (4.109) in its general form, i.e., without the use of initial conditions because we do not want to explain quantum effects by classical mechanics.

The situation without initial conditions

What does equation (4.109) mean without initial condition? It means that there is a relationship between $E, \mathbf{p}$, and $\mathbf{r}$ where two of them, for example E and $\mathbf{p}$, can be chosen arbitrarily. Clearly these values E and $\mathbf{p}$ must be compatible with the shape of the potential $v_{cl}(x,y,z)$. For example, when we choose arbitrarily at time τ' for the energy the value E' and for the momentum $\mathbf{p}'$ the value $\mathbf{r}' = (x', y', z')$ is determined over the classical potential $v_{cl}(x', y', z')$ and we have the following relationship:

$$\tau' : E' - \frac{\mathbf{p}'^2}{2m_0} = v_{cl}(x', y', z'). \qquad (4.111)$$

---

[57] For example, Newton's equations of motion.

On the other hand, when we choose, also arbitrarily, at time τ'' with $\tau' < \tau''$ for the energy the value E'' and for the momentum the value $\mathbf{p}''$ the position $\mathbf{r}'' = (x'', y'', z'')$ is determined by the classical potential $v_{cl}(x'', y'', z'')$ and the following relationship is valid:

$$\tau'' : E'' - \frac{\mathbf{p}''^2}{2m_0} = v_{cl}(x'', y'', z''). \qquad (4.112)$$

However, we know that the transition from the state $(E', \mathbf{p}', \mathbf{r}')$ to the other state $(E'', \mathbf{p}'', \mathbf{r}'')$ is not possible on the basis of equation (4.109) alone because the laws for the conservation of momentum and energy are in general not fulfilled as is required in connection with closed systems. Within conventional physics, we work with equation (4.109) in connection with initial conditions and the equations of motion, and the equations of motion are based on the imagination that there are localized material objects with a certain mass which are embedded in space carrying energy and momentum. Within this picture,[58] the conservation laws for the momentum and the energy are fulfilled.

Situation in projection theory

Within projection theory and also in usual quantum theory, we have no classical equations of motion but the values $\mathbf{p}$ and E appear in a certain sense arbitrarily but with a certain probability which is determined by the probability density $\Psi^*(\mathbf{p}, E)\Psi(\mathbf{p}, E)$, quite similar to the situation we have just discussed in connection with the classical equation $E = \mathbf{p}^2/(2m_0) + v_{cl}(x, y, z)$. Also the variables $\mathbf{r}$ and t appear with a certain arbitrariness which is explained here also by probability arguments using the probability density $\Psi^*(\mathbf{r}, t)\Psi(\mathbf{r}, t)$.

Within projection theory, we work on the basis of two spaces, i.e., on the basis of $(\mathbf{p}, E)$-space and $(\mathbf{r}, t)$-space, and the scenario is organized as follows: "reality", described by $\mathbf{p}, E$-fluctuations in $(\mathbf{p}, E)$-space, is projected onto $(\mathbf{r}, t)$-space and produces in this way "pictures of reality",[59] that is, only geometrical structures appear in $(\mathbf{r}, t)$-space.

---

[58] Using equation (4.109), initial conditions and equations of motion.
[59] See Chapter 2.

Thus, within projection theory there are no localized material objects which are carrying E and $\mathbf{p}$ and that are embedded in space-time, i.e., in $(\mathbf{r}, t)$-space at the position $\mathbf{r}, t$.

What does classical mean in projection theory?

Thus, within projection theory, "classical" means that there exist, instead of probability densities, definite values for $E, \mathbf{p}, \mathbf{r}$ at time τ, but it does not mean that there are localized material objects with mass m_0 which are carrying energy E and momentum $\mathbf{p}$ and that are embedded in space at definite positions at time τ. As a starting point for the quantum-theoretical formulation of the basic equations, we will also use here, as in Section 2.3.7.1, a classical equation which is quite similar to $E = \mathbf{p}^2/(2m_0) + v_{cl}(x, y, z)$.[60]

4.6.4. Introduction of Pair Potentials for Certain Configurations

The aim is to find an adequate classical formulation with respect to the variables that appear in equation (4.109) in connection with a system consisting of two subsystems. These variables are $E, \mathbf{p}, \mathbf{r}$ and t and appear as variables of the wave functions $\Psi(\mathbf{p}, E)$ and $\Psi(\mathbf{r}, t)$ which are connected by a Fourier transform. First, we have to state that the system-specific t has no place within such a classical formulation and, therefore only the quantities $E, \mathbf{p}, \mathbf{r}$ can be considered and the fact that the system consists of two subsystems.

Let us consider such a classical two-particle system. This system should be identical with the complex consisting of subsystem k and subsystem m of the system with N subsystems. Here we first consider this complex k, m in classical approximation. If the two particles (subsystems) have the positions $\mathbf{r}_k, \mathbf{r}_m$, the masses m_{0k}, m_{0m}, the momenta $\mathbf{p}_k, \mathbf{p}_m$ and the energy E_{km}, belonging to this complex, the

---

[60] See equation (4.109).

equation is expressed within conventional classical physics by

$$E_{km} = \frac{\mathbf{p}_k^2}{2m_{0k}} + \frac{\mathbf{p}_m^2}{2m_{0m}} + v_{cl}(x_{km}, y_{km}, z_{km}, \alpha_{env}), \quad (4.113)$$

where $v_{cl}(x_{km}, y_{km}, z_{km}, \alpha_{env})$ is a classical form of the quantum-theoretical interaction potential $v(x_{km}, y_{km}, z_{km}, t_{km}, \alpha_{env})$, introduced in Sec. 4.6.1 and which is responsible for the two subsystems. In order to find the basic equations for a certain quantum-theoretical configuration as, for example, for configuration $\mathbf{r}_{k \bullet m \bullet} t_{k \bullet m \bullet}$,[61] we have to base our considerations on a classical expression similar to that of equation (4.113).

The variables that have to be treated classically in connection with our system are defined by the functions $\Psi(\mathbf{p}, E)$ and $\Psi(\mathbf{r}_{km}, t_{km})$. Therefore, we may write $E_{12} = E$. Due to the conditions formulated in Section 4.4.4, we in particular have $\mathbf{p}_1 = \mathbf{p}, \mathbf{p}_2 = -\mathbf{p}$. Then, instead of equation (4.113), we obtain the following classical equation:

$$E = \frac{\mathbf{p}^2}{2\mu_{km}} + v_{cl}(x_{km}, y_{km}, z_{km}, \alpha_{env}), \quad (4.114)$$

with

$$\mu_{km} = \frac{m_{0k} m_{0m}}{m_{0k} + m_{0m}}. \quad (4.115)$$

The values $\mathbf{p}, E, \mathbf{r}_{km}$, which appear in the classical equation (4.114), take definite values at time τ. This is just the definition of "classical" within projection theory. The variable t_{km} first comes into play through the quantum-theoretical treatment, and the classical potential $v_{cl}(x_{km}, y_{km}, z_{km}, \alpha_{env})$ becomes a quantum-theoretical function $v(x_{km}, y_{km}, z_{km}, t_{km}, \alpha_{env})$.[62]

On the basis of (4.114) and the rules for operators, which have been derived in Section 2.3.5.1, it is straightforward to formulate the quantum-theoretical equations for the determination of the wave functions from the pair potential. In the following, we give the

---

[61] See Section 4.5.3.
[62] See Section 2.3.7.

equivalent representations for $(\mathbf{r}, t)$-space, $(\mathbf{p}, E)$-space as well as for $(\mathbf{r}, E)$-space:

$(\mathbf{r}, t)$-space:

$$i\hbar \frac{\partial}{\partial t_{km}} \Psi(\mathbf{r}_{km}, t_{km}) = -\frac{\hbar^2}{2\mu_{km}} \Delta_{km} \Psi(\mathbf{r}_{km}, t_{km})$$
$$+ \upsilon(x_{km}, y_{km}, z_{km}, t_{km}, \alpha_{env}) \Psi(\mathbf{r}_{km}, t_{km}).$$
(4.116)

$(\mathbf{p}, E)$-space:

$$E\Psi(\mathbf{p}, E) = \frac{\mathbf{p}^2}{2\mu_{km}} \Psi(\mathbf{p}, E) + \upsilon[i\hbar \partial/\partial p_x, i\hbar \partial/\partial p_y, i\hbar \partial/\partial p_z,$$
$$- i\hbar \partial/\partial E, \alpha_{env}] \Psi(\mathbf{p}, E).$$
(4.117)

$(\mathbf{r}, E)$-space:

$$E\Psi(\mathbf{r}_{km}, E) = -\frac{\hbar^2}{2\mu_{km}} \Delta_{km} \Psi(\mathbf{r}_{km}, E) + \upsilon[x_{km}, y_{km}, z_{km},$$
$$- i\hbar \partial/\partial E, \alpha_{env}] \Psi(\mathbf{r}_{km}, E),$$
(4.118)

with $k, m = 1, \ldots, N$, $k \neq m$ and

$$\Delta_{km} = \frac{\partial^2}{\partial x_{km}^2} + \frac{\partial^2}{\partial y_{km}^2} + \frac{\partial^2}{\partial z_{km}^2}; \quad \mathbf{r}_{km} = (x_{km}, y_{km}, z_{km}). \quad (4.119)$$

All three equations, i.e., (4.116), (4.117) and (4.118), are completely equivalent concerning their physical information; they only differ in the frame of representation.

Discussion

The function $\Psi(\mathbf{r}_{km}, t_{km})$ is a solution of equation (4.116) and is fixed by the pair potential $\upsilon(x_{km}, y_{km}, z_{km}, t_{km}, \alpha_{env})$ and the parameter μ_{km}. $\Psi(\mathbf{r}_{km}, t_{km})$ is not only dependent on $\mathbf{r}_{km}$ and t_{km} but also on μ_{km} which is a function of the masses m_{0k} and m_{0m}.[63] Since the masses of the $(N-1)N$ pairs of the ensemble can be different from each other,

---

[63] See equation (4.115).

also the wave functions $\Psi(\mathbf{r}_{km}, t_{km})$ with $k, m = 1, \ldots, N, k \neq m$ are in general different from each other. For example, the wave function of the complex k, j with $j \neq m$ and the masses m_{0k} and m_{0j} with $m_{0j} \neq m_{0m}$ must lead to

$$\Psi(\mathbf{r}_{kj}, t_{kj}) \neq \Psi(\mathbf{r}_{km}, t_{km}), \qquad (4.120)$$

because of[63]

$$\mu_{km} \neq \mu_{kj}. \qquad (4.121)$$

Thus, the inverse transformation of (4.92)

$$\Psi(\mathbf{p}, E)_{ab} = \frac{1}{(2\pi\hbar)^2} \int_{-\infty}^{\infty} \Psi(\mathbf{r}_{ab}, t_{ab})$$
$$\times \exp\left\{-\frac{i}{\hbar}[\mathbf{p} \cdot \mathbf{r}_{ab} - Et_{ab}]\right\} d\mathbf{r}_{ab}\, dt_{ab}, \quad (4.122)$$

must lead to

$$\Psi(\mathbf{p}, E)_{kj} \neq \Psi(\mathbf{p}, E)_{km}, \qquad (4.123)$$

for $a = k, b = j$ and $a = k, b = m$, respectively. However, equation (4.123) is in contradiction to condition (4.52). The problem can only be solved if

$$\mu_{km} = \mu_{kj}, \qquad (4.124)$$

and this equation can be satisfied if all N subsystems have exactly the same mass:

$$m_{0i} = m_0, \; i = 1, \ldots, N, \qquad (4.125)$$

that is, equation (4.116) can only be valid if the parameter μ_{km} is given by

$$\mu_{km} = \frac{m_0}{2}, \; k, m = 1, \ldots, N, k \neq m, \qquad (4.126)$$

and not by equation (4.115). Clearly, the same argument holds for equations (4.117) and (4.118) which allow the determination of $\Psi(\mathbf{p}, E)$ and $\Psi(\mathbf{r}_{km}, E)$.

System with two subsystems

The mass-problem only appears for $N > 2$. Therefore, in the case of $N = 2$ we may work with different masses for the subsystems. That is, for a system that consists of two subsystems, say 1 and 2, we may have

$$m_{01} \neq m_{02}. \tag{4.127}$$

Then, instead of equations (4.116)–(4.118), we immediately get the following relationships:

$(\mathbf{r}, t)$-space:

$$i\hbar \frac{\partial}{\partial t_{12}} \Psi(\mathbf{r}_{12}, t_{12}) = -\frac{\hbar^2}{2\mu_{12}} \Delta_{12} \Psi(\mathbf{r}_{12}, t_{12}) \\ + v(x_{12}, y_{12}, z_{12}, t_{12}) \Psi(\mathbf{r}_{12}, t_{12}). \tag{4.128}$$

$(\mathbf{p}, E)$-space:

$$E\Psi(\mathbf{p}, E) = \frac{\mathbf{p}^2}{2\mu_{12}} \Psi(\mathbf{p}, E) + v[i\hbar \partial/\partial p_x, i\hbar \partial/\partial p_y, i\hbar \partial/\partial p_z, \\ -i\hbar \partial/\partial E, \alpha_{env}] \Psi(\mathbf{p}, E). \tag{4.129}$$

$(\mathbf{r}, E)$-space:

$$E\Psi(\mathbf{r}_{12}, E) = -\frac{\hbar^2}{2\mu_{12}} \Delta_{12} \Psi(\mathbf{r}_{12}, E) + v[x_{12}, y_{12}, z_{12}, \\ -i\hbar \partial/\partial E, \alpha_{env}] \Psi(\mathbf{r}_{12}, E), \tag{4.130}$$

with

$$\mu_{12} = \frac{m_{01} m_{02}}{m_{01} + m_{02}}. \tag{4.131}$$

The parameter α_{env} is not defined in the case of $N = 2$ because the two systems do not interact with other systems, i.e., the pair potential, which appears in equations (4.128), (4.129) and (4.130), is simply given by $v(\mathbf{r}_{km}, t_{km})$ and not by $v(\mathbf{r}_{km}, t_{km}, \alpha_{env})$.

Systems with two types of subsystems or more

We assumed that the $\mathbf{p}, E$-fluctuations between all pairs (complexes) of the ensemble with the N subsystems are exactly the same. Then, all pairs must equally interact by the same pair potential $v(\mathbf{r}_{km}, t_{km}, \alpha_{env})$. In this case the $\mathbf{p}, E$-fluctuations are described in $(\mathbf{p}, E)$-space on the basis of only two variables, $\mathbf{p}, E$, and the interaction processes are characterized by the wave function $\Psi(\mathbf{p}, E)$ and the probability density $\Psi^*(\mathbf{p}, E)\Psi(\mathbf{p}, E)$, respectively. However, if the subsystems are different from each other the description by $\Psi(\mathbf{p}, E)$ is no longer possible because the $\mathbf{p}, E$-fluctuations between different subsystems are by definition different from each other. For example, if the system with N subsystems consists of two types of subsystems, say a and b, we have three types of $(\mathbf{p}, E)$-fluctuations:

1. $\mathbf{p}_{aa}, E_{aa}$-fluctuations (interaction between a subsystem of type a and another subsystem of type a);
2. $\mathbf{p}_{bb}, E_{bb}$-fluctuations (interaction between a subsystem of type b and another subsystem of type b);
3. $\mathbf{p}_{ab}, E_{ab}$-fluctuations (interaction between a subsystem of type a and another subsystem of type b).

It is easy to recognize that we must have in this case the more complex wave function $\Psi(\mathbf{p}_{aa}, E_{aa}, \mathbf{p}_{bb}, E_{bb}, \mathbf{p}_{ab}, E_{ab})$ instead of $\Psi(\mathbf{p}, E)$. In other words, we have the transition

$$\Psi(\mathbf{p}, E) \to \Psi(\mathbf{p}_{aa}, E_{aa}, \mathbf{p}_{bb}, E_{bb}, \mathbf{p}_{ab}, E_{ab})$$

when there are two types of subsystems. It is also easy to show that under certain conditions three types of pair potentials appear when there are two types of subsystems:

$$v_{aa}(x_{ij}, y_{ij}, z_{ij}, t_{ij}, \alpha_{env}), v_{bb}(x_{kl}, y_{kl}, z_{kl}, t_{kl}, \alpha_{env}),$$
$$v_{ab}(x_{mn}, y_{mn}, z_{mn}, t_{mn}, \alpha_{env}).$$

That is, here we have the transition

$$v(x_{km}, y_{km}, z_{km}, t_{km}, \alpha_{env}) \to v_{aa}(x_{ij}, y_{ij}, z_{ij}, t_{ij}, \alpha_{env}),$$

$$v_{bb}(x_{kl}, y_{kl}, z_{kl}, t_{kl}, \alpha_{env}), v_{ab}(x_{mn}, y_{mn}, z_{mn}, t_{mn}, \alpha_{env}).$$

The treatment of a system that consists of two types of subsystems or more is straightforward, but we do not want to outline the details in this monograph.

4.6.5. Interaction Effects

Each of the N non-interacting subsystems occupy the whole space-time and each subsystem, as, for example, system k, is described by[64]

$$\Psi(\mathbf{r}_k, t_k) = C_k \exp\left\{\frac{i}{\hbar}(\mathbf{p}_k \cdot \mathbf{r}_k - E_k t_k)\right\}, \quad -\mathbf{r}_a, t_a \leq \mathbf{r}_k, t_k \leq \mathbf{r}_a, t_a.$$

The pictures of these N subsystems are supposed to be projected onto $(\mathbf{r}, t)$-space with an extension of $(\pm \mathbf{r}_a, \pm t_a)$ which corresponds to an interaction in $(\mathbf{p}, E)$-space.[65] However, in this case there is no mutual interaction between the N systems. In other words, there are no $(\mathbf{p}, E)$-fluctuations between the N subsystems and we have in $(\mathbf{p}, E)$-space,[66]

$$\Delta \mathbf{p}_{km} = 0, \quad k, m = 1, \ldots, N, k \neq m,$$

$$\Delta E_{km} = 0, \quad k, m = 1, \ldots, N, k \neq m.$$

In this case the system k occupies the whole $(\mathbf{r}, t)$-space which is limited by $\pm \mathbf{r}_a, \pm t_a$ where the parameters $\mathbf{r}_a$ and t_a can be arbitrarily large but have to be different from infinity. However, when the N subsystems start to interact we have

$$\Delta \mathbf{p}_{km} \neq 0, \quad k, m = 1, \ldots, N, k \neq m,$$

$$\Delta E_{km} \neq 0, \quad k, m = 1, \ldots, N, k \neq m,$$

leading to the wave functions $\Psi(\mathbf{p}, E)$ and $\Psi(\mathbf{r}_{12}, \mathbf{r}_{13}, \ldots, \mathbf{r}_{(N-1)N}, t_{12}, t_{13}, \ldots, t_{(N-1)N})$ which are defined in Section 4.4.3.

---

[64] See equation (4.8).
[65] See also Section 4.4.1.
[66] See also Section 4.4.3.

It is characteristic for a Fourier transform that if the function $\Psi(\mathbf{p}, E)$ and the probability density $\Psi^*(\mathbf{p}, E)\Psi(\mathbf{p}, E)$, respectively, becomes broader and broader in $(\mathbf{p}, E)$-space (the $\mathbf{p}, E$-fluctuations increase) the wave function

$$\Psi(\mathbf{r}_{12}, \mathbf{r}_{13}, \ldots, \mathbf{r}_{(N-1)N}, t_{12}, t_{13}, \ldots, t_{(N-1)N}),$$
$$[\Psi^\bullet(\mathbf{r}_{12}, \mathbf{r}_{13}, \ldots, \mathbf{r}_{(N-1)N}, t_{12}, t_{13}, \ldots, t_{(N-1)N})$$
$$\times \Psi(\mathbf{r}_{12}, \mathbf{r}_{13}, \ldots, \mathbf{r}_{(N-1)N}, t_{12}, t_{13}, \ldots, t_{(N-1)N})]$$

in $(\mathbf{r}, t)$-space becomes smaller and smaller. In other words, with increasing $\mathbf{p}, E$-fluctuations, the attraction between the N subsystems seems to increase because the distances $\mathbf{r}_{12}, \mathbf{r}_{13}, \ldots, \mathbf{r}_{(N-1)N}$, $t_{12}, t_{13}, \ldots, t_{(N-1)N}$ in $(\mathbf{r}, t)$-space become smaller. But this picture is utterly wrong from the point of view of projection theory.

The various probabilities for the possible $\mathbf{p}, E$-values described by the probability density $\Psi^*(\mathbf{p}, E)\Psi(\mathbf{p}, E)$ leads to the effect that the various space-time points are correlated. The physically real processes take place in $(\mathbf{p}, E)$-space and these processes in $(\mathbf{p}, E)$-space are projected onto $(\mathbf{r}, t)$-space and we obtain a picture of this reality. The imagination that the N systems, as, for example, system i and system j, influence each other in $(\mathbf{r}, t)$-space is wrong within projection theory. There is no exchange of real quantities and/or information through $(\mathbf{r}, t)$-space. There is no exchange of certain things between the systems i and j. The complete space-time structure is the result of a projection and, generally speaking, we have the following situation in $(\mathbf{r}, t)$-space.[67] We observe in $(\mathbf{r}, t)$-space global, instantaneous changes in the $(\mathbf{r}, t)$-structure, and it is a feature of the projection that all space-time points $\mathbf{r}, t$ ($-\mathbf{r}_a \leq \mathbf{r} \leq \mathbf{r}_a, -t_a \leq t \leq t_a$) are simultaneously involved. Again, the parameters $\mathbf{r}_a$ and t_a can be arbitrarily large but have to be different from infinity. This effect is comparable with a flashlight on a screen. In other words, there are non-local effects.

In summary, the space-time structure is projected on $(\mathbf{r}, t)$-space and does not come into existence by an exchange of a certain kind of information through $(\mathbf{r}, t)$-space. Therefore, the specific law $\Psi(\mathbf{p}, E)$

---

[67] See also Section 2.3.6.

[$\Psi^*(\mathbf{p}, E)\Psi(\mathbf{p}, E)$], which reflects $\mathbf{p}, E$-fluctuations (interactions) in ($\mathbf{p}, E$)-space, leads to correlated space-time configurations, i.e., ($\mathbf{r}, t$)-correlations; there is nothing else.

Remark

We have discussed in Section 4.4.2 that such a construction seems to be artificial because it is problematic to distinguish between non-interacting and interacting subsystems. If the subsystems start to interact at time τ_{INT}, the subsystems are in a non-interacting state at $\tau < \tau_{INT}$ and they are in an interacting state at $\tau > \tau_{INT}$. However, who switches the interaction at time τ_{INT} on? The answering of such metaphysical questions lies out of the scope of physics, and we should try to find another imagination.

In Section 4.4.2 we stated the following:

In order to avoid such questions, it is more realistic to assume that the N subsystems, which are under mutual interaction, simply exist (emerge) in this form, i.e., in the form of one system consisting of N interacting subsystems, and we should assume that this system is not divisible into two parts, the N systems without mutual interaction and with interaction.

Then, we may assume that the entire system, i.e., its geometrical structure, is positioned in an infinite space-time with $-\infty \leq \mathbf{r}, t \leq \infty$ instead of $-\mathbf{r}_a \leq \mathbf{r} \leq \mathbf{r}_a, -t_a \leq t \leq t_a$. The dynamical behaviour of such a complex system, i.e., the system consisting of N subsystems, is discussed in Section 4.5.

However, the tendencies and relations in connection with the $\mathbf{p}, E$-fluctuations in ($\mathbf{p}, E$)-space and the $\mathbf{r}, t$-correlations in ($\mathbf{r}, t$)-space, which we have discussed above, are not affected by this question. (Who switches the interaction at time τ_{INT} on?)

4.7. Energy Levels

In principle, the system consisting of N subsystems can have certain constant energy levels, say $E_\alpha, E_\beta, \ldots$; such energies do not vary with time τ. We want to assume that our system is permanently in the state E_α. This is fulfilled if the system does not interact with external systems. The $\mathbf{p}, E$-fluctuations in connection with N subsystems are described by $\Psi^*(\mathbf{p}, E)\Psi(\mathbf{p}, E)$. Then, the energy values E defined by $\Psi^*(\mathbf{p}, E)\Psi(\mathbf{p}, E)$ fluctuate in a certain sense around E_α. In the following, we will investigate the details.

The energy E_α is a system-specific quantity and belongs to the solution of the equations for the determination of the wave functions $\Psi(\mathbf{p}, E)$ and $\Psi(\mathbf{r}, E)$.[68] E_α is a constant quantity because the system does not interact with the outside world, i.e., external systems; there is merely an interaction or exchange of momentum and energy between the N subsystems. In particular, E_α belongs to the whole system consisting of N subsystems and is not a property of any specific subsystem.

Within projection theory, the possible values $\mathbf{p}$ and E of $(\mathbf{p}, E)$-space are correlated with certain space-time positions $\mathbf{r}$ and t of $(\mathbf{r}, t)$-space. In Section 2.3.8.3, we have pointed out that the measurement at time τ of one of the possible values for $\mathbf{p}$ and for E is done in the space-time intervals $\mathbf{r}, \mathbf{r} + \Delta \mathbf{r}$ and $t, t + \Delta t$ with the probability density of $\Psi^*(\mathbf{r}, t)\Psi(\mathbf{r}, t)$. The possible values for $\mathbf{p}$ and for E are determined by a probability density too and is given by $\Psi^*(\mathbf{p}, E)\Psi(\mathbf{p}, E)$.

This can be applied to our system consisting of N subsystems, and the general statement about measurements at time τ in connection with the variables $\mathbf{r}$ and t can be transferred to all N space-time positions $\mathbf{r}_1, t_1, \mathbf{r}_2, t_2, \ldots, \mathbf{r}_N, t_N$ and, as we have stated in Section 4.4.5, to each subsystem with $\mathbf{r}_k, t_k, k = 1, \ldots, N$, the same function $\Psi^*(\mathbf{p}, E)\Psi(\mathbf{p}, E)$ has to be assigned. In other words, there are N values for the momentum $\mathbf{p}$ and also N values for the energy E at time τ, but there can exist only one value for the energy E_α at time τ.

Again, the energy E_α belongs to the whole system with N subsystems and is not a property of any specific subsystem, and we are not able to connect E_α with a specific space-time point, as, for example, $\mathbf{r}_i, t_j$, of the ensemble with $\mathbf{r}_1, t_1, \mathbf{r}_2, t_2, \ldots, \mathbf{r}_N, t_N$. E_α is a property of all N subsystems, that is, all N subsystems must be equally involved.

---

[68] See equations (4.129) and (4.130).

Since a measurement of **p**, E is always connected to specific space-time positions, the following question arises: At what space-time position $\mathbf{r}_{E_\alpha}, t_{E_\alpha} \neq \mathbf{r}_i, t_i, i = 1, \ldots, N$, of $(\mathbf{r}, t)$-space can we observe the energy E_α? The only thing we have in connection with $(\mathbf{r}, t)$-space at time τ are the space-time positions $\mathbf{r}_1, t_1, \mathbf{r}_2, t_2, \ldots, \mathbf{r}_N, t_N$ of the N subsystems. That is, such a specific space-time position $\mathbf{r}_{E_\alpha}, t_{E_\alpha}$ obviously does not exist.

Since the energy E_α is not a property of any specific subsystem, we are not able to connect E_α with a specific space-time point of the ensemble $\mathbf{r}_1, t_1, \mathbf{r}_2, t_2, \ldots, \mathbf{r}_N, t_N$. E_α comes into play through all N subsystems. We have to solve the problem with the help of other arguments. The question "At what position $\mathbf{r}_{E_\alpha}, t_{E_\alpha} \neq \mathbf{r}_i, t_i, i = 1, \ldots, N$, of $(\mathbf{r}, t)$-space can we observe the energy E_α?" makes no sense because the values $\mathbf{r}_{E_\alpha}$ and t_{E_α} are obviously not definable. On the other hand, there should be a clear assignment of E_α and the space-time positions $\mathbf{r}_1, t_1, \mathbf{r}_2, t_2, \ldots, \mathbf{r}_N, t_N$ and, as we have already remarked above, it is a general feature of projection theory that the values of $(\mathbf{p}, E)$-space are correlated to the elements of $(\mathbf{r}, t)$-space.[69]

4.7.1. Treatment of the Problem

For simplicity, let us consider a system consisting of two subsystems 1 and 2 instead of N, and let us work within $(\mathbf{r}, t)$-space and $(\mathbf{p}, E)$-space. However, it is more easy to investigate the role of the constant energy E_α when we use a form for $\Psi(\mathbf{p}, E)$ which is not dependent of E_α. In other words, E_α has been eliminated from the original form of $\Psi(\mathbf{p}, E)$.[70] Then, the Fourier transform

$$\Psi(\mathbf{r}_{12}, t_{12}) = \frac{1}{(2\pi\hbar)^2} \int_{-\infty}^{\infty} \Psi(\mathbf{p}, E) \exp\left\{\frac{i}{\hbar}(\mathbf{p}\cdot\mathbf{r}_{12} - Et_{12})\right\} d\mathbf{p}\, dE, \tag{4.132}$$

defines the same situation in $(\mathbf{r}, t)$-space. Instead of $\mathbf{r}, t$ we have used the marking $\mathbf{r}_{12}, t_{12}$ with $\mathbf{r}_{12} = \mathbf{r}_1 - \mathbf{r}_2$ and $t_{12} = t_1 - t_2$.[71]

---

[69] See, in particular, Chapter 2.
[70] This is a solution of equation (4.129).
[71] See equations (4.35) and (4.36).

The wave function $\Psi(\mathbf{r}_{12}, t_{12})$ just reflects the properties of the re-defined function $\Psi(\mathbf{p}, E)$ that is by definition not dependent on E_α. Whereas the energy values E of the solutions of (4.129) fluctuate around E_α, the energy values E of the re-formulated function $\Psi(\mathbf{p}, E)$ fluctuate around zero.

Let us introduce the energy E_α by the following step. The function $\Psi(\mathbf{r}_{12}, t_{12})$ and the probability density $\Psi^*(\mathbf{r}_{12}, t_{12})\Psi(\mathbf{r}_{12}, t_{12})$ remain unchanged when we replace in equation (4.132) the variable E by $E + E_\alpha$:

$$\Psi(\mathbf{r}_{12}, t_{12}) = \frac{1}{(2\pi\hbar)^2} \int_{-\infty}^{\infty} \left[\Psi(\mathbf{p}, E + E_\alpha)\exp\left(-\frac{i}{\hbar}E_\alpha t_{12}\right)\right]$$

$$\times \exp\left\{\frac{i}{\hbar}(\mathbf{p}\cdot\mathbf{r}_{12} - Et_{12})\right\} d\mathbf{p}\, dE. \quad (4.133)$$

That is, the wave function $\Psi(\mathbf{r}_{12}, t_{12})$ in $(\mathbf{r}, t)$-space remains unchanged when we use instead of $\Psi(\mathbf{p}, E)$ the function

$$\left[\Psi(\mathbf{p}, E + E_\alpha)\exp\left(-\frac{i}{\hbar}E_\alpha t_{12}\right)\right]. \quad (4.134)$$

Thus, we have the following correspondence:

$$\Psi(\mathbf{p}, E) \leftrightarrow \left[\Psi(\mathbf{p}, E + E_\alpha)\exp\left(-\frac{i}{\hbar}E_\alpha t_{12}\right)\right]. \quad (4.135)$$

The measurements at time τ lead to events at the space-time positions $\mathbf{r}_1, t_1$ and $\mathbf{r}_2, t_2$, no other space-time positions are defined at time τ for the two subsystems 1 and 2. If we measure for system 1 at position $\mathbf{r}_1, t_1$ the momentum $\mathbf{p}$ and the energy E, then we simultaneously measure with certainty for system 2 at position $\mathbf{r}_2, t_2$ the momentum $-\mathbf{p}$ and the energy $-E$. This statement can be done on the basis of $\Psi(\mathbf{p}, E)$ and $\Psi^*(\mathbf{p}, E)\Psi(\mathbf{p}, E)$, respectively. But what about E_α? In answering this question, we have to apply equation (4.134): Not only $\Psi(\mathbf{p}, E)$ is a solution for $\Psi(\mathbf{r}_{12}, t_{12})$[72] but also the function

$$\Psi(\mathbf{p}, E + E_\alpha)\exp\left(-\frac{i}{\hbar}E_\alpha t_{12}\right),$$

---

[72] See equation (4.132).

and we have the equivalences[73]

$$\Psi(\mathbf{p}, E) \to \Psi(\mathbf{r}_{12}, t_{12}), \tag{4.136}$$

$$\Psi(\mathbf{p}, E + E_\alpha) \exp\left(-\frac{i}{\hbar} E_\alpha t_{12}\right) \to \Psi(\mathbf{r}_{12}, t_{12}). \tag{4.137}$$

Equations (4.134) and (4.135) are the basic expressions for the analysis in connection with the energy E_α. Because both functions, $\Psi(\mathbf{p}, E)$ and $\Psi(\mathbf{p}, E + E_\alpha) \exp(-iE_\alpha t_{12})/\hbar$, are equivalent, we can directly investigate the role of E_α.

4.7.2. Specific Properties

We have outlined in Section 4.4.3 that the wave function in $(\mathbf{r}, t)$-space in not dependent on the space-time positions $\mathbf{r}_1, t_1$ and $\mathbf{r}_2, t_2$ but only on the differences $\mathbf{r}_{12}, t_{12}$, that is, we have $\Psi(\mathbf{r}_{12}, t_{12})$ instead of $\Psi(\mathbf{r}_1, \mathbf{r}_2, t_1, t_2)$. This is the reason why we have at time τ for the whole system consisting of subsystem 1 and subsystem 2 only one variable for the momentum $\mathbf{p}$ and only one variable for the energy E, i.e., the variables $\mathbf{p}_1, E_1$ and $\mathbf{p}_2, E_2$ for each system are not explicitly expressed within this $\Psi(\mathbf{p}, E)$-formulation. Because only two variables $\mathbf{p}$ and E can be realized at time τ, the normalization integral in connection with $\Psi(\mathbf{p}, E)$ must necessarily be one:

$$\int_{-\infty}^{\infty} \Psi^*(\mathbf{p}, E) \Psi(\mathbf{p}, E) d\mathbf{p}\, dE = 1. \tag{4.138}$$

At time τ, one of the possible values for $\mathbf{p}$ and for E are present, and these two values $\mathbf{p}, E$ are given with a certain probability which is determined by the probability density $\Psi^*(\mathbf{p}, E)\Psi(\mathbf{p}, E)$ and this inevitably leads to equation (4.138).

However, we know that the system consists of two subsystems 1 and 2 and the interaction mechanism discussed in Section 4.5 demands that *two* values for $\mathbf{p}$ of $\Psi(\mathbf{p}, E)$ exist at time τ and of course two values for E of $\Psi(\mathbf{p}, E)$ at the same time τ. However, these two values for $\mathbf{p}$ and E are correlated. Due to the conservation of momentum and energy, system 1 has the values $\mathbf{p}, E$ at time τ if system 2 is in the state

---

[73] See equation (4.133).

$-\mathbf{p}, -E$ at the same time τ. Clearly, another combination is possible: system 1 has the values $-\mathbf{p}, E$ at time τ if system 2 is in the state $\mathbf{p}, -E$ at the same time τ. In other words, there is a strong correlation between the two subsystems.

4.7.3. Conditional Wave Functions

Nevertheless, we may define a wave function for each subsystem because both subsystems behave statistically. However, due to the correlation between the $\mathbf{p}, E$-processes, we would like to denote this kind of wave function by "conditional wave function" since we have the following situation. A measurement of $\mathbf{p}_a, E_a$ at the space-time position $\mathbf{r}_1, t_1$ would show that the variables $\mathbf{p}_a, E_a$ behave strictly statistically. On the other hand, a measurement of $\mathbf{p}_b, E_b$ at the space-time position $\mathbf{r}_2, t_2$ would show that also the variables $\mathbf{p}_b, E_b$ behave strictly statistically. Nevertheless, as we have outlined above, both sets $\mathbf{p}_a, E_a$ and $\mathbf{p}_b, E_b$ are strongly correlated because have $\mathbf{p}_a = -\mathbf{p}_b = \mathbf{p}$, $E_a = -E_b = E$. Therefore, the notion "conditional wave function" in connection with the two subsystems 1 and 2 is appropriate. Therefore, instead of $\Psi(\mathbf{p}, E)$ we will use the marking $\Psi_{cond}(\mathbf{p}, E)$ in the case of a conditional wave function, i.e., for system 1 we have $\Psi_{1,cond}(\mathbf{p}, E)$ and for system 2 we have the conditional wave function $\Psi_{2,cond}(\mathbf{p}, E)$.

In conclusion, we can only say something about $\mathbf{p}$ and E in connection with the probability distribution $\Psi^*(\mathbf{p}, E)\Psi(\mathbf{p}, E)$. We can say nothing about the basic mechanisms (processes) that take place between system 1 and system 2 on the basis of $\Psi(\mathbf{p}, E)$ and $\Psi^*(\mathbf{p}, E)\Psi(\mathbf{p}, E)$ alone. Nevertheless, this information, i.e., $\Psi^*(\mathbf{p}, E)\Psi(\mathbf{p}, E)$, is sufficient because the conservation laws for momentum and energy dictate these basic mechanisms. Therefore, it is sufficient to know what values $\mathbf{p}$ and E are different from zero and what is the probability density $\Psi^*(\mathbf{p}, E)\Psi(\mathbf{p}, E)$ for them. The basic processes that take place between system 1 and system 2 are determined by $\Psi^*(\mathbf{p}, E)\Psi(\mathbf{p}, E)$ and the conservation laws for momentum and energy.[74]

---

[74] See also Section 4.7.2.

Therefore, we are able to study the basic interaction mechanism between system 1 and system 2 on the basis of $\Psi(\mathbf{p}, E)$ and $\Psi^*(\mathbf{p}, E)\Psi(\mathbf{p}, E)$, respectively. This can be done by means of the conditional probability densities $\Psi^*_{1,cond}(\mathbf{p}, E)\Psi_{1,cond}(\mathbf{p}, E)$ and $\Psi^*_{2,cond}(\mathbf{p}, E)\Psi_{2,cond}(\mathbf{p}, E)$ for the two systems 1 and 2, where $\Psi_{1,cond}(\mathbf{p}, E)$ and $\Psi_{2,cond}(\mathbf{p}, E)$ have been introduced above. Both functions $\Psi^*_{1,cond}(\mathbf{p}, E)\Psi_{1,cond}(\mathbf{p}, E)$ and $\Psi^*_{2,cond}(\mathbf{p}, E)\Psi_{2,cond}(\mathbf{p}, E)$ can be constructed by means of $\Psi(\mathbf{p}, E)$ and $\Psi^*(\mathbf{p}, E)\Psi(\mathbf{p}, E)$, respectively, and the conservation laws for momentum and energy.

That is, the transition

$$\Psi^*(\mathbf{p}, E)\Psi(\mathbf{p}, E) \rightarrow \Psi^*_{1,cond}(\mathbf{p}, E)\Psi_{1,cond}(\mathbf{p}, E) \\ + \Psi^*_{2,cond}(\mathbf{p}, E)\Psi_{2,cond}(\mathbf{p}, E), \quad (4.139)$$

is possible, and we have the following conditions[75]

$$\Psi^*_{1,cond}(\mathbf{p}', E')\Psi_{1,cond}(\mathbf{p}', E') = \Psi^*_{2,cond}(\mathbf{p}', E')\Psi_{2,cond}(\mathbf{p}', E') \\ = \Psi^*(\mathbf{p}', E')\Psi(\mathbf{p}', E'), \quad (4.140)$$

for $\mathbf{p}', E'$ at time τ' and

$$\Psi^*_{1,cond}(\mathbf{p}'', E'')\Psi_{1,cond}(\mathbf{p}'', E'') = \Psi^*_{2,cond}(\mathbf{p}'', E'')\Psi_{2,cond}(\mathbf{p}'', E'') \\ = \Psi^*(\mathbf{p}'', E'')\Psi(\mathbf{p}'', E''), \quad (4.141)$$

for $\mathbf{p}'', E''$ at time τ'', etc.

Thus, we have for all times τ

$$2\Psi^*(\mathbf{p}, E)\Psi(\mathbf{p}, E) = \Psi^*_{1,cond}(\mathbf{p}, E)\Psi_{1,cond}(\mathbf{p}, E) \\ + \Psi^*_{2,cond}(\mathbf{p}, E)\Psi_{2,cond}(\mathbf{p}, E). \quad (4.142)$$

---

[75] See also the discussion above.

Since the interaction mechanisms are symmetrical with respect to the momentum and the energy we must have[76]

$$\Psi^*(\mathbf{p}, E)\Psi(\mathbf{p}, E) = \Psi^*(-\mathbf{p}, -E)\Psi(-\mathbf{p}, -E), \quad (4.143)$$

$$\Psi^*_{1,cond}(\mathbf{p}, E)\Psi_{1,cond}(\mathbf{p}, E) = \Psi^*_{1,cond}(-\mathbf{p}, -E)$$
$$\times \Psi_{1,cond}(-\mathbf{p}, -E), \quad (4.144)$$

$$\Psi^*_{2,cond}(\mathbf{p}, E)\Psi_{2,cond}(\mathbf{p}, E) = \Psi^*_{2,cond}(-\mathbf{p}, -E)$$
$$\times \Psi_{2,cond}(-\mathbf{p}, -E). \quad (4.145)$$

Each of the two characteristic functions $\Psi^*_{1,cond}(\mathbf{p}, E)\Psi_{1,cond}(\mathbf{p}, E)$ and $\Psi^*_{2,cond}(\mathbf{p}, E)\Psi_{2,cond}(\mathbf{p}, E)$ behaves strictly statistically but there is a strong correlation between them that is dictated by the conservation laws for momentum and energy. If we observe, for example, for subsystem 1 with $\Psi^*_{1,cond}(\mathbf{p}, E)\Psi_{1,cond}(\mathbf{p}, E)$ the values $\mathbf{p}$ and E at time τ, the values for subsystem 2 with $\Psi^*_{2,cond}(\mathbf{p}, E)\Psi_{2,cond}(\mathbf{p}, E)$ are given by $-\mathbf{p}$ and $-E$ at the same time τ. Both functions $\Psi^*_{1,cond}(\mathbf{p}, E)\Psi_{1,cond}(\mathbf{p}, E)$ and $\Psi^*_{2,cond}(\mathbf{p}, E)\Psi_{2,cond}(\mathbf{p}, E)$ produce random sequences but there are cross correlations.

At time τ we have two events: one in connection with system 1 another in connection with system 2. Thus, both functions $\Psi^*_{1,cond}(\mathbf{p}, E)\Psi_{1,cond}(\mathbf{p}, E)$ and $\Psi^*_{2,cond}(\mathbf{p}, E)\Psi_{2,cond}(\mathbf{p}, E)$ must be normalized to unity:

$$\int_{-\infty}^{\infty} \Psi^*_{1,cond}(\mathbf{p}, E)\Psi_{1,cond}(\mathbf{p}, E) d\mathbf{p}\, dE = 1, \quad (4.146)$$

and

$$\int_{-\infty}^{\infty} \Psi^*_{2,cond}(\mathbf{p}, E)\Psi_{2,cond}(\mathbf{p}, E) d\mathbf{p}\, dE = 1. \quad (4.147)$$

Again, both probability densities $\Psi^*_{1,cond}(\mathbf{p}, E)\Psi_{1,cond}(\mathbf{p}, E)$ and $\Psi^*_{2,cond}(\mathbf{p}, E)\Psi_{2,cond}(\mathbf{p}, E)$ are strongly correlated.

---

[76]See Section 4.4.5.

4.7.4. E_α-Fluctuations

General remarks

In the Sections 4.7.1–4.7.3 we have studied the interaction mechanisms in connection with the function $\Psi(\mathbf{p}, E)$ where the energy E_α was not involved. In order to be able to study the role of the energy E_α we have to investigate the expression[77]

$$[\Psi(\mathbf{p}, E + E_\alpha)\phi(E_\alpha)], \qquad (4.148)$$

instead of $\Psi(\mathbf{p}, E)$ where the wave function $\phi(E_\alpha)$ is given by

$$\phi(E_\alpha) = \exp\left(-\frac{i}{\hbar} E_\alpha t_{12}\right). \qquad (4.149)$$

With equation (4.148) we are able to connect E_α with the interaction mechanisms described by $\Psi(\mathbf{p}, E)$. For the explanation of the interaction mechanisms described by $\Psi(\mathbf{p}, E)$, without E_α, we needed two systems, subsystem 1 and subsystem 2, in order to satisfy the conservation laws for the momentum and the energy: Due to these conservation laws, system 1 has the values $\mathbf{p}, E$ at time τ if system 2 is in the state $-\mathbf{p}, -E$ at the same time τ.

The E_α-process

1. *Energy shift*

The transition from $\Psi(\mathbf{p}, E)$ to $\Psi(\mathbf{p}, E + E_\alpha)$ only means that the function is shifted by the constant energy E_α, i.e., the $\mathbf{p}, E$-processes cannot be influenced by this transition and we have for $\Psi(\mathbf{p}, E + E_\alpha)$ the same $\mathbf{p}, E$-processes as in the case of $\Psi(\mathbf{p}, E)$. The $\mathbf{p}, E$-processes are not dependent on the absolute values of $\mathbf{p}$ and E.[78]

Due to the interaction at time τ, the system, as, for example, system 1, changes the state at time τ and there is a transition from one $\mathbf{p}, E$-state to another, i.e., from $\mathbf{p}', E' + E_\alpha$ to $\mathbf{p}'', E'' + E_\alpha$, and there is an exchange of momentum and energy with a second system, as, for example, system 2, and the interaction process (mechanism) is

---

[77] See equation (4.134).
[78] See Section 4.4.2.

only dependent on the differences $\Delta \mathbf{p}$ and ΔE with respect to the two $\mathbf{p}, E$-states, i.e., the interaction mechanism is dependent on

$$\begin{aligned}\Delta \mathbf{p} &= \mathbf{p}'' - \mathbf{p}', \\ \Delta E &= (E'' + E_\alpha) - (E' + E_\alpha) = E'' - E',\end{aligned} \quad (4.150)$$

but not on E_α. That is, the probability densities

$$\Psi^*(\mathbf{p}, E)\Psi(\mathbf{p}, E),$$

and

$$\Psi^*(\mathbf{p}, E + E_\alpha)\Psi(\mathbf{p}, E + E_\alpha),$$

are completely equivalent:

$$\Psi^*(\mathbf{p}, E)\Psi(\mathbf{p}, E) \leftrightarrow \Psi^*(\mathbf{p}, E + E_\alpha)\Psi(\mathbf{p}, E + E_\alpha). \quad (4.151)$$

Therefore, what we have said about $\Psi^*(\mathbf{p}, E)\Psi(\mathbf{p}, E)$ in Section 4.7.3 can be applied on $\Psi^*(\mathbf{p}, E + E_\alpha)\Psi(\mathbf{p}, E + E_\alpha)$. In particular, we have to introduce the conditional probability densities

$$\Psi^*_{1,cond}(\mathbf{p}, E + E_\alpha)\Psi_{1,cond}(\mathbf{p}, E + E_\alpha), \quad (4.152)$$

and

$$\Psi^*_{2,cond}(\mathbf{p}, E + E_\alpha)\Psi_{2,cond}(\mathbf{p}, E + E_\alpha). \quad (4.153)$$

In analogy to equations (4.139)–(4.145), we immediately get

$$\begin{aligned}\Psi^*(\mathbf{p}, E + E_\alpha)&\Psi(\mathbf{p}, E + E_\alpha) \\ &\to \Psi^*_{1,cond}(\mathbf{p}, E + E_\alpha)\Psi_{1,cond}(\mathbf{p}, E + E_\alpha) \\ &\quad + \Psi^*_{2,cond}(\mathbf{p}, E + E_\alpha)\Psi_{2,cond}(\mathbf{p}, E + E_\alpha),\end{aligned} \quad (4.154)$$

$$\begin{aligned}\Psi^*_{1,cond}(\mathbf{p}', E' + E_\alpha)&\Psi_{1,cond}(\mathbf{p}', E' + E_\alpha) \\ &= \Psi^*_{2,cond}(\mathbf{p}', E' + E_\alpha)\Psi_{2,cond}(\mathbf{p}', E' + E_\alpha) \\ &= \Psi^*(\mathbf{p}', E' + E_\alpha)\Psi(\mathbf{p}', E' + E_\alpha),\end{aligned} \quad (4.155)$$

for $\mathbf{p}', E'$ at time τ' and

$$\Psi_{1,cond}^*(\mathbf{p}'', E'' + E_\alpha)\Psi_{1,cond}(\mathbf{p}'', E'' + E_\alpha)$$
$$= \Psi_{2,cond}^*(\mathbf{p}'', E'' + E_\alpha)\Psi_{2,cond}(\mathbf{p}'', E'' + E_\alpha)$$
$$= \Psi^*(\mathbf{p}'', E'' + E_\alpha)\Psi(\mathbf{p}'', E'' + E_\alpha), \qquad (4.156)$$

for $\mathbf{p}'', E''$ at time τ'' etc.

Thus, we have for all times τ

$$2\Psi^*(\mathbf{p}, E + E_\alpha)\Psi(\mathbf{p}, E + E_\alpha)$$
$$= \Psi_{1,cond}^*(\mathbf{p}, E + E_\alpha)\Psi_{1,cond}(\mathbf{p}, E + E_\alpha)$$
$$+ \Psi_{2,cond}^*(\mathbf{p}, E + E_\alpha)\Psi_{2,cond}(\mathbf{p}, E + E_\alpha). \qquad (4.157)$$

In particular, the interaction mechanism requires

$$\Psi^*(\mathbf{p}, E + E_\alpha)\Psi(\mathbf{p}, E + E_\alpha)$$
$$= \Psi^*(-\mathbf{p}, -E + E_\alpha)\Psi(-\mathbf{p}, -E + E_\alpha), \qquad (4.158)$$
$$\Psi_{1,cond}^*(\mathbf{p}, E + E_\alpha)\Psi_{1,cond}(\mathbf{p}, E + E_\alpha)$$
$$= \Psi_{1,cond}^*(-\mathbf{p}, -E + E_\alpha)\Psi_{1,cond}(-\mathbf{p}, -E + E_\alpha), \qquad (4.159)$$
$$\Psi_{2,cond}^*(\mathbf{p}, E + E_\alpha)\Psi_{2,cond}(\mathbf{p}, E + E_\alpha)$$
$$= \Psi_{2,cond}^*(-\mathbf{p}, -E + E_\alpha)\Psi_{2,cond}(-\mathbf{p}, -E + E_\alpha). \qquad (4.160)$$

In the next section, we will make statements about the location of E_α in connection with measurements in $(\mathbf{r}, t)$-space.

2. Specific interaction mechanisms

In Section 4.7.1, we have stated that the probability densities $\Psi^*(\mathbf{p}, E)\Psi(\mathbf{p}, E)$ and $\Psi^*(\mathbf{p}, E + E_\alpha)\Psi(\mathbf{p}, E + E_\alpha)\phi^*(E_\alpha)\phi(E_\alpha)$ have to be considered as completely equivalent. Furthermore, the interaction processes, i.e., the $\mathbf{p}, E$-fluctuations, between subsystem 1 and subsystem 2 are not dependent on E_α.

The next step in our analysis is to answer the following question. What is the interaction mechanism behind $\Psi^*(\mathbf{p}, E + E_\alpha)\Psi(\mathbf{p}, E + $

$E_\alpha)\phi^*(E_\alpha)\phi(E_\alpha)$? We will answer this question by means of conditional wave functions.

The only thing we can say on the basis of $\Psi^*(\mathbf{p}, E + E_\alpha)\Psi(\mathbf{p}, E + E_\alpha)\phi^*(E_\alpha)\phi(E_\alpha)$ is that there are $\mathbf{p}, E$-fluctuations (interactions) and that the whole system, consisting of a certain number of subsystems, is permanently in the state E_α. The last point follows directly from the fact that the probability density $\phi^*(E_\alpha)\phi(E_\alpha)$, with $\phi(E_\alpha) = \exp(-iE_\alpha t_{12}/\hbar)$, for E_α is one:

$$\phi^*(E_\alpha)\phi(E_\alpha) = 1. \qquad (4.161)$$

How are the subsystems involved? This question becomes relevant because E_α reflects the total energy of the system consisting of two subsystems. The value E_α may only exist once at time τ, but we consider here at least two subsystems. Hence, we have to investigate the corresponding conditional wave functions that we have already introduced above. In the following we would like to investigate this point in somewhat more detail.

Also here the principal features can already be discussed on the basis of two interacting systems, that is, the whole system consists of two subsystems: subsystem 1 and subsystem 2. The interaction takes place between these two subsystems and is completely described by $\Psi^*(\mathbf{p}, E)\Psi(\mathbf{p}, E)$ and alternatively by $\Psi^*(\mathbf{p}, E + E_\alpha)\Psi(\mathbf{p}, E + E_\alpha)\phi^*(E_\alpha)\phi(E_\alpha)$ (Section 4.7.1). Since we are interested in the role of the constant energy E_α in connection with the interaction mechanism we will use $\Psi^*(\mathbf{p}, E + E_\alpha)\Psi(\mathbf{p}, E + E_\alpha)\phi^*(E_\alpha)\phi(E_\alpha)$.

We have analyzed above that the interaction takes place between two identical systems (subsystem 1 and subsystem 2) having exactly the same wave functions $\Psi(\mathbf{p}, E)$ and $\Psi(\mathbf{p}, E + E_\alpha)$, respectively.[79] The function $\Psi^\bullet(\mathbf{p}, E + E_\alpha)\Psi(\mathbf{p}, E + E_\alpha)$ is symmetrical, that is, we have $\Psi^\bullet(\mathbf{p}, E + E_\alpha)\Psi(\mathbf{p}, E + E_\alpha) = \Psi^\bullet(-\mathbf{p}, -E + E_\alpha)\Psi(-\mathbf{p}, -E + E_\alpha)$, and the $\mathbf{p}, E$-states of both subsystems are strongly correlated and expressed by the conditional wave functions $\Psi_{1,cond}(\mathbf{p}, E + E_\alpha)$ and $\Psi_{2,cond}(\mathbf{p}, E + E_\alpha)$.

---

[79] See Section 4.7.1.

In conclusion, for the description of $\Psi^*(\mathbf{p}, E + E_\alpha)\Psi(\mathbf{p}, E + E_\alpha)\phi^*(E_\alpha)\phi(E_\alpha)$ in terms of interaction mechanisms, we need two systems with exactly the same (conditional) wave function and (conditional) probability density, respectively. In other words, in the expression $\Psi^*(\mathbf{p}, E + E_\alpha)\Psi(\mathbf{p}, E + E_\alpha)\phi^*(E_\alpha)\phi(E_\alpha)$, we have to replace $\Psi^*(\mathbf{p}, E + E_\alpha)\Psi(\mathbf{p}, E + E_\alpha)$ by $2\Psi^*(\mathbf{p}, E + E_\alpha)\Psi(\mathbf{p}, E + E_\alpha)$:

$$\Psi^*(\mathbf{p}, E + E_\alpha)\Psi(\mathbf{p}, E + E_\alpha) \to 2\Psi^*(\mathbf{p}, E + E_\alpha)\Psi(\mathbf{p}, E + E_\alpha). \quad (4.162)$$

Because the expression $\Psi^*(\mathbf{p}, E + E_\alpha)\Psi(\mathbf{p}, E + E_\alpha)\phi^*(E_\alpha)\phi(E_\alpha)$ may not be changed, we necessarily have to replace the function $\phi^*(E_\alpha)\phi(E_\alpha)$ by $\phi^*(E_\alpha)\phi(E_\alpha)/2$:

$$\phi^*(E_\alpha)\phi(E_\alpha) \to \frac{1}{2}\phi^*(E_\alpha)\phi(E_\alpha). \quad (4.163)$$

Then, we obtain with equation (4.157)

$$\Psi^*(\mathbf{p}, E + E_\alpha)\Psi(\mathbf{p}, E + E_\alpha)\phi^*(E_\alpha)\phi(E_\alpha)$$
$$= 2\Psi^*(\mathbf{p}, E + E_\alpha)\Psi(\mathbf{p}, E + E_\alpha)\frac{1}{2}\phi^*(E_\alpha)\phi(E_\alpha)$$
$$= \Psi^*_{1,cond}(\mathbf{p}, E + E_\alpha)\Psi_{1,cond}(\mathbf{p}, E + E_\alpha)\frac{1}{2}\phi^*(E_\alpha)\phi(E_\alpha)$$
$$+ \Psi^*_{2,cond}(\mathbf{p}, E + E_\alpha)\Psi_{2,cond}(\mathbf{p}, E + E_\alpha)\frac{1}{2}\phi^*(E_\alpha)\phi(E_\alpha). \quad (4.164)$$

We have two terms: one for subsystem 1 at position $\mathbf{r}_1, t_1$ and another for subsystem 2 at position $\mathbf{r}_2, t_2$. Each term is weighted with $\phi^*(E_\alpha)\phi(E_\alpha)/2$, and determines at what space-time position ($\mathbf{r}_1, t_1$ or $\mathbf{r}_2, t_2$) the constant energy E_α is realized. In the next step, we would like to analyze this point.

3. Discussion

The terms $\phi^*(E_\alpha)\phi(E_\alpha)/2$ that appear in equation (4.164) have the following meaning: At time τ we observe the energy E_α with a probability of 50% at position $\mathbf{r}_1, t_1$ (subsystem 1). Exactly the same statement is true in connection with subsystem 2 at position $\mathbf{r}_2, t_2$. Also here we observe at time τ the energy E_α with a probability of 50%. In other words, at time τ there is always one value E_α existent, and both subsystems are involved. The whole system is permanently in the state E_α and this is correct because E_α reflects the total energy of the system, as we have already discussed above. E_α fluctuates statistically between subsystem 1 ($\mathbf{r}_1, t_1$) and subsystem 2 ($\mathbf{r}_2, t_2$). There is no specific space-time position $\mathbf{r}_{E_\alpha}, t_{E_\alpha}$ for the observation of E_α necessary.[80] In fact, $\phi(E_\alpha)$, of $\phi^*(E_\alpha)\phi(E_\alpha)/2$, is given by $\exp(-iE_\alpha t_{12}/\hbar)$[81] and is connected to t_1 and t_2, that is, both subsystems with $\mathbf{r}_1, t_1$ and $\mathbf{r}_2, t_2$ are equally involved. Nevertheless, it is surprising that the total energy E_α is at each time τ only connected to one of the subsystems. However, the two subsystems are completely equivalent leading to uniform E_α-fluctuations whereby the two subsystems are equally involved. This feature indicates that E_α belongs to the whole system even when E_α is realized at time τ only in connection with one of the subsystems.

In connection with equation (4.164), the interaction mechanism, i.e., the specific $\mathbf{p}, E$-fluctuations, behind $\Psi^*(\mathbf{p}, E + E_\alpha)\Psi(\mathbf{p}, E + E_\alpha)\phi^*(E_\alpha)\phi(E_\alpha)$ is the following. If, for example, E_α is localized at subsystem 1 ($\mathbf{r}_1, t_1$) we have, for example, at time τ the following distribution:

$$\mathbf{p}, E + E_\alpha \text{ (subsystem 1)},$$

$$-\mathbf{p}, -E \text{ (subsystem 2)}.$$

On the other hand, if E_α is localized at subsystem 2 ($\mathbf{r}_2, t_2$) we have, for example, at time τ' ($\tau' \neq \tau$) the distribution,

$$\mathbf{p}, -E \text{ (subsystem 1)},$$

$$-\mathbf{p}, E + E_\alpha \text{ (subsystem 2)}.$$

This is the meaning of $\Psi^*(\mathbf{p}, E + E_\alpha)\Psi(\mathbf{p}, E + E_\alpha)\phi^*(E_\alpha)\phi(E_\alpha)$. This equation only makes statements about the probability distribution for $\mathbf{p}, E$ and E_α but not about the specific interaction mechanism in connection with subsystem 1 and subsystem 2. With our construction given by equation (4.164), the interaction mechanisms, i.e., the specific

---
[80] See, in particular, the discussion at the beginning of Section 4.7.
[81] See equation (4.149).

p, E-fluctuations, behind $\Psi^*(\mathbf{p}, E+E_\alpha)\Psi(\mathbf{p}, E+E_\alpha)\phi^*(E_\alpha)\phi(E_\alpha)$ could be found. Only this interpretation of $\Psi(\mathbf{p}, E+E_\alpha)\exp(-iE_\alpha t_{12}/\hbar)$ fulfils all the required features.

4. Conditional wave functions for $\phi(E_\alpha)$

As in the case of the **p**, E-fluctuations where we introduced conditional wave functions for both systems, i.e., $\Psi_{1,cond}(\mathbf{p}, E)$ for system 1 and for system 2 the conditional wave function $\Psi_{2,cond}(\mathbf{p}, E)$; see in particular Section 4.7.3, also for the description of the E_α-fluctuations we may define for both subsystems "conditional wave functions" $\phi_{1,cond}(E_\alpha)$ and $\phi_{2,cond}(E_\alpha)$ which behave both statistically but are, in analogy to $\Psi_{1,cond}(\mathbf{p}, E)$ and $\Psi_{2,cond}(\mathbf{p}, E)$, strictly correlated. Then, the two identical terms $\phi^*(E_\alpha)\phi(E_\alpha)/2$ that appear in equation (4.164) have to be replaced as follows:

$$\text{system 1: } \frac{1}{2}\phi^*(E_\alpha)\phi(E_\alpha) \to \phi^*_{1,cond}(E_\alpha)\phi_{1,cond}(E_\alpha), \quad (4.165)$$

$$\text{system 2: } \frac{1}{2}\phi^*(E_\alpha)\phi(E_\alpha) \to \phi^*_{2,cond}(E_\alpha)\phi_{2,cond}(E_\alpha), \quad (4.166)$$

with

$$\phi^*_{1,cond}(E_\alpha)\phi_{1,cond}(E_\alpha) = \phi^*_{2,cond}(E_\alpha)\phi_{2,cond}(E_\alpha), \quad (4.167)$$

and

$$\phi^*(E_\alpha)\phi(E_\alpha) = \phi^*_{1,cond}(E_\alpha)\phi_{1,cond}(E_\alpha) + \phi^*_{2,cond}(E_\alpha)\phi_{2,cond}(E_\alpha). \quad (4.168)$$

Then, we get instead of equation (4.164)

$$\begin{aligned}\Psi^*(\mathbf{p}, E+E_\alpha)&\Psi(\mathbf{p}, E+E_\alpha)\phi^*(E_\alpha)\phi(E_\alpha)\\ &= \Psi^*_{1,cond}(\mathbf{p}, E+E_\alpha)\Psi_{1,cond}(\mathbf{p}, E+E_\alpha)\phi^*_{1,cond}(E_\alpha)\phi_{1,cond}(E_\alpha)\\ &+ \Psi^*_{2,cond}(\mathbf{p}, E+E_\alpha)\Psi_{2,cond}(\mathbf{p}, E+E_\alpha)\\ &\times \phi^*_{2,cond}(E_\alpha)\phi_{2,cond}(E_\alpha).\end{aligned} \quad (4.169)$$

The energy E_α is not simply a static constant but E_α fluctuates between subsystem 1 (space-time position $\mathbf{r}_1, t_1$) and subsystem 2 (space-time position $\mathbf{r}_2, t_2$). There are two distributions $\phi^*_{1,cond}(E_\alpha)\phi_{1,cond}(E_\alpha)$ and $\phi^*_{2,cond}(E_\alpha)\phi_{2,cond}(E_\alpha)$. The random sequence in connection with subsystem 1 is described by $\phi^*_{1,cond}(E_\alpha)\phi_{1,cond}(E_\alpha)$, and the second random sequence we observe simultaneously is $\phi^*_{2,cond}(E_\alpha)\phi_{2,cond}(E_\alpha)$ and describes subsystem 2 with respect to E_α. For example, if we measure E_α at $\mathbf{r}_1, t_1$ for subsystem 1 with $\phi^*_{1,cond}(E_\alpha)\phi_{1,cond}(E_\alpha)$ the sequence

$$0, 1, 0, 0, 1, 1, 1, \qquad (4.170)$$

we must register for E_α at $\mathbf{r}_2, t_2$ the sequence

$$1, 0, 1, 1, 0, 0, 0. \qquad (4.171)$$

Here 0 means no event, and 1 means the registration of E_α. The addition of both sequences (4.170) and (4.171) leads to

$$1, 1, 1, 1, 1, 1, 1. \qquad (4.172)$$

In other words, the sequences (4.170) and (4.171) behave strictly statistically but they are cross-correlated, which leads to the non-statistical sequence (4.172). In summary, both distributions $\phi^*_{1,cond}(E_\alpha)\phi_{1,cond}(E_\alpha)$ and $\phi^*_{2,cond}(E_\alpha)\phi_{2,cond}(E_\alpha)$ behave strictly statistically but are strongly correlated, and we have exactly the same situation as in the case of $\Psi^*(\mathbf{p}, E)\Psi(\mathbf{p}, E)$ which we have also split into two conditional probability distributions $\Psi^*_{1,cond}(\mathbf{p}, E)\Psi_{1,cond}(\mathbf{p}, E)$ and $\Psi^*_{2,cond}(\mathbf{p}, E)\Psi_{2,cond}(\mathbf{p}, E)$.[82]

4.7.5. Extension to N Subsystems

We may extend the system from 2 to N subsystems without modifying the basic features. Since every subsystem interacts by $\mathbf{p}, E$-fluctuations with all the $(N-1)$ subsystems, we have $(N-1)N$ identical density distributions,[83] and for this case we have treated the $\mathbf{p}, E$-fluctuations

---

[82] See, in particular, Section 4.7.3.
[83] See Section 4.4.

in Section 4.5. Then, instead of equation (4.157) we obtain

$$\begin{aligned} N\Psi^*(\mathbf{p}, E + E_\alpha)&\Psi(\mathbf{p}, E + E_\alpha) \\ = \Psi^*_{1,cond}&(\mathbf{p}, E + E_\alpha)\Psi_{1,cond}(\mathbf{p}, E + E_\alpha) \\ + \Psi^*_{2,cond}&(\mathbf{p}, E + E_\alpha)\Psi_{2,cond}(\mathbf{p}, E + E_\alpha) \\ + \cdots + \Psi^*_{N,cond}&(\mathbf{p}, E + E_\alpha)\Psi_{N,cond}(\mathbf{p}, E + E_\alpha). \end{aligned}$$
(4.173)

Furthermore, we have[84]

$$\begin{aligned} \Psi^*(\mathbf{p}, E + E_\alpha)&\Psi(\mathbf{p}, E + E_\alpha) \\ &\to N\Psi^*(\mathbf{p}, E + E_\alpha)\Psi(\mathbf{p}, E + E_\alpha), \end{aligned}$$
(4.174)

and

$$\phi^*(E_\alpha)\phi(E_\alpha) \to \frac{1}{N}\phi^*(E_\alpha)\phi(E_\alpha).$$
(4.175)

Using these equations, we obtain in analogy to equation (4.164) the following expression:

$$\begin{aligned} \Psi^*(\mathbf{p}, E + E_\alpha)&\Psi(\mathbf{p}, E + E_\alpha)\phi^*(E_\alpha)\phi(E_\alpha) \\ = N\Psi^*(\mathbf{p},& E + E_\alpha)\Psi(\mathbf{p}, E + E_\alpha) \\ &\times \frac{1}{N}\phi^*(E_\alpha)\phi(E_\alpha) \\ = \Psi^*_{1,cond}&(\mathbf{p}, E + E_\alpha)\Psi_{1,cond}(\mathbf{p}, E + E_\alpha) \\ &\times \frac{1}{N)}\phi^*(E_\alpha)\phi(E_\alpha) \\ + \Psi^*_{2,cond}&(\mathbf{p}, E + E_\alpha)\Psi_{2,cond}(\mathbf{p}, E + E_\alpha) \\ &\times \frac{1}{N}\phi^*(E_\alpha)\phi(E_\alpha) \\ + \cdots + \Psi^*_{N,cond}&(\mathbf{p}, E + E_\alpha)\Psi_{N,cond}(\mathbf{p}, E + E_\alpha) \\ &\times \frac{1}{N}\phi^*(E_\alpha)\phi(E_\alpha). \end{aligned}$$
(4.176)

---

[84]This is in analogy to equation (4.162).

In this case E_α fluctuates statistically between all N subsystems having the space-time positions $\mathbf{r}_1, t_1, \mathbf{r}_2, t_2, \ldots, \mathbf{r}_N, t_N$. All N subsystems are equally involved, as in the case of the two subsystems. Also here E_α belongs to the whole system even when E_α is realized at time τ only in connection with one of the N subsystems.

4.7.6. *Summary*

In the case of two subsystems, the general form of the wave function $\Psi(\mathbf{p}, E)$ can be determined by equation (4.129). However, for the investigation of the mechanism in connection with the energy E_α we used a form for $\Psi(\mathbf{p}, E)$ which is not dependent of E_α. In other words, E_α has been eliminated from the original form of $\Psi(\mathbf{p}, E)$.[85] Whereas the energy values E of the solutions of equation (4.129) fluctuate around E_α, the energy values E of the re-formulated function $\Psi(\mathbf{p}, E)$ fluctuate around zero. The energy E_α can be introduced over the re-defined function $\Psi(\mathbf{p}, E)$ by replacing the variable E by $E + E_\alpha$ leading to the function $\Psi(\mathbf{p}, E + E_\alpha) \exp(-iE_\alpha t_{12})/\hbar$. Because both functions, $\Psi(\mathbf{p}, E + E_\alpha) \exp(-iE_\alpha t_{12})/\hbar$ and $\Psi(\mathbf{p}, E)$, are equivalent, we were able to investigate directly the role of E_α.

Clearly, only the solutions of (4.129) for $\Psi(\mathbf{p}, E)$ are relevant and not its re-defined form. The energy values E of this general function $\Psi(\mathbf{p}, E)$ fluctuate around E_α. However, due to our analysis we can make additional statements about the mechanisms with respect to the subsystems. In the case of two subsystems the following is valid in connection with $\Psi(\mathbf{p}, E)$ whose energy values E fluctuate around E_α. If, for example, E_α is localized at subsystem 1 $(\mathbf{r}_1, t_1)$ we have, for example, at time τ the following distribution: $\mathbf{p}, E + E_\alpha$ (subsystem 1), $-\mathbf{p}, -E$ (subsystem 2). On the other hand, if E_α is localized at subsystem 2 $(\mathbf{r}_2, t_2)$ we have, for example, at time τ' ($\tau' \neq \tau$) the distribution $\mathbf{p}, -E$ (subsystem 1), and $-\mathbf{p}, E + E_\alpha$ (subsystem 2).

4.8. Distance-Independent Interactions

In the last section we have treated a system consisting of N interacting subsystems, and we assumed that a distance-dependent interaction

---
[85] See solution of (4.129).

between the subsystems is effective, that is, the mutual interactions between the subsystems are dependent on the space-time distances $\mathbf{r}_{ki} - \mathbf{r}_l$, $t_k - t_l$ with $k, l = 1, \ldots, N$, $k \neq l$ where $\mathbf{r}_1, \mathbf{r}_2, \ldots, \mathbf{r}_N, t_1, t_2, \ldots, t_N$ are the positions and system-specific times in $(\mathbf{r}, t)$-space at time τ. However, projection theory opens the possibility for another kind of interaction that is not dependent on the space-time distances and, therefore, we would like to call it "distance-independent interactions". In other words, there can be interactions between two systems, say i and j, whose strength is not dependent on the space-time distances $\mathbf{r}_i - \mathbf{r}_j$, $t_i - t_j$.

This type of interaction reflects an interesting physical situation, not merely as a formal aspect but in particular with respect to many scientific real effects as, for example, non-local events with respect to the various $\mathbf{r}, t$-states. However, the projection principle not only emerges new aspects in connection with the notion "interaction" but it influences the entire range of physical reality. Most of the new physical aspects have already been quoted and discussed in the preceding chapters, but we will continue this discussion in the forthcoming sections.

4.8.1. *Principal Remarks*

Does such kind of interaction make sense? Yes, it definitely does. These distance-independent interactions furnish a system with a specific form (shape). Therefore, in the following we will call this kind of interaction also "form interaction". If the systems (particles) i and j are elementary in character this kind of interaction creates their shape. In conventional physics we also work with certain forms for elementary systems. We have point-like particle, strings and branes. However, these specific forms had to be assumed in conventional physics and could not be derived. In contrast to these developments, projection theory opens the possibility to explain (derive) certain elementary forms in nature by means of this new kind of interaction, i.e., by distance-independent (form) interactions. In the next sections, we will discuss some basic features of distance-independent interactions.

4.8.2. *Some Minor Changes*

In Section 4.2.2, we already discussed the notion "interaction" in a new light. Within projection theory all interactions exclusively take place in $(\mathbf{p}, E)$-space and not in $(\mathbf{r}, t)$-space, as in conventional physics. We discussed this specific question by means of two systems i and j which were assumed to be the lonely systems in the universe, that is, we assumed in our example that there are no other systems within the universe. We would like to use this model for the discussion of form interactions. In fact, the functions $\chi_k(\mathbf{r}, t), k = i, j$,[86] already define this kind of interaction because there is no relationship between the space-time positions $\mathbf{r}_i, t_i$ and $\mathbf{r}_j, t_j$. The functions are dependent on $\mathbf{r}_i, \mathbf{r}_j$ and t_i, t_j but there is no term which is dependent on the distances $\mathbf{r}_i - \mathbf{r}_j, t_i - t_j$.

Nevertheless, there must be an interaction between both systems. The $\mathbf{p}, E$-values fluctuate for each system in the range $-\infty \leq \mathbf{p}, E \leq \infty$ and this can only be explained by interaction processes, i.e., $\mathbf{p}, E$-fluctuations, between system i and system j.[87] If both systems would perform independent $\mathbf{p}, E$-fluctuations, the conservation laws for momentum and energy would be violated. These $\mathbf{p}, E$-fluctuations in $(\mathbf{p}, E)$-space just reflect distance-independent interactions in $(\mathbf{r}, t)$-space.

However, the functions $\chi_k(\mathbf{r}, t), k = i, j$ are not quite realistic. The width of the peaks (delta functions) is zero and this implies that the $\mathbf{p}, E$-states, described by the probability densities $\chi_k^*(\mathbf{p}, E)\chi_k(\mathbf{p}, E), k = i, j$, become constants for all possible $\mathbf{p}, E$-states $(-\infty \leq \mathbf{p}, E \leq \infty)$: $\chi_k^*(\mathbf{p}, E)\chi_k(\mathbf{p}, E) = 1/(2\pi\hbar)^4, k = i, j$. We need all values of $\mathbf{p}$ and E in order to guarantee that the Fourier transform of $\chi_k(\mathbf{p}, E)$ leads to delta functions for $\chi_k(\mathbf{r}, t)$.[88] However, the values $\mathbf{p} = \infty, E = \infty$ do not appear in the observable world. Therefore, we have to modify the functions $\chi_k(\mathbf{p}, E), k = i, j$ somewhat, and we would like to

---

[86] See equations (4.1) and (4.2).
[87] See Section 4.2.20.
[88] See equations (4.1) and (4.2).

denote these modified functions by $\Psi_F(\mathbf{p}, E)_k, k = i, j$. The function $\Psi_F(\mathbf{p}, E), k = i, j$ must be constructed in a way that the values $\mathbf{p} = \infty, E = \infty$ are excluded, and we have

$$\chi_k(\mathbf{p}, E), k = i, j \to \Psi_F(\mathbf{p}, E)_k, k = i, j, \quad (4.177)$$

with

$$\lim_{\mathbf{p}, E \to \infty} \Psi_F(\mathbf{p}, E)_k = 0, \quad k = i, j. \quad (4.178)$$

Then, the Fourier transform of $\Psi_F(\mathbf{p}, E)_k, k = i, j$ must lead to modified functions in $(\mathbf{r}, t)$-space and, instead of $\chi_k(\mathbf{r}, t), k = i, j$,[89] we get the functions $\Psi_F(\mathbf{r} - \mathbf{r}_k, t - t_k), k = i, j$, that is, we have the transition

$$\chi_k(\mathbf{r}, t) \to \Psi_F(\mathbf{r} - \mathbf{r}_k, t - t_k), \quad k = i, j, \quad (4.179)$$

where

$$\Psi_F(\mathbf{p}, E)_k = \frac{1}{(2\pi\hbar)^2} \int_{-\infty}^{\infty} \Psi_F(\mathbf{r} - \mathbf{r}_k, t - t_k)$$
$$\times \exp\left\{-i\left[\frac{\mathbf{p}}{\hbar} \cdot \mathbf{r} - \frac{E}{\hbar} t\right]\right\} d\mathbf{r}\, dt, \quad k = i, j. \quad (4.180)$$

Equation (4.180) is completely equivalent to equation (4.3). The differences between the wave functions $\Psi_F(\mathbf{r} - \mathbf{r}_k, t - t_k), k = i, j$ and $\chi_k(\mathbf{r}, t), k = i, j$,[88] are allowed to be (very) small. However, the peak-widths of $\Psi_F(\mathbf{r} - \mathbf{r}_k, t - t_k), k = i, j$ may not be zero as in the case of $\chi_k(\mathbf{r}, t), k = i, j$, but can be infinitesimal.

4.8.3. Some Basic Features of Distance-Independent Interactions

We may shift the functions $\Psi_F(\mathbf{r} - \mathbf{r}_k, t - t_k), k = i, j$ relative to $(\mathbf{r}, t)$-space without changing the interaction, i.e., the law for the $(\mathbf{p}, E)$-fluctuations, which are described by the probability densities $\Psi_F^*(\mathbf{p}, E)_k \Psi_F(\mathbf{p}, E)_k, k = i, j$, is not changed when $\mathbf{r}$ and t are shifted by the quantities $\Delta_{k,\mathbf{r}}$ and $\Delta_{k,t}$:

$$\mathbf{r} \to \mathbf{r} - \Delta_{k,\mathbf{r}}, \quad k = i,j,$$
$$t \to t - \Delta_{k,t}, \quad k = i,j, \qquad (4.181)$$

where the values $\Delta_{k,\mathbf{r}}$ and $\Delta_{k,t}$ are assumed not to be dependent on $\mathbf{r}$ and t. With equations (4.180) and (4.181) we immediately get

$$\Psi_F(\mathbf{p}, E)_k = \frac{1}{(2\pi\hbar)^2} \int_{-\infty}^{\infty} \left\{ \Psi_F(\mathbf{r} - (\mathbf{r}_k + \Delta_{k,\mathbf{r}}), t - (t_k + \Delta_{k,t})) \right.$$
$$\left. \times \exp\left\{ i\left[\frac{\mathbf{p}}{\hbar} \cdot \Delta_{k,\mathbf{r}} - \frac{E}{\hbar}\Delta_{k,t}\right]\right\} \right\}$$
$$\times \exp\left\{ -i\left[\frac{\mathbf{p}}{\hbar} \cdot \mathbf{r} - \frac{E}{\hbar}t\right]\right\} d\mathbf{r}\, dt, \quad k = i,j. \qquad (4.182)$$

The functions

$$\Psi_F(\mathbf{r} - \mathbf{r}_k, t - t_k), \qquad (4.183)$$

and

$$\left\{ \Psi_F(\mathbf{r} - (\mathbf{r}_k + \Delta_{k,\mathbf{r}}), t - (t_k + \Delta_{k,t})) \exp\left\{ i\left[\frac{\mathbf{p}}{\hbar} \cdot \Delta_{k,\mathbf{r}} - \frac{E}{\hbar}\Delta_{k,t}\right]\right\} \right\}, \qquad (4.184)$$

have to considered as equivalent because both expressions[89] equally reproduce $\Psi_F(\mathbf{p}, E)_k$, $k = i,j$.[90] It is therefore indifferent whether we use a shifted function, shifted by $\Delta_{k,\mathbf{r}}$ and $\Delta_{k,t}$ in accordance to equation (4.184), or the function $\Psi_F(\mathbf{r} - \mathbf{r}_k, t - t_k)$. This is an important point because the function $\Psi_F(\mathbf{p}, E)_k$, $k = i,j$, is responsible for the interaction between the systems i and j. Clearly, also here the interaction ($\mathbf{p}$, E-fluctuations) is described by $\Psi_F^*(\mathbf{p}, E)_k \Psi_F(\mathbf{p}, E)_k$ with $k = i,j$. That is, the relevant function is $\Psi_F^*(\mathbf{p}, E)_k \Psi_F(\mathbf{p}, E)_k$ and,[91] in order to obtain $\Psi_F^*(\mathbf{p}, E)_k \Psi_F(\mathbf{p}, E)_k$, equation (4.183) can

---
[89] See equations (4.183) and (4.184).
[90] See equations (4.180) and (4.182).
[91] See Section 2.3.8.2.

be reformulated as follows:

$$\Phi_F(\mathbf{p}, E)_k = \frac{1}{(2\pi\hbar)^2} \int_{-\infty}^{\infty} \Psi_F(\mathbf{r} - (\mathbf{r}_k + \Delta_{k,\mathbf{r}}), t - (t_k + \Delta_{k,t}))$$
$$\times \exp\left\{-i\left[\frac{\mathbf{p}}{\hbar}\cdot\mathbf{r} - \frac{E}{\hbar}t\right]\right\} d\mathbf{r}\, dt, \quad k = i, j, \quad (4.185)$$

with

$$\Phi_F(\mathbf{p}, E)_k = \Psi_F(\mathbf{p}, E)_k \exp\left\{-i\left[\frac{\mathbf{p}}{\hbar}\cdot\Delta_{k,\mathbf{r}} - \frac{E}{\hbar}\Delta_{k,t}\right]\right\}, \quad k = i, j. \quad (4.186)$$

Then, we obtain

$$\Phi_F^*(\mathbf{p}, E)_k \Phi_F(\mathbf{p}, E)_k = \Psi_F^*(\mathbf{p}, E)_k \Psi_F(\mathbf{p}, E)_k, \quad k = i, j. \quad (4.187)$$

Because of this property both functions, $\Psi_F(\mathbf{p}, E)_k$ and $\Phi_F(\mathbf{p}, E)_k$, are equivalent in connection with the description of interaction processes. They are equally suitable to describe $\mathbf{p}, E$-fluctuations.

The function $\Psi_F(\mathbf{r}-\mathbf{r}_k, t-t_k)$ is shifted but its form (shape) remains conserved. Thus, the forms of

$$\Psi_F^*(\mathbf{r} - \mathbf{r}_k, t - t_k)\Psi_F(\mathbf{r} - \mathbf{r}_k, t - t_k), \quad k = i, j, \quad (4.188)$$

and

$$\Psi_F^*(\mathbf{r} - (\mathbf{r}_k + \Delta_{k,\mathbf{r}}), t - (t_k + \Delta_{k,t}))$$
$$\times \Psi_F(\mathbf{r} - (\mathbf{r}_k + \Delta_{k,\mathbf{r}}), t - (t_k + \Delta_{k,t})), \quad k = i, j, \quad (4.189)$$

are exactly the same. They only differ in the space-time positions and, because of equation (4.187), in both cases the probability density $\Psi_F^*(\mathbf{p}, E)_k \Psi_F(\mathbf{p}, E)_k$, $k = i, j$ is relevant.

4.8.4. Absolute Space-Time Positions

Can a human observer distinguish between the absolute space-time positions $\mathbf{r}_k, t_k$ and $\mathbf{r}_k + \Delta_{k,\mathbf{r}}, t_k + \Delta_{k,t}$? No, he cannot. There is no law or phenomenon in projection theory defined that would formulate a relationship between an observer and his absolute space-time positions. We always measure at certain space-time positions certain values of $\mathbf{p}$

and E and never space-time coordinates (instead of $\mathbf{p}$ and E). It makes no sense to define absolute space-time positions within the framework of projection theory because within the basic frame of this theory, space and time exclusively play the role of auxiliary elements for the description of physical phenomena and, therefore, we are principally not able to measure at certain space-time positions specific values for space-time coordinates as, for example, $\mathbf{r}_k, t_k$ or $\mathbf{r}_k + \Delta_{k,\mathbf{r}}, t_k + \Delta_{k,t}$. This would make no sense. This general statement about space and time within projection theory is consistent with the analysis concerning probability densities in $(\mathbf{r}, t)$-space as well as in $(\mathbf{p}, E)$-space and their interconnections.

First, let us repeat the general statement that we have given in Section 2.3.8.3. The measurement of one of the possible values for $\mathbf{p}$ and for E is done in the space-time intervals $\mathbf{r}, \mathbf{r} + \Delta \mathbf{r}$ and $t, t + \Delta t$ with the probability density of $\Psi^*(\mathbf{r}, t)\Psi(\mathbf{r}, t)$. The possible values for $\mathbf{p}$ and for E are given with a certain probability that is determined by the probability density $\Psi^*(\mathbf{p}, E)\Psi(\mathbf{p}, E)$.

Let us apply this general statement on the situation in connection with wave functions described by equations (4.188) and (4.189) where we have to consider that the probability density $\Psi_F^*(\mathbf{p}, E)_k \Psi_F(\mathbf{p}, E)_k$, $k = i, j$, is relevant in both cases. Since the probability densities in $(\mathbf{r}, t)$-space, described by the functions (4.188) and (4.189), have exactly the same form[92] and, furthermore, since both probability densities in $(\mathbf{r}, t)$-space[93] are connected to exactly the same probability density $\Psi_F^*(\mathbf{p}, E)_k \Psi_F(\mathbf{p}, E)_k$, $k = i, j$ in $(\mathbf{p}, E)$-space, there is principally no possibility to distinguish between the space-time positions $\mathbf{r}_k + \Delta_{k,\mathbf{r}}$, $t_k + \Delta_{k,t}$ and $\mathbf{r}_k, t_k$ ($k = i$ or j).

In conclusion, the observer, who is resting relative to the system with $\mathbf{r}_k + \Delta_{k,\mathbf{r}}, t_k + \Delta_{k,t}$ ($k = i$ or j), measures exactly the same $\mathbf{p}, E$-distribution, i.e., $\Psi_F^*(\mathbf{p}, E)_k \Psi_F(\mathbf{p}, E)_k$, as the observer, which is resting with respect to the system with the space-time position $\mathbf{r}_k, t_k$ ($k = i$ or j). This scenario can be summarized schematically as

---

[92] They only differ in the space-time positions.
[93] See equations (4.188) and (4.189).

follows:

$$\Psi_F^*(\mathbf{p}, E)_k \Psi_F(\mathbf{p}, E)_k$$
$$\downarrow$$
$$\Psi_F^*(\mathbf{r} - (\mathbf{r}_k + \Delta_{k,\mathbf{r}}), t - (t_k + \Delta_{k,t})) \tag{4.190}$$
$$\times \Psi_F(\mathbf{r} - (\mathbf{r}_k + \Delta_{k,\mathbf{r}}), t - (t_k + \Delta_{k,t}))$$
$$\Psi_F^*(\mathbf{p}, E)_k \Psi_F(\mathbf{p}, E)_k$$
$$\downarrow \tag{4.191}$$
$$\Psi_F^*(\mathbf{r} - \mathbf{r}_k, t - t_k) \Psi_F(\mathbf{r} - \mathbf{r}_k, t - t_k)$$

4.8.5. Arbitrary Jumps

Both observers can make statements about the form (shape) of the systems in $(\mathbf{r}, t)$-space in connection with a $\mathbf{p}, E$-measurement, but they are principally not be able to say something about the space-time positions $\mathbf{r}_k + \Delta_{k,\mathbf{r}}, t_k + \Delta_{k,t}$ and $\mathbf{r}_k, t_k$ where the distribution functions[93] are positioned. A relationship to these positions is not defined. The reason is the fact that the wave function $\Psi_F(\mathbf{p}, E)_k$ merely describes the shapes of the functions $\Psi_F(\mathbf{r} - \mathbf{r}_k, t - t_k)$ and $\Psi_F(\mathbf{r} - (\mathbf{r}_k + \Delta_{k,\mathbf{r}}), t - (t_k + \Delta_{k,t}))$ in $(\mathbf{r}, t)$-space.

Because there is no relationship between $\mathbf{r}_k + \Delta_{k,\mathbf{r}}, t_k + \Delta_{k,t}$ (or $\mathbf{r}_k, t_k$) and the space-time coordinates of $(\mathbf{r}, t)$-space defined, the quantities $\Delta_{k,\mathbf{r}}$ and $\Delta_{k,t}$ must behave statistically, i.e., the values of $\Delta_{k,\mathbf{r}}$ and $\Delta_{k,t}$ are dependent on time τ, and we have

$$\Delta_{k,\mathbf{r}} = \Delta_{k,\mathbf{r}}(\tau), \quad k = i, j,$$
$$\Delta_{k,t} = \Delta_{k,t}(\tau), \quad k = i, j, \tag{4.192}$$

but both functions $\Delta_{k,\mathbf{r}}$ and $\Delta_{k,t}$ do not vary systematically with time τ because such a law does not exist. If we have certain values $\Delta_{k,\mathbf{r}}(\tau_1)$ and $\Delta_{k,t}(\tau_1)$ at time τ_1, we cannot say something about the values $\Delta_{k,\mathbf{r}}$ and $\Delta_{k,t}$ at time τ_2, even when τ_2 is the next time-value after τ_1, i.e., the quantities $\Delta_{k,\mathbf{r}}(\tau_2)$ and $\Delta_{k,t}(\tau_2)$ can principally not be predicted. In conclusion, the quantities $\Delta_{k,\mathbf{r}}$ and $\Delta_{k,t}$ behave statistically. Thus, the space-time positions of the peaks defined by

$$\Psi_F^*(\mathbf{r} - (\mathbf{r}_k + \Delta_{k,\mathbf{r}}(\tau)), t - (t_k + \Delta_{k,t}(\tau))) \Psi_F(\mathbf{r} - (\mathbf{r}_k + \Delta_{k,\mathbf{r}}(\tau)),$$
$$t - (t_k + \Delta_{k,t}(\tau))), \quad k = i, j, \tag{4.193}$$

must behave statistically, i.e., both peaks jump arbitrarily through space and time. These jumps are independent from each other because the interaction between both systems (system i and system j) is not dependent on their distance but we assumed that this interaction is distance-independent and creates the form of the systems.

On the other hand, if both systems, or one of them, would reside at a certain position in space and time, a relationship between $\mathbf{r}_k + \Delta_{k,\mathbf{r}}, t_k + \Delta_{k,t}$ and certain space-time coordinates of $(\mathbf{r}, t)$-space would be defined, i.e., certain space-time coordinates would be privileged and not all space-time points would be equivalent. However, within projection theory all space-time points $(-\infty \leq \mathbf{r}, t \leq \infty)$ are by definition equivalent; the Fourier transform and its inverse formulation[94] do not allow privileged $\mathbf{r}, t$-points. In other words, because such a relationship between $\mathbf{r}_k + \Delta_{k,\mathbf{r}}, t_k + \Delta_{k,t}$ and certain space-time coordinates of $(\mathbf{r}, t)$-space does not exist, it is strictly forbidden that one of the two systems i and j (or both) is able to reside at a certain position in space and time, i.e., in $(\mathbf{r}, t)$-space. This is a relevant statement just in connection with Mach's principle. We will come back to this principal question below.

Remark

The arbitrary jumps of the geometrical structures in $(\mathbf{r}, t)$-space corresponding to random phases in $(\mathbf{p}, E)$-space where the phases are expressed by[95]

$$\exp\left\{-i\left[\frac{\mathbf{p}}{\hbar} \cdot \Delta_{k,\mathbf{r}} - \frac{E}{\hbar}\Delta_{k,t}\right]\right\}, \qquad (4.194)$$

i.e., the transition from $\mathbf{r}, t$ to the shifted space-time positions $\mathbf{r} - \Delta_{k,\mathbf{r}}$, $t - \Delta_{k,t}$,[96] is given in $(\mathbf{r}, t)$-space by

$$\Psi_F(\mathbf{r} - \mathbf{r}_k, t - t_k) \to \Psi_F(\mathbf{r} - (\mathbf{r}_k + \Delta_{k,\mathbf{r}}), t - (t_k + \Delta_{k,t})), \quad k = i, j,$$

---

[94] See equations (2.4) and (2.5).
[95] See equation (4.186).
[96] See equation (4.181).

and the corresponding situation in $(\mathbf{p}, E)$-space leads to the transition

$$\Psi_F(\mathbf{p}, E)_k \to \Psi_F(\mathbf{p}, E)_k \exp\left\{-i\left[\frac{\mathbf{p}}{\hbar} \cdot \Delta_{k,\mathbf{r}} - \frac{E}{\hbar}\Delta_{k,t}\right]\right\}, \quad k = i, j. \tag{4.195}$$

Reality has to be identified with the states in $(\mathbf{p}, E)$-space and through their projection onto $(\mathbf{r}, t)$-space the arbitrary jumps come into play. Thus, we have the following scheme:

random phases
↓
arbitrary jumps

In the following, we have to analyze these arbitrary jumps in more detail because they reflect an interesting state (solution) with respect to the $\mathbf{p}, E$-states of the system. It will turn out that the arbitrary jumps in $(\mathbf{r}, t)$-space do not influence the $\mathbf{p}, E$-states.

4.8.6. Effective Velocities

In connection with form interactions, i.e., distance-independent interactions, we have developed the following picture in Section 4.8.2. There are $\mathbf{p}, E$-fluctuations (interactions) in $(\mathbf{p}, E)$-space that are described by the probability densities $\Psi_F^*(\mathbf{p}, E)_k \Psi_F(\mathbf{p}, E)_k, k = i, j$. If, for example, system i takes at time τ the values $\mathbf{p}_a, E_a$, system j must simultaneously be in the state of $-\mathbf{p}_a, -E_a$.[97,98] As in the case of distance-dependent interactions, the $\mathbf{p}, E$-states of both systems are strictly correlated, and this is due to the conservation laws for momentum and energy that have to be assumed to be valid here.[99]

---

[97] See Section 4.4.
[98] The total momenta and energies of both systems must be zero; we will recognize that in Section 4.8.9.
[99] We have outlined in Chapter 3 that the total momentum and the total energy must be zero because the systems i and j do not interact with other systems.

The projection of $\Psi_F(\mathbf{p}, E)_k$, $k = i, j$ onto $(\mathbf{r}, t)$-space leads to the wave functions

$$\Psi_F(\mathbf{r} - (\mathbf{r}_k + \Delta_{k,\mathbf{r}}), t - (t_k + \Delta_{k,t})), \quad k = i, j,$$

and the peaks of[100]

$$\Psi_F^*(\mathbf{r} - (\mathbf{r}_k + \Delta_{k,\mathbf{r}}), t - (t_k + \Delta_{k,t}))$$
$$\times \Psi_F(\mathbf{r} - (\mathbf{r}_k + \Delta_{k,\mathbf{r}}), t - (t_k + \Delta_{k,t})), \quad k = i, j,$$

should be considered as a definition of the form (shape) of the systems i and j and, as we have outlined above, these peaks, i.e., geometrical structures, jump arbitrarily relative to $(\mathbf{r}, t)$-space. Let us briefly repeat and deepen the physical content of this effect. We want to assume that system i is at time τ_a at the space-time position $\mathbf{r}_{1a}, t_{1a}$ and that system j is at the same time τ_a at the space-time position $\mathbf{r}_{2a}, t_{2a}$. Furthermore, let us assume that system i is at time $\tau_b \neq \tau_a$ at the space-time position $\mathbf{r}_{1b}, t_{1b}$ and system j at the same time τ_b at $\mathbf{r}_{2b}, t_{2b}$. Then, the following is relevant. There is no law defined that would predict the values $\mathbf{r}_{1b}, t_{1b}$ on the basis of $\mathbf{r}_{1a}, t_{1a}$ and there is of course also no law that would predict the values $\mathbf{r}_{2b}, t_{2b}$ on the basis of $\mathbf{r}_{2a}, t_{2a}$. In other words, both systems take arbitrary space-time positions in the course of time τ.

Because there is no distance-dependent interaction effective between system i and system j, both peaks jump independently from each other through $(\mathbf{r}, t)$-space, i.e., there can also be no relationship between the space-time positions $\mathbf{r}_{1a}, t_{1a}$ and $\mathbf{r}_{2a}, t_{2a}$ and also not between $\mathbf{r}_{1b}, t_{1b}$ and $\mathbf{r}_{2b}, t_{2b}$. In other words, there are no relationships, i.e., physical laws, between the space-time positions $\mathbf{r}_{1a}, t_{1a}, \mathbf{r}_{1b}, t_{1b}$, $\mathbf{r}_{2a}, t_{2a}$ and $\mathbf{r}_{2b}, t_{2b}$, even when τ_b is the next time value after τ_a and, therefore, the values $\mathbf{r}_{2b}, t_{2b}$ and $\mathbf{r}_{2a}, t_{2a}$ can principally not be predicted. That is, all the quantities $\mathbf{r}_{1a}, t_{1a}, \mathbf{r}_{1b}, t_{1b}, \mathbf{r}_{2a}, t_{2a}$ and $\mathbf{r}_{2b}, t_{2b}$ are completely independent from each other and behave strictly statistically.

---

[100] See equation (4.189).

However, we always have to keep in mind that within projection theory the peaks, i.e., geometrical structures, defined by equation (4.189), do not move through space-time, i.e., through $(\mathbf{r}, t)$-space, but are projected on it. Nevertheless, we may define "effective velocities" which we would like to call $v_{m\mathbf{r}}$ and v_{mt}. $v_{m\mathbf{r}}$ is the effective velocity with respect to the variable $\mathbf{r}$ and is given by

$$v_{m\mathbf{r}} = \frac{\mathbf{r}_{mb} - \mathbf{r}_{ma}}{\tau_b - \tau_a}, \quad m = 1, 2. \qquad (4.197)$$

v_{mt} is the effective velocity with respect to the variable t and has the form

$$v_{mt} = \frac{t_{mb} - t_{ma}}{\tau_b - \tau_a}, \quad m = 1, 2. \qquad (4.198)$$

The time interval $\tau_b - \tau_a$ can be close to zero and, on the other hand, $\mathbf{r}_{mb} - \mathbf{r}_{ma}$ and $t_{mb} - t_{ma}$ may be as large as the universe, where large really means with respect to the maximum space-extension as well as with respect to the maximum time-extension.[101] With

$$\tau_b - \tau_a \to 0, \qquad (4.199)$$

we obtain for the velocities

$$\begin{aligned} v_{m\mathbf{r}} &\to \infty, \quad m = 1, 2, \\ v_{mt} &\to \infty, \quad m = 1, 2. \end{aligned} \qquad (4.200)$$

The situation is illustrated in Figure 4.1.

In general, we may express the arbitrary behaviour of the $\mathbf{r}, t$-structures as follows. In equation (4.192) the quantities $\Delta_{k,\mathbf{r}}$ and $\Delta_{k,t}$ are given as a function of time τ. If $\Delta_{k,\mathbf{r}}$ and $\Delta_{k,t}$ take at time τ' the values $\Delta_{k,\mathbf{r}}(\tau')$ and $\Delta_{k,t}(\tau')$ and at time τ'' the values $\Delta_{k,\mathbf{r}}(\tau'')$ and

---

[101] That is, from the beginning to the end of time.

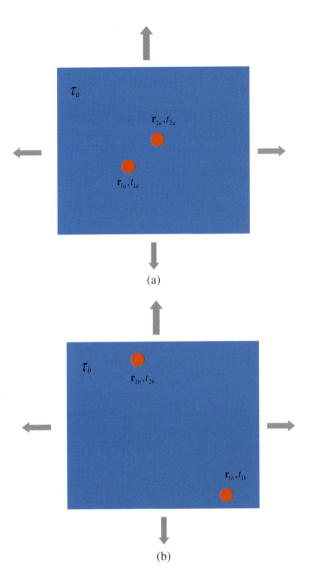

Fig. 4.1. (a) At time τ_a system i is at the space-time position $\mathbf{r}_{1a}, t_{1a}$ and system j is at $\mathbf{r}_{2a}, t_{2a}$. (b) At time τ_b system i is at the space-time position $\mathbf{r}_{1b}, t_{1b}$ and system j is at $\mathbf{r}_{2b}, t_{2b}$. Because of $\tau_b - \tau_a \to 0$ and the fact that the distances $\mathbf{r}_{mb} - \mathbf{r}_{ma}$ and $t_{mb} - t_{ma}$ (with $m = 1, 2$) may be arbitrarily large (in principle, as large as the space-time extension of the universe), the velocities $v_{m\mathbf{r}}$ and v_{mt} may be close to infinity.

$\Delta_{k,t}(\tau'')$, the quantities

$$\Delta_{k,\mathbf{r}}(\tau'') - \Delta_{k,\mathbf{r}}(\tau'), \quad k = i, j, \tag{4.201}$$

and

$$\Delta_{k,t}(\tau'') - \Delta_{k,t}(\tau'), \quad k = i, j, \tag{4.202}$$

may be expressed by

$$\Delta_{k,\mathbf{r}}(\tau'') - \Delta_{k,\mathbf{r}}(\tau') = v_{k\mathbf{r}}(\tau'' - \tau')$$
$$+ b_{k\mathbf{r}}(\tau'' - \tau')^2 + \cdots, \quad k = i, j, \tag{4.203}$$
$$\Delta_{k,t}(\tau'') - \Delta_{k,t}(\tau') = v_{kt}(\tau'' - \tau')$$
$$+ b_{kt}(\tau'' - \tau')^2 + \cdots, \quad k = i, j, \tag{4.204}$$

where $v_{k\mathbf{r}}$ and v_{kt} are again effective velocities and $b_{k\mathbf{r}}$ and b_{kt} are effective accelerations. With $\tau'' - \tau' \to 0$ and arbitrary values for $\Delta_{k,\mathbf{r}}(\tau'') - \Delta_{k,\mathbf{r}}(\tau')$ and $\Delta_{k,t}(\tau'') - \Delta_{k,t}(\tau')$, the values for $v_{k\mathbf{r}}$ and v_{kt} and for $b_{k\mathbf{r}}$ and b_{kt} may take arbitrary values but must fulfil the relations (4.203) and (4.204). In particular, they may be close to infinity[102]:

$$v_{k\mathbf{r}}, v_{kt}, b_{k\mathbf{r}}, b_{kt} \to \infty, \quad k = 1, 2. \tag{4.205}$$

In conclusion, the arbitrary jumps of the geometrical structures contain effective velocities $v_{k\mathbf{r}}$ and v_{kt} and effective accelerations $b_{k\mathbf{r}}$ and b_t and of course other types of motion when we consider more than two terms in the relations (4.203) and (4.204). As we have mentioned several times, these motions have no influence on the $\mathbf{p}, E$-states of both systems i and j.[103]

Remark

We know from Special Theory of Relativity that the velocity of a body having the rest mass of m_0 cannot exceed the velocity of light. However, the property $v_{m\mathbf{r}} \to \infty$, $m = 1, 2$,[104] seems to be in contradiction to this relevant statement of Special Theory of Relativity. But this is not the

---

[102] Here we have assumed that only the first two terms in equations (4.203) and (4.204) are different from zero.
[103] See, in particular, Sections 4.8.3 and 4.8.4.
[104] See equation (4.197).

case. The reason is obvious. The observer, resting in system i, cannot say anything about the space-time positions of system j and vice versa. Therefore, a relative motion of both systems is not definable. This is reason why the laws of Special Theory of Relativity cannot be applied here. In this connection, it should also be mentioned that the effective velocity v_{mt} is not defined in Special Theory of Relativity because the system-specific time t is not defined here. In Section 4.12, we will discuss possible relativistic effects within projection theory.

4.8.7. Space-Effects

The arbitrary jumps within $(\mathbf{r}, t)$-space, which we have analyzed in the last sections, reflect a fundamental effect. In particular, the role of the "absolute observer" who is resting relative to $(\mathbf{r}, t)$-space and, on the other hand, that of the observer, who is resting relative to arbitrarily jumping system, are of basic relevance. In this section, we would like to discuss typical space-effects in conventional physics as well as within projection theory. In this connection, Mach's principle inevitably comes into play.

Conventional physics

Within conventional physics, reality is embedded in space, i.e., in $(\mathbf{r}, \tau)$-space. Here the notion "space" is often considered as an element with absolute properties. However, in accordance to Ernst Mach such an absolute space can be no basis for a physical theory.[105] There is no instruction defined for the measurement of the coordinates x, y, z and also not for the time τ.[106] This is an important principal point because Mach's principle could not be realized in conventional physics. For example, Newton's theory is based on an absolute space and on an absolute time too. Therefore, Mach rejected radically Newton's theory. After Mach, the motion of an object with mass m_0 may not be influenced by space but exclusively by the other bodies around it. This requirement is not fulfilled within Newton's theory, but also not in

---

[105] Known as Mach's principle.
[106] The variable t is not defined here.

the Theory of Relativity. This is a serious problem, and we will discuss this point in more detail below.

Projection theory

Within projection theory Mach's principle is definitely fulfilled. Because space and time, i.e., the element of $(\mathbf{r}, \tau)$-space, have to be considered as auxiliary elements for the representation of geometrical structures,[107] $(\mathbf{r}, t)$-space can principally not have physically real effects on other systems and, on the other hand, it cannot be influenced by certain physical conditions. However, due to the interaction processes, i.e., due to the $\mathbf{p}, E$-fluctuations in $(\mathbf{p}, E)$-space, there could nevertheless be a relationship between the geometrical structures of the projected information and the absolute elements of space and time, that is, there would be a relationship between the coordinates x, y, z and the time t of $(\mathbf{r}, t)$-space and the geometrical structure. Due to the basic features of space-time, outlined in Chapter 2, such an effect has to be excluded. Nevertheless, such an effect would not be in conflict with Mach's principle because space and time would remain auxiliary elements without being able to act on things and, in turn, to react on them. Our analysis in connection with the probability densities in $(\mathbf{r}, t)$-space as well as those for $(\mathbf{p}, E)$-space[108] showed however that there is no possibility for an observer to make statements about the absolute coordinates x, y, z and the absolute time t.

Conclusion

In summary, we can fulfil Mach's principle when we apply the projection principle. This is obviously not possible within conventional physics, and the reason for this drawback is probably due to the basic conception of conventional physics: Reality is embedded in space; space plays the role of a container here. This container-picture corresponds to our impressions we have in front of us within the observations in

---

[107] Reality is projected onto $(\mathbf{r}, t)$-space and is not embedded in it.
[108] See Section 4.8.3.

everyday life. As we have outlined,[109] these everyday-life impressions are obviously a fallacy when we analyze the situation in terms of an "absolute reality". The events in everyday life do not take place within a frame that is often called "absolute reality".

Within projection theory there are arbitrary jumps of geometrical structures,[110] i.e., there is no relationship between the elements of $(\mathbf{r}, t)$-space, i.e., coordinates x, y, z and time t and these geometrical structures, which are defined by probability densities. As we have outlined in Section 4.8.5, these arbitrary jumps in $(\mathbf{r}, t)$-space correspond to random phases in $(\mathbf{p}, E)$-space.

The arbitrary jumps that appear in connection with projection theory would also be expected within conventional physics. The real bodies, embedded in space-time, that do not interact with other bodies in the universe, should move arbitrarily through space-time because Mach's principle requires that space-time cannot act on these real bodies and, on the other hand, the real bodies should not be able to act on space-time.

4.8.8. Arbitrary Jumps and p, E-States

Motion relative to nothing

The arbitrary jumps in connection with the quantities $\Delta_{k,\mathbf{r}} = \Delta_{k,\mathbf{r}}(\tau)$ and $\Delta_{k,t} = \Delta_{k,t}(\tau)$,[111] reflect a certain dynamics of the systems i and j relative to that what we have called "absolute space-time", which is identical with $(\mathbf{r}, t)$-space. Just the existence of the effective velocities $v_{m\mathbf{r}}$ and v_{mt}[112] and their possible changes in the course of time τ suggest that the $\mathbf{p}, E$-states of both systems, that are seemingly moving relative to $(\mathbf{r}, t)$-space, are changed. However, this is not the case as we have discussed in connection with the functions $\Phi_F^*(\mathbf{p}, E)_k \Phi_F(\mathbf{p}, E)_k$ and $\Psi_F^*(\mathbf{p}, E)_k \Psi_F(\mathbf{p}, E)_k$ with $k = i, j$.[113] This property is confirmed when we investigate the mean values for the momentum $\bar{p}_k$ and the

---

[109] Refer to Schommers, W., Cosmic secrets: World Scientific, in preparation.
[110] See Section 4.8.3.
[111] See equation (4.192).
[112] See equations (4.197) and (4.198).
[113] See equation (4.187).

mean energy $\bar{E}_k$. Both quantities, that is, $\bar{\mathbf{p}}_k$ and $\bar{E}_k$, are completely independent on the (arbitrary) motion of the systems in $(\mathbf{r}, t)$-space.[114] In fact, we have

$$\bar{\mathbf{p}}_k = \int_{-\infty}^{\infty} \mathbf{p} \Psi_F^*(\mathbf{p}, E)_k \Psi_F(\mathbf{p}, E)_k d\mathbf{p}\, dE$$

$$= \int_{-\infty}^{\infty} \mathbf{p} \Phi_F^*(\mathbf{p}, E)_k \Phi_F(\mathbf{p}, E)_k d\mathbf{p}\, dE, \quad k = i, j, \quad (4.206)$$

and

$$\bar{E}_k = \int_{-\infty}^{\infty} E \Psi_F^*(\mathbf{p}, E)_k \Psi_F(\mathbf{p}, E)_k d\mathbf{p}\, dE$$

$$= \int_{-\infty}^{\infty} E \Phi_F^*(\mathbf{p}, E)_k \Phi_F(\mathbf{p}, E)_k d\mathbf{p}\, dE, \quad k = i, j, \quad (4.207)$$

with the condition

$$\int_{-\infty}^{\infty} \Psi_F^*(\mathbf{p}, E)_k \Psi_F(\mathbf{p}, E)_k d\mathbf{p}\, dE$$

$$= \int_{-\infty}^{\infty} \Phi_F^*(\mathbf{p}, E)_k \Phi_F(\mathbf{p}, E)_k d\mathbf{p}\, dE = 1, \quad k = i, j. \quad (4.208)$$

In other words, the arbitrary jumps have no influence on the real properties of the systems. Then, we may conclude that it is a "motion relative to nothing".

Remarks in connection with Newton's theory

The motion of the geometrical structures in $(\mathbf{r}, t)$-space with certain effective velocities $v_{m\mathbf{r}}$ and v_{mt}[115] does not lead to changes in the $\mathbf{p}, E$-states of the systems i and j. We have characterized this situation by the statement "motion relative to nothing". This is in contrast to the facts of conventional physics. Let us briefly discuss this point in connection with Newton's theory.

The physical picture within Newton's theory is quite different from that of projection theory. In classical mechanics the $\mathbf{p}, E$-states

---

[114]This result also supports the use of the product $\Psi^*\Psi$ for the description of physical processes instead of the function Ψ.
[115]See Section 4.8.6.

are definitely changed when a body moves through space, i.e., relative to space, with varying velocities, and this is explained in Newton's mechanics by space-effect. However, such kind of space-effect has to be considered as unphysical and has been strongly criticized by Ernst Mach and, in particular, also by Albert Einstein. Nevertheless, also within Special Theory of Relativity and General Theory of Relativity, the situation concerning space-effects could not really improve. More details concerning Theory of Relativity will be given in Section 4.11. The situation within Newton's theory can be summarized as follows:

Within Newton's theory, the notion "inertia" is of particular relevance. Here all real bodies are embedded in space. Even when a body does not interact with other bodies, the effect of inertia is effective, i.e., the body moves through space with constant velocity v where the velocity of zero is included, and the effect of inertia is entirely due the interaction of the body with space. The body moves relative to space, and its $\mathbf{p}$, E-state is dependent on its velocity relative to space, i.e., the variation of v leads to changes of the bodies' $\mathbf{p}$, E-states. In particular, we have the following situation.

Let us consider an object having the mass m_0 that is resting relative to space, and we would like to assume that this body does not interact with any other body in the cosmos. Now we want to apply an accelerating force $\mathbf{K}$ on this object during the time interval $\Delta \tau = \tau_2 - \tau_1$, i.e., the effect of the force starts at τ_1 and stops at τ_2. During that time interval the work

$$A = \int_{\tau_1}^{\tau_2} \mathbf{K} \cdot d\mathbf{s} \tag{4.209}$$

is done on the body along the path $\mathbf{s}$, that is, due to the force $\mathbf{K}$ the object will be accelerated from the velocity $\mathbf{v}_1 = \mathbf{v}_1(\tau_1) = 0$ to the final velocity $\mathbf{v}_2 = \mathbf{v}_2(\tau_2) = \mathbf{v}$. While the body has no energy E and no momentum $\mathbf{p}$ up to τ_1, it has, as it is well-known, at time τ_2 the energy

$$E = m_0 \mathbf{v}^2 / 2, \tag{4.210}$$

and the momentum

$$\mathbf{p} = m_0 \mathbf{v}. \tag{4.211}$$

In other words, due to the motion relative to space there is a change of the **p**, E-state of the object. This result is in distinct contrast to that which we have found in connection within projection theory where the motion through (**r**, t)-space has definitely no influence on the **p**, E-states.

4.8.9. Resting and Moving Frames

The quantities $\Delta_{k,\mathbf{r}}$ and $\Delta_{k,t}$[116] can be treated as the variables of a free, i.e., non-interacting, system in (**r**, t)-space, where $\Delta_{k,\mathbf{r}}$ is a space position and $\Delta_{k,t}$ is a position with respect to time. With equation (4.182) we can construct the following identity:

$$\Psi_F(\mathbf{p}, E)_k = \frac{1}{(2\pi\hbar)^2} \int_{-\infty}^{\infty} \{\Psi_F(\mathbf{r} - (\mathbf{r}_k + \Delta_{k,\mathbf{r}}), t - (t_k + \Delta_{k,t}))\}$$

$$\times \exp\left\{i\left[\frac{\mathbf{p}_{k,\mathbf{r}}}{\hbar}\cdot\Delta_{k,\mathbf{r}} - \frac{E_{k,t}}{\hbar}\Delta_{k,t}\right]\right\}$$

$$\times \exp\left\{i\left[\frac{\mathbf{p}-\mathbf{p}_{k,\mathbf{r}}}{\hbar}\cdot\Delta_{k,\mathbf{r}} - \frac{E-E_{k,t}}{\hbar}\Delta_{k,t}\right]\right\}$$

$$\times \exp\left\{-i\left[\frac{\mathbf{p}}{\hbar}\cdot\mathbf{r} - \frac{E}{\hbar}t\right]\right\} d\mathbf{r}\,dt, \quad k = i,j, \quad (4.212)$$

where $\mathbf{p}_{k,\mathbf{r}}$ and $E_{k,t}$, $k = i,j$, are constant values and have to be considered as the momenta and the energies of free systems with the variables $\Delta_{k,\mathbf{r}}$ and $\Delta_{k,t}$, $k = i,j$.

Then, the terms $\exp\{i[\mathbf{p}_{k,\mathbf{r}}\cdot\Delta_{k,\mathbf{r}} - E_{k,t}\Delta_{k,t}]/\hbar\}$, $k = i,j$, which appear in equation (4.212), define wave functions of the form

$$\Psi(\Delta_{k,\mathbf{r}}, \Delta_{k,t}) = \exp\left\{\frac{i}{\hbar}[\mathbf{p}_{k,\mathbf{r}}\cdot\Delta_{k,\mathbf{r}} - E_{k,t}\Delta_{k,t}]\right\}, \quad k = i,j.$$

(4.213)

The corresponding probability densities for the variables $\Delta_{k,\mathbf{r}}$ and $\Delta_{k,t}$ are expressed by

$$\Psi^*(\Delta_{k,\mathbf{r}}, \Delta_{k,t})\Psi(\Delta_{k,\mathbf{r}}, \Delta_{k,t}) = 1, \quad k = i,j.$$

---

[116] See equation (4.181).

This result has to be interpreted as follows. The equations are expressions for an observer resting relative to system i (frame of reference S') and for another observer who is resting relative to system j (frame of reference S''). These systems (frames of reference S' and S'') are moving geometrical structures. In the course of time τ they jump arbitrarily relative to the resting frame of reference S which defines an absolute space-time, which is identical with $(\mathbf{r}, t)$-space, and the quantities $\Delta_{k,\mathbf{r}}(\tau)$ and $\Delta_{k,t}(\tau)$,[117] which characterize the arbitrary jumps, are the space-time positions with respect to the frame of reference S. However, the moving observers, who are resting relative to S' and S'', jump arbitrarily as well, that is, the statistical fluctuations due to $\Delta_{k,\mathbf{r}}(\tau)$ and $\Delta_{k,t}(\tau)$ are eliminated for the observers who are resting relative to S' and S''.

We have called the frame of reference S "absolute space-time". However, here we have to be careful. It is not the sole, absolute space of conventional physics but it is the absolute space-time of projection theory, and this is nothing other than the space-time frame that is positioned within the brain of an observer. Because there are a lot of observers, there are also a lot of these absolute space-time frames.

Equation (4.212) are expressions for observers which are resting relative to the frames S' and S''. But what about an observer who is resting in the frame of reference S? Clearly, here the quantities $\Delta_{k,\mathbf{r}}$ and $\Delta_{k,t}$[118] have to be treated as the variables of a free (non-interacting) system in $(\mathbf{r}, t)$-space. However, in contrast to the moving observers (who are resting relative to S' and S''), the resting observer in S registers the fluctuations with respect to $\Delta_{k,\mathbf{r}}$ and $\Delta_{k,t}$ of both systems. Because $\Delta_{k,\mathbf{r}}$ and $\Delta_{k,t}$ behave like the space-time points of free (non-interaction) systems that we have treated in Chapter 3 and, therefore, we have to replace the terms $\exp\{i[\mathbf{p}_{k,\mathbf{r}} \cdot \Delta_{k,\mathbf{r}} - E_{k,t}\Delta_{k,t}]/\hbar\}$ which appear in equation (4.212) by

$$C\Psi(\mathbf{p}_{k,\mathbf{r}}, E_{k,t}) \exp\left\{\frac{i}{\hbar}[\mathbf{p}_{k,\mathbf{r}} \cdot \Delta_{k,\mathbf{r}} - E_{k,t}\Delta_{k,t}]\right\},$$

---

[117] See equation (4.192).
[118] See equation (4.181).

and, furthermore, we have to replace the functions $\Psi_F(\mathbf{p}, E)_k$ on the left hand side of equation (4.212) by

$$C\Psi(\mathbf{p}_{k,\mathbf{r}}, E_{k,t})\Psi_F(\mathbf{p}, E)_k.$$

Thus, instead of equation (4.213) we get for $\Psi(\Delta_{k,\mathbf{r}}, \Delta_{k,t})$ the expression

$$\Psi(\Delta_{k,\mathbf{r}}, \Delta_{k,t}) = C\Psi(\mathbf{p}_{k,\mathbf{r}}, E_{k,t}) \exp\left\{\frac{i}{\hbar}[\mathbf{p}_{k,\mathbf{r}} \cdot \Delta_{k,\mathbf{r}} - E_{k,t}\Delta_{k,t}]\right\}. \tag{4.214}$$

An equation of this type has been found in Section 3.4 for a free, non-interacting system. Then, equation (4.212) takes the form

$$C\Psi(\mathbf{p}_{k,\mathbf{r}}, E_{k,t})\Psi_F(\mathbf{p}, E)_k$$
$$= \frac{1}{(2\pi\hbar)^2} \int_{-\infty}^{\infty} \{\Psi_F(\mathbf{r} - (\mathbf{r}_k + \Delta_{k,\mathbf{r}}), t - (t_k + \Delta_{k,t}))\}$$
$$\times C\Psi(\mathbf{p}_{k,\mathbf{r}}, E_{k,t}) \exp\left\{i\left[\frac{\mathbf{p}_{k,\mathbf{r}}}{\hbar} \cdot \Delta_{k,\mathbf{r}} - \frac{E_{k,t}}{\hbar}\Delta_{k,t}\right]\right\}$$
$$\times \exp\left\{i\left[\frac{\mathbf{p} - \mathbf{p}_{k,\mathbf{r}}}{\hbar} \cdot \Delta_{k,\mathbf{r}} - \frac{E - E_{k,t}}{\hbar}\Delta_{k,t}\right]\right\}$$
$$\times \exp\left\{-i\left[\frac{\mathbf{p}}{\hbar} \cdot \mathbf{r} - \frac{E}{\hbar}t\right]\right\} d\mathbf{r}\, dt, \quad k = i, j.$$

The function $\Psi(\mathbf{p}_{k,\mathbf{r}}, E_{k,t})$ is the Fourier transform of $\Psi(\Delta_{k,\mathbf{r}}, \Delta_{k,t})$, i.e., we have

$$\Psi(\mathbf{p}_{k,\mathbf{r}}, E_{k,t}) = \frac{1}{(2\pi\hbar)^2} \int_{-\infty}^{\infty} d\Delta_{k,\mathbf{r}} d\Delta_{k,t} \Psi(\Delta_{k,\mathbf{r}}, \Delta_{k,t})$$
$$\times \exp\left\{-i\left[\frac{\mathbf{p}_{k,\mathbf{r}}}{\hbar} \cdot \Delta_{k,\mathbf{r}} - \frac{E_{k,t}}{\hbar}\Delta_{k,t}\right]\right\}, \quad k = i, j.$$

The variables $\mathbf{p}_{k,\mathbf{r}}, E_{k,t}$ are the Fourier variables with respect to $\Delta_{k,\mathbf{r}}$ and $\Delta_{k,t}$; $\mathbf{p}_{k,\mathbf{r}}, E_{k,t}$ belong to $(\mathbf{p}, E)$-space and $\Delta_{k,\mathbf{r}}, \Delta_{k,t}$ to $(\mathbf{r}, t)$-space.

After some simple mathematical manipulations we obtain

$$\Psi(\mathbf{p}'_{k,\mathbf{r}}, E'_{k,t})\Psi_F(\mathbf{p}, E)_k$$
$$= \frac{1}{(2\pi\hbar)^4} \int_{-\infty}^{\infty} \{\Psi_F(\mathbf{r} - (\mathbf{r}_k + \Delta_{k,\mathbf{r}}), t - (t_k + \Delta_{k,t}))\}$$
$$\times C\Psi(\mathbf{p}_{k,\mathbf{r}}, E_{k,t}) \exp\left\{i\left[\frac{\mathbf{p}_{k,\mathbf{r}}}{\hbar} \cdot \Delta_{k,\mathbf{r}} - \frac{E_{k,t}}{\hbar}\Delta_{k,t}\right]\right\}$$
$$\times \Psi(\Delta_{k,\mathbf{r}}, \Delta_{k,t}) \exp\left\{i\left[\frac{\mathbf{p}}{\hbar} \cdot \Delta_{k,\mathbf{r}} - \frac{E}{\hbar}\Delta_{k,t}\right]\right\}$$
$$\times \exp\left\{-i\left[\frac{\mathbf{p}'_{k,\mathbf{r}}}{\hbar} \cdot \Delta_{k,\mathbf{r}} - \frac{E'_{k,t}}{\hbar}\Delta_{k,t}\right]\right\}$$
$$\times \exp\left\{-i\left[\frac{\mathbf{p}}{\hbar} \cdot \mathbf{r} - \frac{E}{\hbar}t\right]\right\} d\Delta_{k,\mathbf{r}} d\Delta_{k,t} d\mathbf{r}\, dt, k = i,j$$

and with (4.214)

$$\Psi(\mathbf{p}'_{k,\mathbf{r}}, E'_{k,t})\Psi_F(\mathbf{p}, E)_k$$
$$= \frac{1}{(2\pi\hbar)^4} \int_{-\infty}^{\infty} \Psi_F(\mathbf{r} - (\mathbf{r}_k + \Delta_{k,\mathbf{r}}), t - (t_k + \Delta_{k,t}))\Psi(\Delta_{k,\mathbf{r}}, \Delta_{k,t})$$
$$\times \exp\left\{-i\left[\frac{\mathbf{p}'_{k,\mathbf{r}}}{\hbar} \cdot \Delta_{k,\mathbf{r}} - \frac{E'_{k,t}}{\hbar}\Delta_{k,t}\right]\right\}$$
$$\times \exp\left\{i\left[\frac{\mathbf{p}}{\hbar} \cdot \Delta_{k,\mathbf{r}} - \frac{E}{\hbar}\Delta_{k,t}\right]\right\}$$
$$\times \exp\left\{-i\left[\frac{\mathbf{p}}{\hbar} \cdot \mathbf{r} - \frac{E}{\hbar}t\right]\right\} d\Delta_{k,\mathbf{r}} d\Delta_{k,t} d\mathbf{r}\, dt, k = i,j$$

With the Fourier representation introduced above for $\Psi(\mathbf{p}'_{k,\mathbf{r}}, E'_{k,t})$ and the use of equation (4.214) we get

$$\Psi(\mathbf{p}'_{k,\mathbf{r}}, E'_{k,t}) = \frac{C}{(2\pi\hbar)^2} \int_{-\infty}^{\infty} d\Delta_{k,\mathbf{r}} d\Delta_{k,t} \Psi(\mathbf{p}_{k,\mathbf{r}}, E_{k,t})$$
$$\times \exp\left\{\frac{i}{\hbar}[(\mathbf{p}_{k,\mathbf{r}} - \mathbf{p}'_{k,\mathbf{r}}) \cdot \Delta_{k,\mathbf{r}} - (E_{k,t} - E'_{k,t})\Delta_{k,t}]\right\},$$
$$k = i,j.$$

The evaluation of this type of equation has been done in Chapter 3[119] and leads to

$$C = \frac{\hbar^2}{(2\pi)^2[\delta(0)]^4},$$
$$C = 0.$$

In Chapter 3 we came to the conclusion that the wave functions for a free (non-interacting) system must be zero. If we allow that the position $\Delta_{k,\mathbf{r}}$ and the time $\Delta_{k,t}$ take all possible values, i.e., $-\infty \leq \Delta_{k,\mathbf{r}}, \Delta_{k,t} \leq \infty$, we have just the situation which we have treated and discussed in Chapter 3. Thus, the functions $\Psi(\Delta_{k,\mathbf{r}}, \Delta_{k,t})$ and $\Psi(\mathbf{p}'_{k,\mathbf{r}}, E'_{k,t})$ must be zero if we allow that the position $\Delta_{k,\mathbf{r}}$ and the time $\Delta_{k,t}$ take all possible values, i.e., $-\infty \leq \Delta_{k,\mathbf{r}}, \Delta_{k,t} \leq \infty$.[120] The values for $\mathbf{p}'_{k,\mathbf{r}}$ and $E'_{k,t}$ must be zero too. In summary, we have

$$\Psi(\Delta_{k,\mathbf{r}}, \Delta_{k,t}) = 0,$$
$$\Psi(\mathbf{p}'_{k,\mathbf{r}}, E'_{k,t}) = 0,$$
$$\mathbf{p}'_{k,\mathbf{r}} = 0, E'_{k,t} = 0,$$

with $k = i, j$. More details concerning the treatment of free systems are given in Chapter 3.

In conclusion, an observer, who is resting in the frame S, is not able to observe the systems i and j. The situation is illustrated in Figure 4.2 and Figure 4.3. Only the observers in the moving frames S' and S'' can give experimental statements about the systems i and j.

4.8.10. *Arbitrary Jumps within Single Systems*

The entire geometrical structures (systems) i and j jump arbitrarily relative to $(\mathbf{r}, t)$-space, and the effective velocities $v_{m\mathbf{r}}$ and v_{mt} for these arbitrary processes may be close to infinity: $v_{m\mathbf{r}} \to \infty, v_{mt} \to \infty$,

---

[119] See Section 3.4.
[120] See also Figures 4.2 and 4.3.

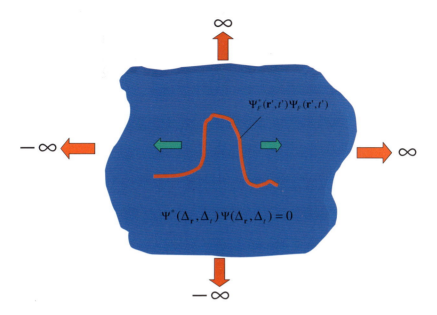

Fig. 4.2. The geometrical structure $\Psi_F^*(\mathbf{r}',t')\Psi_F(\mathbf{r}',t')$ moves arbitrarily relative to $(\mathbf{r},t)$-space and can take space-time positions in the range of $-\infty \leq \mathbf{r}, t \leq \infty$. The observer who is resting relative to the geometrical structure $\Psi_F^*(\mathbf{r}',t')\Psi_F(\mathbf{r}',t')$ can detect this structure. On the other hand, an observer who is resting relative to $(\mathbf{r},t)$-space is not able to observe this system because we have $\Psi^*(\Delta_{\mathbf{r}}, \Delta_t)\Psi(\Delta_{\mathbf{r}}, \Delta_t) = 0$. These statements are equally valid for both systems, system i and system j.

$m = 1, 2$.[121] Such an effect not only appears in connection with arbitrary jumps of the *whole* structures i and j but also *within* each of these structures, i.e., within the $\mathbf{r}, t$-range where the probability density of a single system (i and/or j) is different from zero: $\Psi_F^*(\mathbf{r}, t)_i \Psi_F(\mathbf{r}, t)_i \neq 0$ and/or $\Psi_F^*(\mathbf{r}, t)_j \Psi_F(\mathbf{r}, t)_j \neq 0$. In the following, we will analyze this point in more detail, also in connection with usual quantum theory.

Situation within projection theory

Let us consider an observer who is resting within the frame S' (or S'') and, furthermore, let us put the origin of the coordinate system at the

---

[121] See equations (4.199) and (4.200).

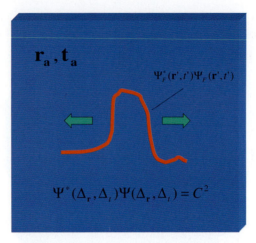

Fig. 4.3. The geometrical structure $\Psi_F^*(\mathbf{r}',t')\Psi_F(\mathbf{r}',t')$ moves arbitrarily relative to $(\mathbf{r},t)$-space and, due to a space-time limiting interaction, $\Psi_F^*(\mathbf{r}',t')\Psi_F(\mathbf{r}',t')$ can only take space-time positions in the range of $(\pm \mathbf{r}_a, \pm t_a)$. The observer who is resting relative to the geometrical structure $\Psi_F^*(\mathbf{r}',t')\Psi_F(\mathbf{r}',t')$ can detect this structure. Due to the space-time limiting interaction $(\pm \mathbf{r}_a, \pm t_a)$, an observer who is resting relative to $(\mathbf{r},t)$-space is able to observe this system because we have $\Psi^*(\Delta_\mathbf{r}, \Delta_t)\Psi(\Delta_\mathbf{r}, \Delta_t) = C^2$.[122] These statements are equally valid for both systems, system i and system j.

space-time positions $\mathbf{r}_k + \Delta_{k,\mathbf{r}}, t_k + \Delta_{k,t}$ ($k = i$ or j). Then, the observer in S' makes observations in connection with $\Psi_F^*(\mathbf{r},t)_i \Psi_F(\mathbf{r},t)_i$, and he can give the following statements:

1. *Measurement at time τ_a*:

The observer measures within a small space-time interval around $\mathbf{r}_{ia}, t_{ia}$ one of the possible values for $\mathbf{p}$ and for E, where the corresponding probability density must be different from zero: $\Psi_F^*(\mathbf{r}_{ia}, t_{ia})_i \Psi_F(\mathbf{r}_{ia}, t_{ia})_i \neq 0$.

2. *Measurement at time $\tau_b > \tau_a$*:

The observer also measures within a small space-time interval around $\mathbf{r}_{ib}, t_{ib}$ one of the possible values for $\mathbf{p}$ and for E, where the corresponding probability density must be different from zero: $\Psi_F^*(\mathbf{r}_{ib}, t_{ib})_i \Psi_F(\mathbf{r}_{ib}, t_{ib})_i \neq 0$.

---

[122] See Section 4.4.1, equation (4.8).

Clearly, in both cases the values for **p** and for E are given with a certain probability that is determined by the probability density $\Psi_F^*(\mathbf{p}, E)_i \Psi_F(\mathbf{p}, E)_i$.

Then, the information transfer between the space-time positions $\mathbf{r}_{ia}, t_{ia}$ and $\mathbf{r}_{ib}, t_{ib}$ seems to take place with the effective velocities $v_{i\mathbf{r}} = (\mathbf{r}_{ib} - \mathbf{r}_{ia})/(\tau_b - \tau_a)$ and $v_{it} = (t_{ib} - t_{ia})/(\tau_b - \tau_a)$.[123,124]

Since $\tau_b - \tau_a$ can be close to zero and we get $\tau_b - \tau_a \to 0$ when τ_b is the next time-value after τ_a. On the other hand, if the function $\Psi_F^*(\mathbf{r}, t)_i \Psi_F(\mathbf{r}, t)_i$ is different from zero, i.e., $\Psi_F^*(\mathbf{r}, t)_i \Psi_F(\mathbf{r}, t)_i \neq 0$, in the range $\Delta \mathbf{r}, \Delta t$ and if $\mathbf{r}_{ia}, t_{ia}$ and $\mathbf{r}_{ib}, t_{ib}$ are two space-time positions within the intervals $\Delta \mathbf{r}$ and Δt, with $\Delta \mathbf{r} \neq 0$ and $\Delta t \neq 0$, we obtain for the effective velocities $v_{i\mathbf{r}}$ and v_{it} *within* a geometrical structure the following properties:

$$v_{i\mathbf{r}} = (\mathbf{r}_{ib} - \mathbf{r}_{ia})/(\tau_b - \tau_a) \to \infty, \qquad (4.215)$$

$$v_{it} = (t_{ib} - t_{ia})/(\tau_b - \tau_a) \to \infty. \qquad (4.216)$$

While $\tau_b - \tau_a$ can be close to zero, i.e., $\tau_b - \tau_a \to 0$, the quantities $\mathbf{r}_{ib} - \mathbf{r}_{ia}$ and $t_{ib} - t_{ia}$ are in general not close to zero.[125] This situation can lead to the effect that the velocities $v_{i\mathbf{r}}$ and v_{it} may become infinity, as is expressed by equations (4.215) and (4.216).

In conclusion, both velocities can be close to infinity even when the intervals $\mathbf{r}_{ib} - \mathbf{r}_{ia}$ and $t_{ib} - t_{ia}$ are small as, for example, 10^{-8} cm and 10^{-10} s (or even smaller). However, the time interval $\tau_b - \tau_a$ must be close to zero, i.e., $\tau_b - \tau_a \to 0$, and this is the case when τ_b is the next time-value after τ_a. As we have seen in connection with the space-limiting interaction expressed by $(\pm \mathbf{r}_a, \pm t_a)$,[126] a wave function may be macroscopic in character.[127] In such a case it is admitted to measure at time τ_a at the space-time position $\mathbf{r}_a, t_a$ and to measure at time τ_b at the opposite space-time position $-\mathbf{r}_a, -t_a$ where the parameters $|\mathbf{r}_a|, |t_a|$

---

[123] See also equations (4.197) and (4.198).
[124] There is of course no systematic information transfer between the two events $\mathbf{r}_{ia}, t_{ia}$ and $\mathbf{r}_{ib}, t_{ib}$ because we deal here with statistical processes and the event $\mathbf{r}_{ia}, t_{ia}$ at time τ_a is completely independent of the event $\mathbf{r}_{ib}, t_{ib}$ at time τ_b.
[125] For example, $\Delta \mathbf{r}$ can be of the order of a typical atomic diameter.
[126] See Section 4.4.1.
[127] See equation (4.8).

may be arbitrarily large but must be different from infinity. Also τ_b can be the next time-value after τ_a.

Clearly, all the features that we have developed in connection with observer S' must also be valid for the observer in S'', i.e., he must come to exactly the same conclusions in connection with $\Psi_F^*(\mathbf{r},t)_j \Psi_F(\mathbf{r},t)_j$.

Situation within usual quantum theory

As we have already outlined in Chapter 1, within usual quantum theory we have a particle that is embedded in space but it moves through space without trajectory, i.e., the classical concept of velocity cannot be used here and the particle motion must be accidental leading to the probability interpretation for the wave function $\psi_n(\mathbf{r})$ where n denotes a certain energy state. In other words, there is no longer a physical law that tells us *where* a particle jumps. If the particle is at time τ_A at position $\mathbf{r}_A$ we cannot say something about the position of the same particle at time τ_B. Within classical mechanics motion is a

> *"continuous blend of changing positions. The object moves in a flow from one point to another. Quantum mechanics failed to reinforce that picture. In fact, it indicated that motion could not take place in that way. They jumped from one place to another, seemingly without effort and without bothering to go between two places"*.[128]

This statement by Fred Alan Wolf suggests that within usual quantum theory the particle moves without inertia. Let us briefly investigate this point.

This argument, i.e., that within usual quantum theory the particle moves without inertia, becomes more accessible when we consider the particle-problem in terms of typical observations in connection with the probability density $\psi_n^*(\mathbf{r})\psi_n(\mathbf{r})$. Let $\psi_n^*(\mathbf{r})\psi_n(\mathbf{r}) \neq 0$ be in the range $\Delta \mathbf{r}$ and let A and B be two positions within $\Delta \mathbf{r}$; at time τ_A we register the particle with a detector at position A, while at time τ_B we register the same particle at position B. Because there is no

---

[128] Refer to Wolf, F.A., (1981) Taking the quantum leap: Harper and Row.

physical law that tells us *when* the real particle with mass m_0 arrives at position B after it has left position A, the time interval $\Delta \tau = \tau_B - \tau_A$ can take any value. For example, it may be infinitesimal but different from zero. If, for example, τ_B is the next time value after τ_A, we have $\Delta \tau = \tau_B - \tau_A \to 0$. On the other hand, if $\Delta \mathbf{r}$ takes a value of the order of a typical atomic diameter, the seeming velocity $v = \Delta \mathbf{r}/\Delta \tau$ can take any value ($v \to \infty$). In particular, it may be larger than the velocity of light c. The property $v \to \infty$ suggests that it is a motion without inertia.

Conclusion

In summary, also within usual quantum theory a motion without inertia can take place within the volume where $\psi_n^*(\mathbf{r})\psi_n(\mathbf{r})$ is different from zero. However, the mass m_0 is nevertheless defined, and this is because $\psi_n(\mathbf{r})$ and $\psi_n^*(\mathbf{r})\psi_n(\mathbf{r})$, respectively, is based on Schrödinger's equation of usual quantum theory where the mass $m_0 \neq 0$ appears as a parameter. The "motion without inertia in connection with a non-vanishing mass m_0" has to be considered as a contradiction within usual quantum theory because within this approach the mass m_0 reflects the effect of inertia and nothing else.

The situation is different in projection theory. Here the mass m_0 comes into play through the basic equations as, for example, through equation (4.218), which exclusively determines $\Psi_F(\mathbf{r}, t)_k$, i.e., the form $\Psi_F^*(\mathbf{r}, t)_k \Psi_F(\mathbf{r}, t)_k$ of a system. The form is dependent on the potential $V_F(\mathbf{r}, t)_k$ and the mass m_0. In other words, here the mass m_0 plays the role of a form-determining parameter and has nothing to do with the notion "inertia". This is consistent with the fact that the geometrical structures[129]

$$\Psi_F^*(\mathbf{r} - (\mathbf{r}_k + \Delta_{k,\mathbf{r}}), t - (t_k + \Delta_{k,t}))$$
$$\times \Psi_F(\mathbf{r} - (\mathbf{r}_k + \Delta_{k,\mathbf{r}}), t - (t_k + \Delta_{k,t})), \quad k = i, j,$$

jump arbitrarily within $(\mathbf{r}, t)$-space, that is, without any inertia. This picture cannot be used in usual quantum theory because a description

---

[129] See equation (4.189).

of the form is not defined here and, therefore, "motion without inertia in connection with a non-vanishing mass m_0" has to be interpreted as a contradiction. Within usual quantum theory, the real particle with mass m_0 is embedded in space and it seems to be problematic when the accidental motion within the volume, where $\psi_n^*(\mathbf{r})\psi_n(\mathbf{r})$ is different from zero, can lead to velocities that are larger than the velocity of light c, or even close to infinity.

4.9. The Meaning of the Potential Functions

Again, both systems i and j jump arbitrarily in $(\mathbf{r}, t)$-space and their space-time distance at a certain time τ may be as large as the space-extension of the universe, where large really means with respect to the maximum space-extension as well as with respect to the maximum time-extension, that is, from the beginning to the end of time. Nevertheless, both systems i and j interact with a constant strength even when the space-time distance takes the largest possible value. In other words, both systems interact, but this interaction is independent on the actual space-time positions of both systems.

This property reflects the non-local character of projection theory. Because the interaction processes exclusively take place in $(\mathbf{p}, E)$-space and the structures in $(\mathbf{r}, t)$-space are projections from $(\mathbf{p}, E)$-space onto space-time, a position-independent interaction becomes possible in a quite natural way. Note that within $(\mathbf{r}, t)$-space no signals are exchanged between the two systems; within $(\mathbf{r}, t)$-space there are "only" geometrical structures, i.e., $\mathbf{r}, t$-correlations, and nothing else.

In this section, we will formally define a potential function that describes these geometrical structures (form of the systems) in $(\mathbf{r}, t)$-space, which reflect $\mathbf{r}, t$-correlations and correspond to $\mathbf{p}, E$-fluctuations in $(\mathbf{p}, E)$-space, i.e., $\mathbf{r}, t$-correlations and $\mathbf{p}, E$-fluctuations take place simultaneously. After that, we will discuss the physical meaning of the notion "interaction" within projection theory but also within conventional physics where the original thoughts by Newton will be of particular interest. The result can be anticipated as follows. Within projection theory, the interaction potential can be interpreted at best as an auxiliary element without any imaginable background.[130] On the other hand, we will recognize that within conventional physics the notion "interaction" takes in its original form a conception with an occult quality.[131]

4.9.1. Introduction of a Potential Function in the Case of Distance-Independent Interactions (Form Interactions)

What does distance-independent interaction mean? As we have outlined above, this interaction exclusively describes the *form* of a system, and it is characterized in $(\mathbf{r}, t)$-space in the case of system i and system j by the wave function $\Psi_F(\mathbf{r} - (\mathbf{r}_k + \Delta_{k,\mathbf{r}}), t - (t_k + \Delta_{k,t}))$,[132] and in $(\mathbf{p}, E)$-space by $\Psi_F(\mathbf{p}, E)_k$ with $k = i, j$.

Since the form of the wave function in $(\mathbf{r}, t)$-space is independent on the position in space-time, we may put the origin of the coordinate system at the space-time positions $\mathbf{r}_k + \Delta_{k,\mathbf{r}}, t_k + \Delta_{k,t}$ ($k = i$ or j) and, instead of $\Psi_F(\mathbf{r} - (\mathbf{r}_k + \Delta_{k,\mathbf{r}}), t - (t_k + \Delta_{k,t}))$ we have the function $\Psi_F(\mathbf{r}, t)_k$:

$$\Psi_F(\mathbf{r} - (\mathbf{r}_k + \Delta_{k,\mathbf{r}}), t - (t_k + \Delta_{k,t})) \to \Psi_F(\mathbf{r}, t)_k. \quad (4.217)$$

In accordance with equations (4.190) and (4.191), the wave function in $(\mathbf{p}, E)$-space is not changed by transition (4.217) and is given by $\Psi_F(\mathbf{p}, E)_k$. As we have outlined in Section 2.3.7, the wave functions $\Psi_F(\mathbf{r}, t)_k$ may be expressed in terms of an interaction potential, say $V_F(x, y, z, t)_k$ with $k = i, j$, which must be independent on the space-time distance of both systems i and j. Then, it is straightforward to show that the equation for $(\mathbf{r}, t)$-space is expressed by[133]

$$i\hbar \frac{\partial}{\partial t} \Psi_F(\mathbf{r}, t)_k = -\frac{\hbar^2}{2m_0} \Delta \Psi_F(\mathbf{r}, t)_k$$
$$+ V_F(x, y, z, t)_k \Psi_F(\mathbf{r}, t)_k, \quad k = i, j. \quad (4.218)$$

---

[130] There is no information or energy transfer through $(\mathbf{r}, t)$-space.
[131] Interested readers might refer to Kuhn, T.S., (1970) The structure of Scientific revolution: University of Chicago Press.
[132] See equation (4.182).
[133] See Section 2.3.7.

The corresponding equation for $\Psi_F(\mathbf{p}, E)_k$ in $(\mathbf{p}, E)$-space has the form[133]

$$E\Psi_F(\mathbf{p}, E)_k = \frac{\mathbf{p}^2}{2m_0} \Psi_F(\mathbf{p}, E)_k + V_F\left(i\hbar\frac{\partial}{\partial p_x}, i\hbar\frac{\partial}{\partial p_y}, i\hbar\frac{\partial}{\partial p_z}, -i\hbar\frac{\partial}{\partial E}\right)_k$$
$$\times \Psi_F(\mathbf{p}, E)_k, \quad k = i, j. \tag{4.219}$$

Again, the potential $V_F(\mathbf{r}, t)_k$ describes the form interaction and is, at any time τ, independent of the actual space-time distances $\mathbf{r}_{12} = \mathbf{r}_1 - \mathbf{r}_2$ and $t_{12} = t_1 - t_2$ of both systems. In principle, both systems could additionally interact by a distance-dependent potential but this kind of interaction is assumed to be not effective in connection with our specific study given in this section.

It is important to note that within conventional physics, only distance-dependent interactions are known, that is, form interactions are not defined here. This is a very principal point, in particular with respect to the notion "interaction". The source for the notion "interaction" is an invention of classical physics and goes back to Newton. Let us briefly discuss once more the main facts in connection with distance-dependent interactions as they are defined and used in conventional physics.

4.9.2. Interaction within Conventional Physics

Let us consider two systems, say i and j, that are embedded as real bodies in space. The variable for the system-specific time t is not defined in conventional physics but only the reference time τ. Also the real interactions processes exclusively take place in space, and we would like to assume that these two systems interact with each other, but none of them interact with other bodies.

The effect of an interaction between the systems, having at time τ the sharp space positions $\mathbf{r}_i$ and $\mathbf{r}_j$, is that the space positions are changed.[134]

---

[134] In contrast to projection theory, the form (shape) of the two systems i and j have to be assumed within conventional physics and has to be considered as a pre-requisite here, and there is no conception to deduce the form of systems in conventional physics.

That is, within conventional physics, i.e., Newton's mechanics, there is the following mechanism: The space positions $\mathbf{r}_i$ and $\mathbf{r}_j$ are changed by the interaction process:

$$\mathbf{r}_i, \mathbf{r}_j \to \mathbf{r}_i + \Delta\mathbf{r}_i, \mathbf{r}_j + \Delta\mathbf{r}_j. \tag{4.220}$$

In the case of point-like systems, the bodies take the forms $\delta(\mathbf{r} - \mathbf{r}_i)$ (system i) and $\delta(\mathbf{r} - \mathbf{r}_j)$ (system j) and we have after the interaction[135]

$$\textit{system } i : \delta(\mathbf{r} - \{\mathbf{r}_i + \Delta\mathbf{r}_i\}), \tag{4.221}$$

$$\textit{system } j : \delta(\mathbf{r} - \{\mathbf{r}_j + \Delta\mathbf{r}_j\}). \tag{4.222}$$

It is normally assumed within usual physics that the interaction strength decreases with increasing distance $\mathbf{r}_i - \mathbf{r}_j$ between the systems, and this is intuitively understandable because it is assumed in usual physics that the interaction processes take place in space and, therefore, the assumption that with decreasing potential function the influence of system i on system j and vice versa decreases becomes intuitively understandable, but we have to be careful. Within projection theory there cannot be such kind space-time connections since the interaction processes do not take place in $(\mathbf{r}, t)$-space. As we have outlined above, within projection theory we have "merely" $\mathbf{r}, t$-correlations in $(\mathbf{r}, t)$-space, and the real interaction processes are identified with $\mathbf{p}, E$-fluctuations in $(\mathbf{p}, E)$-space.

4.9.3. Interaction Potentials are Auxiliary Elements

In the case of form (distance-independent) interactions such a "classical" potential picture is not applicable. Instead of one potential for both systems (distance-dependent interaction) here we have two potentials $V_F(x, y, z, t)_k$, $k = i, j$[136] one for system i and another for system j. Thus, the usual interpretation in terms of more or less classical notions can no longer be applied in projection theory. In fact, the potential

---

[135] See equation (4.220).
[136] See equation (4.218).

takes in $(\mathbf{p}, E)$-space the form of an operator[137]

$$V_F\left(i\hbar\frac{\partial}{\partial p_x}, i\hbar\frac{\partial}{\partial p_y}, i\hbar\frac{\partial}{\partial p_z}, -i\hbar\frac{\partial}{\partial E}\right),$$

and this expression is not accessible to visualizable pictures. Therefore, interaction potentials should be considered as abstract quantities for the determination of the $\mathbf{p}, E$-fluctuation in $(\mathbf{p}, E)$-space which appear in $(\mathbf{r}, t)$-space as correlations between the various space-time positions $\mathbf{r}, t$, i.e., $\mathbf{r}, t$-correlations.

In other words, the potential is not the primary quantity but the $\mathbf{p}, E$-fluctuations and $\mathbf{r}, t$-correlations, respectively. In $(\mathbf{r}, t)$-space, we have exclusively geometrical structures, and no real objects are embedded here. Therefore, it is wrong to assume that there is an exchange of energy $V_F(x, y, z, t)_k$, $k = i, j$ between the (geometrical) structures i and j.

Conclusion: Within projection theory the interaction potential can be at best interpreted as an auxiliary element without any imaginable background.[138] But also within conventional physics, the notion "interaction" becomes questionable when we analyze this term in somewhat more detail.

4.9.4. Conventional Physics: What Mechanism is Behind Interaction?

Again, within conventional physics the real bodies are embedded in space and there is an information (energy) transfer through space. What does "information (energy) transfer through space" mean? How comes the mutual influence between two bodies as, for example, between the earth and the sun, into existence? What mechanism can explain this mutual influence? We have already discussed this point in Section 4.3. In this connection two "pictures" turned out to be of particular relevance: the "proximity effect" and the "action-at-the-distance". In Section 4.3 and we noted the following:

---

[137] See equation (4.219).
[138] There is no information or energy transfer through $(\mathbf{r}, t)$-space.

It is of principal interest to note that the "proximity effect" and the "action-at-the-distance" are merely "expressions" or interpretations of the gravitational law to $m_1 m_2/r^2$ and Coulomb's law qQ/r^2. These force laws cannot, however, be derived from these notions. Many people believe that a mechanism, which is composed of many familiar single processes (preferably from everyday life), can explain the mathematical structure of the force laws. What mechanism is, for example, responsible for the fact that the forces expressed by $m_1 m_2/r^2$ are inversely proportional to the square of the distance between the masses m_1 and m_2. As already mentioned, the ideas "proximity effect" and "action-at-the-distance" cannot give the answer to this question, since they interpret the relation $m_1 m_2/r^2$, but are not able to explain the mathematical structure of this force law.

In summary, there is no possibility to explain by a mechanism how the mutual influence between two bodies as, for example, between the earth and the sun, comes into existence. The notion "interaction" has therefore to be considered as an irreducible primary property of matter.

4.9.5. "Gravity ... an Occult Quality"

After Thomas Kuhn, the fact that there is no possibility to explain the interaction by a mechanism has to be considered as problematic because Newton's gravity becomes after Thomas Kuhn, an occult quality. He wrote:

Yet, though Newton's work was directed to problems and embodied standards derived from the mechanical-corpuscular world view, the effect of the paradigm that resulted from his work was a further and partially destructive change in the problems and standards legitimate for science. Gravity, interpreted as an innate attraction between every pair of particles of matter, was an occult quality in the same sense as the scholastics' "tendencies to fall" had been. Therefore, while the standards of corpuscularism remained in effect, the research for a mechanical explanation of gravity was one of the most challenging problems for those who accepted the

Principia as paradigm. Newton devoted much attention to it and so did many of his eighteenth-century successors. The only apparent opinion was to reject Newton's theory for its failure to explain gravity, and that alternative, too, was widely adopted. Yet neither these views ultimately triumphed. Unable either to practice science without the principia or to make that work conform to the corpuscular standards of the seventeenth century, scientists gradually accepted the view that gravity was indeed innate. By the mid-eighteenth century that interpretation had been almost universally accepted, and the result was a genuine revision (which is not the same as a retrogression) to a scholastic standard. Innate attractions and repulsions joined size, shape, position, and motion as physically irreducible primary properties of matter.[139]

4.9.6. Phenomena in Usual Quantum Theory

We can go a step further and ask for the role of the potential energy $U(r)$ in usual quantum theory. Newton's mechanics is a classical theory but its interaction concept is also used in usual quantum theory. Schrödinger's equation completely contains the notion potential energy $U(r)$ in the classical sense even when electric fields are used here[140] for the description of atoms etc.

It may be doubted whether the classical concept of potential energy can be used in connection with quantum phenomena.[141] Schrödinger's equation could not be derived, so that an interpretation of its elements, like for example $U(r)$, is, strictly speaking, not immediately possible. Thus, the whole concept is dependent on the practical success. In fact, Schrödinger's equation has been tested unusually well, but the classical explanation of $U(r)$ can be considered also here as problematic. On this point we find the following remark in a book by W. Finkelnburg:

---

[139] Refer to Kuhn, T.S., (1970) The structure of scientific revolution: University of Chicago Press.
[140] Not gravitational fields.
[141] In fact, projection theory clearly shows that it should not be possible.

"... While Schrödinger arrived at wave mechanics via de Broglie's matter waves, Heisenberg recognized that the difficulties concerning Bohr's theory were based in particular on the unscrupulous application of such ideas to atomic problems, which — as he realized — were impossible to test experimentally. Heisenberg therefore radically rejected to introduce terms and conceptions into his quantum mechanics of atoms which were not verifiable experimentally. For example, in the alternative theory of Schrödinger the introduction of potential energy is still necessary, and this is characterized by the idea of point-like nuclei and point-like electrons (Coulomb's law) ...".[142]

Quantum phenomena are very different from those in classical physics. Therefore, it must appear as problematic to use classical concepts here as, for example, the classical concept of potential energy in connection with the potential function $U(r)$. Within projection theory there cannot exist potential energies in $(\mathbf{r}, t)$-space because there are no real bodies embedded in $(\mathbf{r}, t)$-space. Furthermore, in contrast to usual quantum theory, the classical function $U(r)$ gets a quantum-theoretical aspect within projection theory, and this is because the system-specific time t has to be considered. $U(r)$ is automatically extended by the variable t and we have $V(r, t)$.[143] In other words, when we go from the classical theory to the quantum-theoretical description, we have within projection theory the transition $U(r) \to V(r, t)$.

4.9.7. Summary

The notion "interaction in space" becomes problematic in conventional physics when we analyze the situation in terms of Newton's original conception but also within usual quantum theory. After Thomas Kuhn gravity has to be considered as an occult quality. Within projection theory the notion "interaction" is a clear conception. In $(\mathbf{r}, t)$-space, there are no interactions possible and defined, respectively.

---

[142] Refer to Finkelnburg, W., (1951) Einführung in die theoretische physic: Walfer de Gruyted.
[143] See, in particular, Section 2.3.7.

Here we merely have certain $\mathbf{r}, t$-correlations[144] and there can be no information or energy transfer thorough $(\mathbf{r}, t)$-space. Within projection theory, the interaction between individual systems exclusively takes place in $(\mathbf{p}, E)$-space in the form of $\mathbf{p}, E$-fluctuations.

The $\mathbf{p}, E$-fluctuations and $\mathbf{r}, t$-correlations are the primary effects and not the interaction potentials which play at best the role of auxiliary elements for the description of $\mathbf{p}, E$-fluctuations and $\mathbf{r}, t$-correlations and, therefore, the interaction potentials have to be considered as secondary in character.

4.10. Further Basic Features

In Section 4.2.2, we have considered for the two systems i and j sharp space-time positions $\mathbf{r}_i, \mathbf{r}_j$ and t_i, t_j in $(\mathbf{r}, t)$-space, and we have characterized these structures by delta-functions[145]

$$\text{system } i : \chi_i(\mathbf{r}, t) = \delta(\mathbf{r} - \mathbf{r}_i)\delta(t - t_j),$$
$$\text{system } j : \chi_j(\mathbf{r}, t) = \delta(\mathbf{r} - \mathbf{r}_j)\delta(t - t_j).$$

The projection onto $(\mathbf{p}, E)$-space has led to[146]

$$\chi_k(\mathbf{p}, E) = \frac{1}{(2\pi\hbar)^2} \exp\left[-\frac{i}{\hbar}(\mathbf{p} \cdot \mathbf{r}_k - Et_k)\right], \quad k = i, j.$$

Let us use the functions $\chi_j(\mathbf{r}, t)$ and $\chi_k(\mathbf{p}, E)$ with $k = i, j$ for the investigation of two basic questions: 1. Can systems be elementary in character? 2. What is the cause for the existence of certain systems?

4.10.1. Can Systems be Elementary in Character?

Can the systems i and j exist independent of other systems? We already know that this cannot be the case.[147] In this connection the following questions are relevant: How have the systems i and j been created? How did both systems come into existence? Are they already existent

---
[144] Geometrical structures with respect to the variables $\mathbf{r}$ and t.
[145] See equations (4.1) and (4.2).
[146] See equation (4.4).
[147] See, in particular, Section 4.8.6.

as static entities without interaction? In the following we would like to answer these questions.

We have $\mathbf{p}, E$-fluctuations, that is, the values for $\mathbf{p}$ and E vary with time τ and according to the probability densities $\chi_i^*(\mathbf{p}, E)\chi_i(\mathbf{p}, E)$ and $\chi_j^*(\mathbf{p}, E)\chi_j(\mathbf{p}, E)$. Therefore, the conservation laws for the momentum $\mathbf{p}$ and the energy E require an interaction process with another system. This fact is valid for all possible systems and, as we have often remarked, it is therefore a general statement. In other words, within projection theory no system can exist without another system. No system as, for example, system i or system j, is definable without a "partner system". Thus, any system defined within projection theory cannot be *elementary* in character. This is a relevant principle.

How is this principle realizable in the case of system i or system j? There are two possibilities for the realization of partner systems:

1. System i and system j interact independent from each other with their environment.
2. There is a mutual interaction between both systems. In this case we do not need an environment.

The first case needs no further explanations. The second case is of particular interest because we need for the existence of system i and/or system j a certain interaction between both systems that we have called above "distance-independent (form) interaction". We have also mentioned above that this kind of interaction is not known in conventional physics. In this section let us give some additional remarks with respect to the elementary processes in $(\mathbf{p}, E)$-space and their projections onto $(\mathbf{r}, t)$-space.

4.10.2. *Self-Creating Interaction Processes*

For the interaction processes between system i and system j, the probability densities $\chi_i^*(\mathbf{p}, E)\chi_i(\mathbf{p}, E)$ and $\chi_j^*(\mathbf{p}, E)\chi_j(\mathbf{p}, E)$ are relevant, and both expressions take the value $1/(2\pi\hbar)^4$.[148] In contrast to the

---

[148] See equation (4.5).

functions $\chi_i(\mathbf{p}, E)$ and $\chi_j(\mathbf{p}, E)$,[149] note that the probability densities are independent on the space-time points $\mathbf{r}_i, t_i$ and $\mathbf{r}_j, t_j$.

Let us assume that system i is at time τ_1 in the state $\mathbf{p}_1, E_1$ and that system j is at the same time in the state $-\mathbf{p}_1, -E_1$. Both systems interact and interaction means that there are $\mathbf{p}, E$-fluctuations in $(\mathbf{p}, E)$-space, and when we express the fluctuation at a certain time τ by $\Delta \mathbf{p}$ and ΔE, system i and system j changes their states in accordance with the conservation laws of momentum and energy. Thus, instead of the states $\mathbf{p}_1, E_1$ and $-\mathbf{p}_1, -E_1$, we come to $\mathbf{p}_1 \pm \Delta \mathbf{p}, E_1 \pm \Delta E$ and $-\mathbf{p}_1 \mp \Delta \mathbf{p}, -E_1 \mp \Delta E$ at time τ_2 where we assume that τ_2 is the next possible τ value after τ_1, i.e., $\tau_2 - \tau_1$ is close to zero. In summary, we have the following situation in $(\mathbf{p}, E)$-space:

$$\tau_1 : (\mathbf{p}_1, E_1)_i \quad \text{and} \quad (-\mathbf{p}_1, -E_1)_j, \qquad (4.223)$$

$$\tau_2 : (\mathbf{p}_1 \pm \Delta \mathbf{p}, E_1 \pm \Delta E)_i \quad \text{and} \quad (-\mathbf{p}_1 \mp \Delta \mathbf{p}, -E_1 \mp \Delta E)_j, \qquad (4.224)$$

with

$$\Delta \tau = \tau_2 - \tau_1 \to 0. \qquad (4.225)$$

In $(\mathbf{r}, t)$-space, the probability densities are given by $\chi_i^*(\mathbf{r}, t)_i \chi_i^*(\mathbf{r}, t)$ and $\chi_j^*(\mathbf{r}, t) \chi_j(\mathbf{r}, t)$, and we measure in connection with system i at the space-time position $\mathbf{r}_i, t_i$ the values $\mathbf{p}_1, E_1$ at time τ_1 and the values $\mathbf{p}_1 \pm \Delta \mathbf{p}, E_1 \pm \Delta E$ at time τ_2. In connection with system j we have at the space-time position $\mathbf{r}_j, t_j$ the values $-\mathbf{p}_1, -E_1$ at time τ_1 and the values $\mathbf{p}_1 \mp \Delta \mathbf{p}$ and $E_1 \mp \Delta E$ at time τ_2. In summary, we have

$$\tau_1 : \mathbf{r}_i, t_i \leftarrow (\mathbf{p}_1, E_1)_i \atop \mathbf{r}_j, t_j \leftarrow (-\mathbf{p}_1, -E_1)_j, \qquad (4.226)$$

$$\tau_2 : \mathbf{r}_i, t_i \leftarrow (\mathbf{p}_1 \pm \Delta \mathbf{p}, E_1 \pm \Delta E)_i \atop \mathbf{r}_j, t_j \leftarrow (-\mathbf{p}_1 \mp \Delta \mathbf{p}, -E_1 \mp \Delta E)_j, \qquad (4.227)$$

with $\Delta \tau = \tau_2 - \tau_1 \to 0$.[150]

---

[149] See equation (4.4).
[150] See equation (4.225).

It is remarkable that the space-time distance $\mathbf{r}_1 - \mathbf{r}_2, t_2 - t_1$ between both systems may take any value, and may be as large as the space-time extension of the universe. Due to $\Delta\tau = \tau_2 - \tau_1 \to 0$, the registration of the momenta and the energies at the space-time positions $\mathbf{r}_i, t_i$ and $\mathbf{r}_j, t_j$ take place simultaneously although the space-time distance $\mathbf{r}_1 - \mathbf{r}_2$, $t_2 - t_1$ between both system may take any value and, in principle, it can take cosmological space-time extensions. The process takes place without any delay in time because we have $\Delta\tau \to 0$.

This is in fact a non-local effect in $(\mathbf{r}, t)$-space and reflects a projection effect, that is, it is not due to an exchange of information through space-time, i.e., through $(\mathbf{r}, t)$-space. The non-local effect is due to the fact that the functions $\chi_i^*(\mathbf{p}, E)\chi_i(\mathbf{p}, E)$ and $\chi_j^*(\mathbf{p}, E)\chi_j(\mathbf{p}, E)$ are independent of the space-time points $\mathbf{r}_i, t_i$ and $\mathbf{r}_j, t_j$ and their relative distances $\mathbf{r}_1 - \mathbf{r}_2, t_2 - t_1$. In other words, the interaction process, i.e., $\mathbf{p}, E$-fluctuations in $(\mathbf{p}, E)$-space are independent on the specific structure in $(\mathbf{r}, t)$-space, that is, on the space-time points $\mathbf{r}_i, t_i$ and $\mathbf{r}_j, t_j$ and their relative distances $\mathbf{r}_1 - \mathbf{r}_2, t_2 - t_1$. Again, such kind of interaction is not known in conventional physics. This interaction is important for the existence of the systems and this is reflected in the case studied here by their shape (form). Within conventional physics the form of so-called elementary particles or strings have to be assumed.

The cause for the existence of both systems in $(\mathbf{r}, t)$-space are the $\mathbf{p}, E$-fluctuations in $(\mathbf{p}, E)$-space. Again, existence in $(\mathbf{r}, t)$-space does not mean "material existence" in $(\mathbf{r}, t)$-space but "geometrical existence". The geometrical structure is described by the functions $\chi_i^*(\mathbf{r}, t)_i\chi_i^*(\mathbf{r}, t)$ and $\chi_j^*(\mathbf{r}, t)\chi_j(\mathbf{r}, t)$ where the functions $\chi_k(\mathbf{r}, t)$, $k = 1, 2$, are assumed to be delta-functions[151] within our specific considerations. The interaction between system i and system j are the reason for their existence in both spaces, i.e., in $(\mathbf{r}, t)$-space as well as in $(\mathbf{p}, E)$-space. These interactions have to be considered as a self-creating processes. Non-interacting (free) systems cannot exist within projection theory.[152] Hence, a single system (i and/or j) without any interaction cannot exist within projection theory.

---

[151] See equations (4.1) and (4.2).
[152] See Chapter 3.

4.11. Absolute Space-Time Conceptions

Within projection theory, reality is projected onto space-time, and there are merely geometrical structures in $(\mathbf{r}, t)$-space. It turned out[153] that these geometrical structures perform arbitrary jumps in $(\mathbf{r}, t)$-space and contain effective velocities $v_\mathbf{r}$ and v_t and effective accelerations $b_\mathbf{r}$ and b_t etc. However, this kind of motion has no influence on the $\mathbf{p}, E$-states of the arbitrarily jumping systems. In other words, space-time has no physical effects on the real systems; $(\mathbf{r}, t)$-space has no influence on the properties of the systems. We concluded that the arbitrary jumps in $(\mathbf{r}, t)$-space are characterized by a "motion relative to nothing".

4.11.1. Mach's Principle

Mach's principle has already been discussed in Section 4.8.8 but only on the surface. Because this principle is very important and basic, we want to continue and deepen the considerations given in Section 4.8.8.

As we have already pointed out in Section 4.8.8, the physical picture within Newton's theory is quite different from that of projection theory. In classical mechanics the $\mathbf{p}, E$-states are definitely changed when a body moves through space with varying velocities, and this is explained by space-effect, that is, within such a conception the space "interacts" with the body. Or more specifically: The space influences the body but the body does not have a physical effect on space. Therefore, the space plays the role of an absolute element within Newton's mechanics. However, such an absolute space has to be considered as unphysical and has been strongly criticized by Ernst Mach, but also by Einstein. Einstein wrote:

> *The term "absolute" not only means that space is physically real but also [...] independent in its physical properties, having a physical effect, but not itself influenced by physical conditions.*[154]

---

[153] See Section 4.8.6.
[154] Refer to Einstein, A., (1973) Grundzüge der Relativitätstheorie: Verlag Viehweg and Sohn.

In fact, we can never observe such a space (space-time) because its elements[155] are in principle not accessible to empirical tests. In other words, coordinates x, y, z and time τ can never be observed in isolated form. Also isolated distances between two space points and isolated time-intervals are not accessible to empirical tests. We can only say something about *distances in connection with bodies* and time intervals in connection with physical processes. In other words, an *empty* space-time as physical-theoretical conception should not be existent. This is a fundamental point and may not be neglected in the basic formulation of the physical laws.

All that was the reason why Ernst Mach required the following principle: A body does not move in unaccelerated and/or in accelerated motion relative to space, but relative to the centre of all the other bodies (masses) in the universe. This requirement is often discussed in literature under the notion "Mach's Principle". However, as we will recognize below, Mach's principle could not be fulfilled in conventional physics, i.e., within Newton's theory and Theory of Relativity. As we have already remarked and justified in Section 4.8.8, Mach's principle is valid in the framework of projection theory. Let us start our debate with Newton's theory.

4.11.2. *The Effect of Inertia within Newton's Theory*

Within Newton's theory Mach's principle is definitely not fulfilled. In classical mechanics the notion "inertia" is of particular relevance; as we have already remarked several times, here all real bodies are embedded in space. Even when a body does not interact with other systems, the effect of inertia is effective, i.e., also in this case the body moves through space with constant velocity v where the velocity of zero is included, and the effect of inertia is entirely due to the interaction of the body with space. The body moves relative to space, and its **p**, E-state is dependent on its velocity relative to space, i.e., the variation of

---

[155] Coordinates x, y, z and time τ where τ is again the time measured with our clocks used in everyday life.

v leads to changes of the bodies' **p**, E-states. We have the following situation within Newton's theory.

Absolute space was invented by Newton for the explanation of inertia. However, we do not know other phenomena for which absolute space would be responsible. Hence, the hypothesis of absolute space can only be proved by the phenomenon (inertia) for which it has been introduced. This is unsatisfactory and artificial.

Concerning absolute space, Max Born wrote:

Indeed, the concept of absolute space is almost spiritualistic in character. If we ask "What is the cause for the centrifugal forces?", the answer is "absolute space". If, however, we ask what absolute space is and in what other way it expresses itself, no one can furnish an answer other than that absolute space is the cause of centrifugal forces but has no further properties. This consideration shows that space as the cause of physical occurrences must be eliminated from the world picture. [...] Sound epistemological criticism refuses to accept such made-to-order hypothesis. They are too facile and are at odds with the aims of scientific research, which is to determine criteria for distinguishing its results from dreams of fancy. If the sheet of paper on which I have just written suddenly flies up from the table, I should be free to make the hypothesis that a ghost, say the spectre of Newton, had spirited it away. But common sense leads me instead to think of a draft coming from the open window because someone is entering by the door. Even if I do not feel the draft myself, this hypothesis is reasonable because it brings the phenomenon which is to be explained into a relationship with other observable events.[156]

4.11.3. Mach's Principle and Theory of Relativity

Special theory of relativity

In contrast to Newton's theory within Special Theory of Relativity space and time are no longer independent of each other; they are tied together into a space-time. Is Mach's principle fulfilled

---

[156] Refer to Born, M., (1964) Die Relativitätstheorie Einsteins: Springer Verlag.

within the Special Theory of Relativity? Definitely not! Newton's three-dimensional space is merely extended within Special Theory of Relativity to a four-dimensional space-time, without overcoming the absoluteness. In other words, instead of Newton's three-dimensional absolute space, within Special Theory of Relativity we have a four-dimensional absolute space-time (Minkowski's space). Also this space is — as in the case of Newton's three-dimensional space — the seat of the absolute forces of inertia. In conclusion, Mach's principle is not fulfilled at all within the Special Theory of Relativity; Einstein took this point very seriously and developed the General Theory of Relativity.

General theory of relativity

In order to relativize the forces of inertia, Einstein was led to formulate the General Theory of Relativity. In this way he wanted to eliminate the absolute space-time. In accordance to Mach, Einstein argued that the inertia of a body cannot come into play through space-time effects but should be completely due the other matter in the universe. This relativity of inertia was the foundation of his entire considerations.

With the formulation of General Theory of Relativity Einstein took up a completely new direction; this theory represents a magnificent building, and its results are confirmed by many experiments. However, also General Theory of Relativity failed to fulfil its initial goal, namely to eliminate the absoluteness of space-time of Special Theory of Relativity. In the following we would like to discuss this point by means of some solutions that follow from Einstein's field equations.

1. *De Sitter's Solution for a Lone Body*

Again, according to Mach, space-time can never be the source for physically real effects, as, for example, inertia. As in the case of Newton's theory and Special Theory of Relativity, this condition is also not fulfilled within General Theory of Relativity. This has been demonstrated, for example, by de Sitter's solution for a lone body. In a book by Peter and Neal Graneau, we find the following text:

> *In 1917, an eminent Dutch astronomer, Willem de Sitter, pointed out to Einstein that there was a finite valued solution of his field*

equations that gave the inertial mass of a particle even if it was the only one in the universe. In this case, the curved space-time of general relativity would be flat, that is the geodesic line passing through the particle would be straight. The lone particle would be guided along this geodesic line as if it was made of inertial matter. Einstein initially argued strongly against this solution. However eventually he conceded that his interpretation of inertia could not therefore be due to other matter, as required by Mach's principle, because there was no other matter around in the de Sitter's example.[157]

2. Empty Space within General Theory of Relativity

We have already outlined above that we can never observe an empty space (space-time) because its elements, i.e., the coordinates x, y, z and the time τ, are in principle not accessible to empirical tests. Therefore, no realistic space-time theory should contain an empty space-time as a solution. Einstein thought that his field equations would fulfil this important and basic condition. However, in 1917, de Sitter gave a solution to Einstein' field equations which corresponds to an empty universe, i.e., within the framework of this solution, space-time could exist without matter, and this is in obvious contradiction to Mach's principle. This fact was annotated by Einstein's collaborator Banesh Hoffmann as follows:

> *Barely had Einstein taken his pioneering step when in 1917 in neutral Holland de Sitter discovered a different solution to Einstein's cosmological equation. This was embarrassing. It showed that Einstein's equation did not lead to a unique model of the universe at all. Moreover, unlike Einstein's universe, de Sitter's was empty. It thus ran counter to Einstein's belief, an outgrowth of the idea of Mach, that matter and space-time are closely linked and that neither should be able to exist without the other.*[158]

---

[157] Refer to Graneau, P. and Graneau, N., (2006) In the grip of the distance universe: World Scientific.
[158] Refer to Hoffman, B., (1972) Albert Einstein: creator and rebel: The Viking press.

In conclusion, de Sitter's universe, which is based on Einstein's field equations, is empty and exists without any mass, that is, space-time with the elements x, y, z, τ can exist in isolated form. This is however unphysical because we are principally not able to observe the elements x, y, z, τ.[159]

Furthermore, we may state the following. Within General Theory of Relativity the mass m of a body curves space-time. Since both the mass and space-time (x, y, z, τ) can exist independently of each other, the space-time curvature should be produced by an interaction of m with the elements x, y, z, τ, i.e., the elements x, y, z, τ would be the source for physically real effects which is, in accordance to Mach, strictly forbidden. We can only say something about *distances in connection with bodies* and time intervals in connection with physical processes, i.e., the elements x, y, z, τ never appear in isolated form.[159]

In conclusion, de Sitter's universe leads exactly to the same problems as we have outlined above in connection with Newton's inertia. This is not surprising at all because also within General Theory of Relativity a lone mass, which is embedded in space-time, experiences inertia.[160]

But there are further clear indications for the fact that Mach's principle is not fulfilled within General Theory of Relativity. Let us collect some additional comments which are relevant in this connection, but first let us quote a third solution of Einstein's field equations. It is Gödel's solution which also demonstrates that General Theory of Relativity does not contain Mach's principle.

3. *Gödel's Solution*

Within General Theory of Relativity, it is still possible to talk about the rotation of the entire matter of the universe relative to absolute space. Gödel's solution of Einstein's field equations leads to this result. Heckmann remarked:

> *This solution by Gödel describes a model of the world which is uniformly filled with matter. All points in it are equivalent,*

---

[159] See Section 4.11.1 and Section 2.2.
[160] We have outlined this above.

which is therefore homogeneous as in the cases mentioned up to now; it is infinitely large and rotates absolutely, but is not able to expand. At the beginning, in 1916, and still long time after that, Einstein himself believed that his theory would contain the relativity of all motions. Gödel's solution was the first solid evidence that Einstein's belief was an error. [...] The absoluteness of space, which Newton claimed, and Einstein thought to have eliminated, is still contained in Einstein's theory, insofar as in Einstein's theory the concept of an absolute rotation is completely legitimate. Within Einstein' theory it is still possible to talk about the rotation of the entire mass of the world relative to absolute space.[161]

Conclusion

In summary, the concept of absolute space-time could not be eliminated within General Theory of Relativity. We have demonstrated that by means of three examples: 1. A lone mass, which is embedded in space-time, experiences inertia. 2. De Sitter's universe, which is based on Einstein's field equations, is an empty space-time without mass. 3. The rotation of the entire matter of the universe relative to an absolute space is possible within General Theory of Relativity.

The concept of absolute space (Newton's theory) and absolute space-time (Theory of Relativity) are actually ill constructions. We definitely cannot see, hear, smell, or taste space and time, that is, space and time (absolute or non-absolute) are not accessible to our senses. Also measuring instruments for the experimental determination of the space-time points x, y, z, and τ are not known and not even thinkable.

Although Newton's mechanics and the Theory of Relativity were very successful, and they are still used in many calculations, a lot of physicists could not accept such absolute concepts for space because they are unphysical constructions. This is demonstrated by the fact that scientists tried to solve this problem again and again up to the present

---

[161] Refer to Heckmann, O., (1980) Sterne, Kosmos, Weltmodelle: Deutscher Taschenbuch Verlag Gmblt and Co. KG.

day. The elimination of this space-problem is urgent, but the present scientific community often ignores this problem.

Further comments

Although Mach's principle reflects a fundamental and important feature, it could not be realized within General Theory of Relativity up to the present day. Dehnen remarked:

> *In those days Einstein had in mind that the structure of space and time is given completely by the particular distribution of matter in the world in accordance with his field equations of gravitation. As a result of this, Mach's idea would be fulfilled simultaneously, after which the inertia of material bodies is determined by other masses in the world. [...] However, it should be emphasized that Einstein's vision that Mach's principle could be realized within the framework of the general theory of relativity failed, even by additional modifications of the original field equations. [...] The problem in connection with the absolute space-time within the framework of special theory of relativity — a relict from Newton's mechanics — is, in the general theory of relativity, still not solved ...*[162]

Within General Theory of Relativity, there is no clear line between space-time and matter, and this is clearly against the facts. In the course of time, during the development of General Theory of Relativity, space-time took more and more the properties of matter which is in contradiction to what we can observe in connection with space and time. We can put a piece of matter on the table but not a piece of space-time. Kanitscheider remarked:

> *Although Einstein was led in the construction of his theory by Mach's idea of the ontological dominance of matter over space-time, he was even temporarily convinced that his theory had taken space and time the slightest trace of an objective*

---

[162] Refer to Dehnen, H., (1988) in Philosophie und Physik der Raumzeit, J. Audretsch und K. Mainzer (Hrgs.): Wissenschafts verlag.

reality. But the opposed tendency turned out to be the case. A tendency against Mach's principle became effective by a kind of inherent self-dynamics of the theory in the course of time of the development of the conception. Space-time became more and more an ontologically respectable entity, and it took over more and more the properties of material objects.[163]

4.11.4. Final Remarks

In summary, no doubt that General Theory of Relativity has a certain problem with space-time. Mach's principle is not fulfilled within General Theory of Relativity as it should be because this principle has to be considered as an important condition for any space-time theory, and Einstein was well aware of its importance. It is more than unsatisfactory that within General Theory of Relativity, an empty space-time can exist and even an absolute space-time is definable within this theory. In the course of the development of General Theory of Relativity, space-time took more and more the properties of material objects. Also this point reflects a critical situation and can basically not be accepted.

Both Newton's theory and also the Theory of Relativity are obviously not able to treat space (space-time) within a non-absolute framework. Why is it so difficult in conventional physics to construct non-absolute space-time theories? Newton's theory and also the Theory of Relativity have a common point which is probably relevant in this connection. Both developments treat space and space-time, respectively, as a rigid or flexible container in which matter is embedded. This basic feature might be the reason why Mach's principle cannot be fulfilled within the framework of these theories.

As we have already outlined very often, within projection theory the situation is completely different. Nothing real is embedded in space-time, but reality is projected on it and, therefore, it is not surprising that Mach's principle works within projection theory without any problem.

---

[163] Refer to Kanitscheider, B., (1984) Kosmologie: Philipp Reclam Jun.

The **p**, E-states of a system are not dependent on the movements relative to space-time.[164]

4.12. Relativistic Effects
4.12.1. *General Remarks*

We have investigated two systems (i and j) that interact through a distance-independent interaction,[165] and this kind of interaction is responsible for the form (shape) of both systems. This situation leads to arbitrary jumps of both systems, that is, a motion of system i relative to system j and vice versa is not definable. An observer in system i is not able to say something about the position of system j and, on the other hand, an observer in system j cannot say something about the position of system i. Furthermore, an observer in system i (system j) cannot say something about the (absolute) position of system i (system j) in (**r**, t)-space. There does not exist a relationship between certain positions in (**r**, t)-space and the **r**, t-states of the systems i and j, that is, a motion relative to (**r**, t)-space is not observable. As we have outlined above, there is also no motion of both systems relative to each other definable. The statement "motion relative to nothing" can characterize this situation. Therefore, the foundations for a relativistic treatment are not given here. The principles of Theory of Relativity are not applicable in such and similar cases.

As we have analyzed in Section 4.8.6, the arbitrary jumps of the geometrical structures contain effective velocities $v_{k\mathbf{r}}$ and v_{kt} effective accelerations $b_{k\mathbf{r}}$ and b_{kt} and other types of motion, in particular, the values for the quantities $v_{k\mathbf{r}}, v_{kt}, b_{k\mathbf{r}}, b_{kt}$ may be close to infinity: $v_{k\mathbf{r}}, v_{kt}, b_{k\mathbf{r}}, b_{kt} \to \infty$ with $k = 1, 2$. This seems to be in contrast to Special Theory of Relativity. Within this theory the velocity of a body, having the rest mass of m_0, cannot exceed the velocity of light $c : v_\mathbf{r} < c$. However, such a conflict does not exist. Our analysis has led to the result that the principles of Theory of Relativity cannot be applied at the basic quantum level of projection theory.

---

[164] See Section 4.8.8.
[165] See Sections 4.8.1–4.8.6.

In conclusion, projection theory does not allow combining the basic quantum picture with the principles of relativity. Within usual quantum theory such a unification is not forbidden and we come here to "relativistic usual quantum theory".

We also have to consider that the conceptions of projection theory and Theory of Relativity are quite different from each other. Within Theory of Relativity the real bodies are embedded in space-time and the real masses actually move through space; within projection theory we have projections onto space-time, i.e., onto $(\mathbf{r}, t)$-space, and the resulting geometrical structures do not really move through $(\mathbf{r}, t)$-space but the projections appear as jumps between two arbitrary space-time positions, where the $\mathbf{r}, t$-points between these two arbitrary positions are not involved. It is therefore not surprising that the principles of relativity are not applicable in their basic form to projection theory. However, relativistic effects must also be of relevance in connection with projection theory. Let us briefly analyze under what conditions relativistic effects can appear.

4.12.2. Frames of Reference within Projection Theory

So far, we have discussed the situation by means of two systems (i and j), which interact through a distance-independent interaction. Clearly, the statements for two systems must also be valid for N systems. In the following let us discuss, as an example, the situation for $N = 5$ systems that do not only interact by *distance-independent* interactions but also through *distance-dependent* interactions. In other words, we have a superposition of *distance-independent* and *distance-dependent* interactions. Just the switching on of distance-dependent interactions changes the situation fundamentally and, in a certain sense, relativistic effects become possible.

The five systems do not jump arbitrarily through $(\mathbf{r}, t)$-space but, due to the distance-dependent interaction, they form a complex with definite mean distances between the five systems. In Figure 4.4(a) this complex is shown. Now we would like to divide the five systems into two parts, S and S'. S consists of the systems 2, 3, 4 and 5; S' is identical

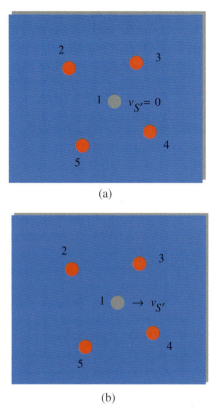

Fig. 4.4. (a) $N = 5$ systems not only interact by *distance-independent* interactions but also *distance-dependent* interactions are assumed to be effective. Through the switching on of distance-dependent interactions relativistic effects become possible. The whole system has been divided into two parts, S and S'. S consists of the systems 2, 3, 4 and 5; S' is identical with system 1. The observer is resting relative to S. S' is able to move relative to S. In (a) also S' (system 1) is at rest, that is, the velocity $v_{S'} = (v_{\mathbf{r}S'}, v_{tS'})$ of S' relative to S is zero: $v_{\mathbf{r}S'} = 0, v_{tS'} = 0$. (b) The same conditions as in (a). However, the velocity of S' relative to S is no longer zero: $v_{S'} \neq 0 \rightarrow v_{\mathbf{r}S'} \neq 0, v_{tS'} \neq 0$. For details, see text.

with system 1. The observer is resting relative to S. It is assumed that S' is able to move relative to S. We may state that S defines a resting frame of reference and S' a moving frame of reference. In Figure 4.4(a) we assumed that S' (system 1) is at rest, that is, the velocity $v_{S'} = (v_{\mathbf{r}S'}, v_{tS'})$ of S' relative to S is zero.

In the course of time τ the relative distances between the five members of the complex vary around their mean distances, and this situation defines a structural equilibrium. In this case no systematic motion of the five subsystems relative to each other is possible, that is, the mean velocities must be zero as, for example, we do not observe a velocity $v_{S'} \neq 0$ of S' relative to the resting frame of reference S.

Now we would like to apply an external interaction on S'. It is effective in the time interval $\Delta\tau = \tau_b - \tau_a$, where we assumed that the external interaction has been switched on at time τ_a and switched off at time τ_b. Furthermore, let us require that this external interaction leads to the effect that system S' (system 1) moves with a constant $v_{S'} = (v_{\mathbf{r}S'} \neq 0, v_{tS'} \neq 0)$ relative to S for times $\tau > \tau_b$. This however provides that S' (system 1) "feels" the same environment in the course of time $\tau > \tau_b$. In our example,[166] the environment is approximated by the subsystems 2–5.

Let us symbolically express the wave function for the complex consisting of five subsystems by $\Psi(\mathbf{r}_1, t_1, \mathbf{r}_2, t_2, \ldots, \mathbf{r}_5, t_5)$ and $\Psi'(\mathbf{r}_1 - v_{\mathbf{r}S'}\tau, t_1 - v_{tS'}\tau, \mathbf{r}_2, t_2, \ldots, \mathbf{r}_5, t_5)$, in particular, we define for

$$\tau < \tau_a : \Psi(\mathbf{r}_1, t_1, \mathbf{r}_2, t_2, \ldots, \mathbf{r}_5, t_5), \qquad (4.228)$$

and for

$$\tau > \tau_b : \Psi'(\mathbf{r}_1 - v_{\mathbf{r}S'}\tau, t_1 - v_{tS'}\tau, \mathbf{r}_2, t_2, \ldots, \mathbf{r}_5, t_5). \qquad (4.229)$$

In other words, between τ_a and τ_b we have the transition

$$\Psi(\mathbf{r}_1, t_1, \mathbf{r}_2, t_2, \ldots, \mathbf{r}_5, t_5)$$
$$\rightarrow \Psi'(\mathbf{r}_1 - v_{\mathbf{r}S'}\tau, t_1 - v_{tS'}\tau, \mathbf{r}_2, t_2, \ldots, \mathbf{r}_5, t_5), \qquad (4.230)$$

with

$$\Psi(\mathbf{r}_1, t_1, \mathbf{r}_2, t_2, \ldots, \mathbf{r}_5, t_5)$$
$$\neq \Psi'(\mathbf{r}_1 - v_{\mathbf{r}S'}\tau, t_1 - v_{tS'}\tau, \mathbf{r}_2, t_2, \ldots, \mathbf{r}_5, t_5). \qquad (4.231)$$

---

[166] See Figure 4.4(b).

This leads to a mean momentum $\bar{\mathbf{p}}_0$ and a mean energy $\bar{E}_0$ for $v_{S'} = 0 = (v_{\mathbf{r}S'} = 0, v_{tS'} = 0)$ that are different from the corresponding values $\bar{\mathbf{p}}_v, \bar{E}_v$ for $v_{S'} = (v_{\mathbf{r}S'}, v_{tS'}) \neq 0$:

$$\bar{\mathbf{p}}_0 \neq \bar{\mathbf{p}}_v, \bar{E}_0 \neq \bar{E}_v. \tag{4.232}$$

4.12.3. Transformation Formulas

In general, the complete geometrical structure $\Psi^*(\mathbf{r},t)\Psi(\mathbf{r},t)$ covers the range $-\infty \leq \mathbf{r},t \leq \infty$. In most cases $\Psi^*(\mathbf{r},t)\Psi(\mathbf{r},t)$ has a certain width Δ_W and we may say that $\Psi^*(\mathbf{r},t)\Psi(\mathbf{r},t)$ is only in the range $-\Delta_W \leq \mathbf{r},t \leq \Delta_W$ essentially different from zero. However, we never observe at a certain time τ the complete geometrical structure $\Psi^*(\mathbf{r},t)\Psi(\mathbf{r},t)$. We know that the function $\Psi^*(\mathbf{r},t)\Psi(\mathbf{r},t)$ is the probability density that we have identified above with the form of the system under investigation. At time τ only a certain couple of values, say $\mathbf{r}_\tau, t_\tau$, becomes real with a definite probability, which is determined by $\Psi^*(\mathbf{r}_\tau, t_\tau)\Psi(\mathbf{r}_\tau, t_\tau)$. The measurement of one of the possible values for $\mathbf{p}$ and for E is done in the space-time intervals $\mathbf{r}_\tau, \mathbf{r}_\tau + \Delta \mathbf{r}_\tau$ and $t_\tau, t_\tau + \Delta t_\tau$ with the probability density of $\Psi^*(\mathbf{r}_\tau, t_\tau)\Psi(\mathbf{r}_\tau, t_\tau)$. The possible values for $\mathbf{p}$ and for E are given with a certain probability which is determined by the probability density $\Psi^*(\mathbf{p}, E)\Psi(\mathbf{p}, E)$.

In summary, we never observe at time τ the complete geometrical structure $\Psi^*(\mathbf{r},t)\Psi(\mathbf{r},t)$, but only a certain couple of values $(\mathbf{r}_\tau, t_\tau)$. We have to consider this probability description in the determination of the transformation formulas which exist between the space-time coordinates of the resting frame of reference S and those of the moving frame of reference S'. Let us discuss this point qualitatively.

We know that the description of quantum-theoretical systems in terms of probabilities leads to uncertainties which must also be reflected in the relative motion of S' relative to S because the velocity of S' relative to S also becomes unavoidably uncertain.[167] This effect must have a serious consequence on the structure of the transformation formulas which exist between space-time coordinates of S and those of

---

[167] All systems 1–5 which are involved are only defined within a certain range $-\Delta_W \leq \mathbf{r},t \leq \Delta_W$; see Figure 4.4.

S'. Within Theory of Relativity, the Lorentz-transformations connect the space-time elements x, y, z, t of S with those of S' (x', y', z', t'), and the velocity $v_{rS'}$ is involved as a parameter together with the velocity of light c.[168] However, the Lorentz-transformations cannot be valid in their original form within projection theory, and this is because the velocity of S' relative to S becomes unavoidably uncertain. Therefore, the velocity $v_{S'} = (v_{rS'}, v_{tS'})$ introduced above can only be a mean velocity. Thus, there can be no transformation formulas in analogy to the Lorentz-transformations. Because $v_{S'} = (v_{rS'}, v_{tS'})$ defines a mean velocity, also the space-time transformations between S and S' can only be defined on the basis of mean space-time coordinates. In other words, within projection theory we should have transformations between the mean space-time elements $\bar{x}, \bar{y}, \bar{z}, \bar{t}$ and $\bar{x}', \bar{y}', \bar{z}', \bar{t}'$ instead of x, y, z, t and x', y', z', t' (Lorentz-transformations). In summary, we come to the following transformation schemes:

$$\text{Projection theory: } \bar{x}, \bar{y}, \bar{z}, \bar{t} \leftrightarrow \bar{x}', \bar{y}', \bar{z}', \bar{t}', \tag{4.233}$$

$$\text{Special Theory of Relativity: } x, y, z, t \leftrightarrow x', y', z', t'. \tag{4.234}$$

Within projection theory it makes not much sense to define transformations laws between x, y, z, t and x', y', z', t'. The reason is simple. Let us denote the form of a system as, for example, that of subsystem 1 of the complex consisting of five subsystems,[169] by $\Psi_F^*(x, y, z, t)_1 \Psi_F(x, y, z, t)_1$. It is the structure with respect to S which is identical with that in S' because the relative velocity is zero.[170] Then, we observe arbitrary jumps *within* the $(\mathbf{r}, t)$-range where the structure $\Psi_F^*(x, y, z, t)_1 \Psi_F(x, y, z, t)_1$ is different from zero,[171] that is, we may have for the effective velocities v_{1r} and v_{1t} *within* the geometrical structure $\Psi_F^*(x, y, z, t)_1 \Psi_F(x, y, z, t)_1 \neq 0$ the values $v_{1r} \to \infty$ and $v_{1t} \to \infty$. Then, it makes no sense to express the structure $\Psi_F^*(x', y', z', t')_1 \Psi_F(x', y', z', t')_1$ in the moving frame of

---

[168] $v_{tS'}$ is not defined within Theory of Relativity.
[169] See Figure 4.4.
[170] $v_{S'} = 0$; see Figure 4.4(a).
[171] See Section 4.8.10.

reference S'[172] because the velocity $v_{S'}$ is not really definable in the case $v_{1r} \to \infty$ and $v_{1t} \to \infty$.

The derivation of the transformation laws within projection theory is an interesting and important topic but we do not want to give more details here. It should however be mentioned that the law for the constancy of the velocity of light can obviously be deduced within the framework of projection theory. Within Special Theory of Relativity, the law for the constancy of the velocity of light is a postulate, that, is, the source for this law remains hidden here but obviously not in projection theory. A more detailed discussion will be given in a forthcoming publication.

4.12.4. Arbitrary Jumps of the Entire Complex in Space-Time

The entire system consisting of the five subsystems 1–5 and the observer (resting in S) jump arbitrarily through $(\mathbf{r}, t)$-space. This can easily be verified if we apply the procedure which we have applied above. If all subsystems 1–5 are equally shifted by $\Delta_\mathbf{r}$ and Δ_t,[173] with $\Delta_\mathbf{r} = \Delta_{1\mathbf{r}} = \cdots = \Delta_{5\mathbf{r}}$ and $\Delta_t = \Delta_{1t} = \cdots = \Delta_{5t}$, nothing is changed in connection with the relative distances of the whole system. Then, the whole system, together with the observer, may perform arbitrary jumps because the quantities $\Delta_\mathbf{r}$ and Δ_t must behave statistically, as we have outlined in Section 4.8.3. In other words, the geometrical structure of the whole system contains effective velocities $v_\mathbf{r}$ and v_t, effective accelerations $b_\mathbf{r}$ and b_t and other types of motion, in particular, the values for $v_\mathbf{r}, v_t, b_\mathbf{r}, b_t$ may be close to infinity: $v_\mathbf{r}, v_t, b_\mathbf{r}, b_t \to \infty$.

Again, the inner structure of the entire system does not change and the observer, resting in S, is not able to detect these arbitrary jumps. In particular, the motion of S' (system 1) relative to S (systems 2–5) with the mean velocity $v_{S'} = (v_{\mathbf{r}S'}, v_{tS'})$ is not influenced by these arbitrary jumps which are characterized by the values for $v_\mathbf{r}, v_t, b_\mathbf{r}, b_t$.

---

[172] $v_{S'} \neq 0$; see Figure 4.4(b).
[173] See, in particular, Section 4.8.3.

Furthermore, we may state the following. An observer who is resting relative to $(\mathbf{r}, t)$-space is not able to observe the arbitrarily jumping complex consisting of five subsystems; a similar situation has been discussed in Section 4.8.5. Only the observer, who is resting relative to the frame of reference S, can make statements about the inner structure of the complex. In general, an observer can only detect the structure of the complex when he is a member of the frame of reference S, that is, when he interacts with it.[174] Otherwise the observer cannot detect the structure of the complex. This is an analogous situation as we have already analyzed in connection with two systems (system i and system j) and a distance-independent interaction.[175]

Remark

An observer cannot find out whether he is resting relative to $(\mathbf{r}, t)$-space or not. Nevertheless, an observer, which is resting relative to $(\mathbf{r}, t)$-space, is imaginable. Why? We have to distinguish between an observer and another observer who develops all these pictures that appear in $(\mathbf{r}, t)$-space; let us call him the "absolute observer". These pictures are positioned in the head of the "absolute observer" together with the $(\mathbf{r}, t)$-space. Then, we may define an observer who is resting relative to $(\mathbf{r}, t)$-space. He is resting relative to the "absolute observer" and, therefore, relative to $(\mathbf{r}, t)$-space. Clearly, observer and "absolute observer" are exchangeable, i.e., an observer is simultaneously an "absolute observer".

4.13. Hierarchy of the Parts in a Part

The principle "hierarchy of the parts in a part" is based on the assumption that there are certain units in the world and levels of description that are more fundamental than others. First, we will discuss this principle within usual physics and after that we discuss the situation within the framework of projection theory.

---

[174] For example, with one of its 4 subsystems.
[175] See Sections 4.8.3–4.8.5.

4.13.1. Conventional Physics

What is the nature of matter? Within conventional physics, this question is normally answered on the basis of

> the assumption that the analysis of matter always leads us to simpler and more fundamental levels, a belief that has been given free reign in twentieth century physics. Complicated objects like typewriters, automobiles, grandfather clocks and radios consist of simple elements, each doing its own job. When these cogs, levers, springs, resistors, and transistors are brought together in the proper way, cogs and springs can tell the time, while transistors, capacitors, and resistors can play music or calculate numbers. [...] Physics pursues this analogy and suggests that the universe itself can be broken down into simpler and smaller parts called elementary particles. Physics, it is hoped, will one day reach the ultimate level of nature in which everything can be described and from which the entire universe develops. The belief could be called the quest for the ultimative.[176]

The concept of elementary particles, and of course that of strings, is based on the conception that matter can be split into smaller and smaller units up to the ultimate level, i.e., we have a hierarchy of parts in the part.

However, such a conception, i.e., the principle of the "hierarchy of the part in a part", has its roots at the macroscopic level and ignores at first that we have fundamentally to change the frame of description when we enter the microscopic level, i.e., that level where quantum phenomena take place.

Our intuitive concepts are based on systems and processes that we observe in everyday life with our sense organs, i.e., they describe reality at the so-called macroscopic level. Classical mechanics is also based on macroscopic observations, and its notions have been chosen with respect to our intuitive demands for visualization. Using more and

---

[176] Refer to Peat, F.D., (1988) Superstrings and the search for the theory of everything: Contemporary Books.

more refined measuring instruments we extend our knowledge about nature, and the macroscopic level is consequently left step by step. At the microscopic level, quantum phenomena are dominant and we have to cope with a conceptual revision of the classical framework. The new (quantum-mechanical) level has led to violations of our intuitive concepts. The framework of quantum theory is in striking conflict with our intuitive demands for visualization.

No doubt, the principle of the "hierarchy of the part in a part" is based on the concepts that we use in everyday life. Therefore, we have to be careful when we apply this principle to levels, which are different from the level of everyday life. It is quite questionable to transfer our intuitive concepts that we use in everyday life to the microscopic level where quantum effects dominate. We have recognized in Sections 4.8–4.12 that non-local effects seem to play an important role at the quantum theoretical level; our intuitive concepts are tailor-made to localized bodies and local interactions and non-local effects remain hidden at the level of everyday life where we observe the world with our sense organs. In the next subsection we will discuss the situation within projection theory.

4.13.2. Is this Principle Realizable within Projection Theory?

Within usual physics in most cases the following concept of matter is used. There exist elementary particles (strings, branes), i.e., particles without substructure. Then, it is assumed that all matter in the cosmos is made from these basic building blocks (elementary particles). In particular, it is assumed within this concept that each of these elementary units can exist independently from other things in the cosmos. Matter is formed by putting the basic building blocks systematically together, and we have that which we have called "hierarchy of the parts in a part".

In the standard theory quarks, leptons and gauge bosons are elementary particles. From quarks and leptons we get atoms and molecules; from atoms and molecules we obtain matter in the various forms. There are certain interaction types, defined in connection with gauge bosons, that are qualitatively different from each other and they enable the possibility for the hierarchical structure. Here it is essential that certain forms of matter, i.e., elementary particles or strings, branes or other units, are more fundamental than other forms of matter as, for example, atoms and molecules. However, we know that this hierarchical scenario is not yet complete because gravitation is not involved; only the string theory is discussed as a theory for everything.

Such a conception, i.e., "hierarchy of the parts in a part", is obviously not possible within projection theory. Let us briefly give the main arguments why such a principle should not be realizable within projection theory. In general, the reason for this situation is obviously due to the fact that certain notions of conventional physics cannot be applied in projection theory, and it is relatively easy to recognize that. In the following let us discuss this point in more detail.

4.13.2.1. *Pictures and* $\mathbf{p}, E$*-fluctuations*

Within projection theory the material world is not embedded in space-time, i.e., in $(\mathbf{r}, t)$-space. Instead of this concept (usual physics) we have here a "reality", embedded in $(\mathbf{p}, E)$-space, and a "picture of this reality", and these pictures are represented in $(\mathbf{r}, t)$-space; both spaces are interconnected by a Fourier transform. Within projection theory the following point is important: "*Within the memory of man all essential things are represented within the frame of pictures*". Therefore, we considered the pictures, i.e., certain facts from basic reality that appear spontaneously in front of us, as the most basic facts.[177]

In other words, the description of reality makes no sense without the existence of a picture. On the other hand, a picture can only come into existence if there are $\mathbf{p}, E$-fluctuations and, as we have outlined several times, these $\mathbf{p}, E$-fluctuations mean nothing else than "interaction" in $(\mathbf{p}, E)$-space. The $\mathbf{p}, E$-fluctuations, expressed by the probability density $\Psi^*(\mathbf{p}, E)\Psi(\mathbf{p}, E)$, are of course influenced by other system-specific features so that we have not an infinite number of fluctuation curves, but we should have only a selection of specific $\mathbf{p}, E$-laws.

---

[177] See Section 2.2.

However, the existence of various specific p, E-curves does not mean that there are interactions which are qualitatively different from each other. There are no p, E-fluctuations which are more fundamental than others. All p, E-values are equivalent and there is no p, E-value which would be more fundamental than another (p, E)-value. Therefore, when we talk about the "hierarchy of the parts in a part", only the existence of p, E-fluctuations are necessary but not the specific p, E-law because all p, E-fluctuations (interactions) are equivalent. Then, the following question arises: Is the construction of such a hierarchy within projection theory rich in meaning under these circumstances, i.e., the equivalence of all kind of interactions?

4.13.2.2. *No static building blocks*

Within projection theory, it is problematic to assume that matter is formed by putting certain basic units systematically together. In particular, within projection theory it is not possible to work with the concept that each of these elementary units can exist independently from the others and of other things in the cosmos. Hence, the concept of conventional physics, that matter is formed by putting certain static basic building blocks systematically together, is possibly not realizable within projection theory. Let us briefly explain the situation by means of a simple example.

For simplicity, let us consider a cosmos with only two subsystems. The two subsystems cannot exist independently from each other. In other words, this simple universe cannot be formed by putting two static units (the two subsystems) systematically together, and this is because the two subsystems cannot be thought as independent from each other. Each of these units needs the other unit in order to be able to exist,[178] that is, both units come into existence in a self-creating process; they interact by p, E-fluctuations and these interaction processes are a necessary condition for their existence. This situation is independent of the type of interaction. The two systems may interact by a form interaction and/or by a distance-dependent interaction.

---

[178] See Section 4.10.

Again, the universe, here defined by two subsystems that are under mutual influence, is not formed by putting the two independent subsystems systematically together but it emerges with two subsystems.

The features discussed in connection with the simple universe consisting of two units have to be taken into account, and we have to analyze the situation for each case carefully. Here the following question is relevant. What parts of a system can be treated as a "unit construction system", and what parts cannot be analyzed in this way but simply emerge as a definite structure consisting of a certain number of subsystems (units)?

In other words, such a structure emerges as an ensemble and comes not into existence through basic building blocks, which are put together for the creation of the ensemble consisting of a certain number of subsystems. The reason is given in Section 4.10: there cannot exist independent (isolated) units in projection theory but each unit is defined by specific interactions $\mathbf{p}, E$-fluctuations. In other words, the subsystems of the ensemble can only be existent within the ensemble. Such units cannot be called "basic". Clearly, a certain subsystem (unit) may appear in connection with various ensembles that are different from each other. In contrast to conventional physics, we have no static units in projection theory but all comes into existence by fluctuations, i.e., by interactions defined by $\mathbf{p}, E$-fluctuations.

Let us underline the main facts by the following arguments. If a certain unit is characterized by the certain features, say $\alpha_1, \alpha_2, \ldots$, we may characterize the differences between usual physics and projection theory as follows:

$$\alpha_1, \alpha_2, \ldots, \mathbf{p}_0, E_0 \rightarrow \alpha_1, \alpha_2, \ldots, \Delta\mathbf{p}, \Delta E, \qquad (4.235)$$

with $\mathbf{p}_0 = $ constant and $E_0 = $ constant. Let us explain the scheme (4.235). The state $\alpha_1, \alpha_2, \ldots, \mathbf{p}_0, E_0$ reflects the situation in usual physics where a separate unit is static in character. In projection theory we have no static units and their state is characterized by $\alpha_1, \alpha_2, \ldots, \Delta\mathbf{p}, \Delta E$ and, instead of $\alpha_1, \alpha_2, \ldots, \mathbf{p}_0, E_0$, we come to $\alpha_1, \alpha_2, \ldots, \Delta\mathbf{p}, \Delta E$, as is symbolized in equation (4.235).

We know that a state $\alpha_1, \alpha_2, \ldots, \mathbf{p}_0, E_0$ is not possible in projection theory. In Chapter 3, we have outlined that a free (non-interacting) system with $\mathbf{p}_0 = $ constant and $E_0 = $ constant cannot exist. For such systems we found $\mathbf{p}_0 = 0$, $E_0 = 0$ and $\Psi(\mathbf{p}_0, E_0) = 0$, $\Psi(\mathbf{r}, t) = 0$. Only in the case of $\mathbf{p}, E$-fluctuations ($\Delta \mathbf{p} \neq 0, \Delta E \neq 0$) we have $\Psi(\mathbf{p}_0, E_0) \neq 0$, $\Psi(\mathbf{r}, t) \neq 0$ where the quantities $\mathbf{p}_0$ and E_0 can no longer be treated as constants: $\mathbf{p}_0 \neq $ constant and $E_0 \neq $ constant. In other words, a certain unit can only exist in connection with other physical systems because such $\mathbf{p}, E$-fluctuations can only be real if the conservation laws for the momentum and the energy are fulfilled.

This situation makes it impossible or at least difficult to base the matter of the cosmos on that what we have called "hierarchy of the parts in a part". In this connection, the following point has to be considered. Within projection theory, the interactions by $\mathbf{p}, E$-fluctuations are relevant and, as we have outlined above, no $\mathbf{p}, E$-fluctuation is more fundamental than others. Also this feature should be of particular relevance in connection with the realization of a "hierarchy of the parts in a part".

Summary

In projection theory a system is defined as follows. A system normally consists of N subsystems (units) and these units can be different from each other. The subsystems interact with each other and there can be two kinds of interactions: 1. The form interaction which is responsible for the shape (form) of each subsystem, and this kind of interaction is not dependent of the space-time distances of the subsystems. 2. In addition to the form interaction there can be an interaction between the N subsystems that is dependent on their relative space-time distances. Furthermore, within projection theory the N units cannot exist independently from each other, that is, if the whole system is closed, each subsystem can only be defined on the basis of the other $(N - 1)$ subsystems or a fraction of them.

In other words, the system cannot come into existence by putting the N subsystems together because the N subsystems do not form an ensemble of N independent units, i.e., the system is not formed by putting the N subsystems systematically together. In conclusion, the system is not composed of N independent units but it emerges with N subsystems. All these points have to be considered when we want to describe a system within projection theory.

Concerning the two kinds of interaction, i.e., form interaction and distance-dependent interaction, there are two interesting limiting cases. Let us briefly discuss them:

1. When we assume that the distance-dependent interaction is zero, the N subsystems only interact with each other by the form interaction, that is, the interaction is not dependent on the space-time distances between the N subsystems and each subsystem may take arbitrary positions. This in particular means that each subsystem (its space-time position) seemingly moves arbitrarily through space-time with arbitrary velocity and arbitrary acceleration varying in the course of time τ. Strictly speaking the space-time position does not move through $(\mathbf{r}, t)$-space but it is projected on it within projection theory. The feature that a subsystem can take arbitrary positions in $(\mathbf{r}, t)$-space means absence of inertia, and this behaviour corresponds to random phases in $(\mathbf{p}, E)$-space.[179]
2. In principle, the distance-dependent interaction can be defined without the existence of the form interaction. That is, the N subsystems may emerge without shape, i.e., point-like or with the shape of a string or any other form.

4.13.2.3. No fluctuations of fluctuations!

If a "hierarchy of the parts in a part" is defined, the following property should be realizable. When we are at a certain level of matter organization, say level 1 with $\Psi_1(\mathbf{p}, E)$, another level of matter organization, say level 2 with $\Psi_2(\mathbf{p}, E)$, should be attainable on the basis of $\Psi_1(\mathbf{p}, E)$, that is, there should be a connection between level 1 and level 2, and the features of $\Psi_2(\mathbf{p}, E)$ have to be a function

---

[179] See, for example, Section 4.8.5.

of $\Psi_1(\mathbf{p}, E)$: $\Psi_2(\mathbf{p}, E) = \Psi_2(\Psi_1(\mathbf{p}, E))$. From level 2, a third level, level 3, should be attainable from level 2 and, therefore, we would have $\Psi_3(\mathbf{p}, E) = \Psi_3(\Psi_2(\Psi_1(\mathbf{p}, E)))$. If the hierarchy consists of n levels we can continue this procedure and we get the following hierarchical scheme:

$$\Psi_1(\mathbf{p}, E),$$
$$\Psi_2(\mathbf{p}, E) = \Psi_2(\Psi_1(\mathbf{p}, E)),$$
$$\Psi_3(\mathbf{p}, E) = \Psi_3(\Psi_2(\Psi_1(\mathbf{p}, E))),$$
$$\vdots$$
$$\Psi_n(\mathbf{p}, E) = \Psi_n(\Psi_{n-1}(\cdots \Psi_2(\Psi_1(\mathbf{p}, E)) \cdots)). \quad (4.236)$$

Is such a scenario[180] possible within projection theory? No, it is not. Why? The reason is relatively simple. In order to reach a new level of matter as, for example, level 2, the structure of the subsystem at level 1 should be involved, i.e., there must be a certain interaction of a new subsystem with the structure of the system defined at level 1. This structure at level 1 is defined without the new subsystem and is given by $\Psi_1(\mathbf{p}, E)$, $\Psi_1^*(\mathbf{p}, E)\Psi_1(\mathbf{p}, E)$ and $\Psi_1(\mathbf{r}, t)$, $\Psi_1^*(\mathbf{r}, t)\Psi_1(\mathbf{r}, t)$, respectively. That is, the structure of the system at level 1 is defined by $\mathbf{p}, E$-fluctuations.[181] In other words, the new level of matter organization (level 2) can only be reached if the new subsystem interacts with the $\mathbf{p}, E$-fluctuations in connection with the system defined at level 1, i.e., we would have $\mathbf{p}, E$-fluctuations of $\mathbf{p}, E$-fluctuations which can be symbolically expressed by $\Delta\Delta\mathbf{p}_1, \Delta\Delta E_1$. A situation without $\Delta\Delta\mathbf{p}_1, \Delta\Delta E_1$ cannot lead to the effect that the structure of the system at level 1 is also reflected at level 2. However, $\mathbf{p}, E$-fluctuations of $\mathbf{p}, E$-fluctuations cannot be defined within projection theory because such a mechanism would definitely lead to contradictions. This must of course also be the case if the hierarchy consists of n levels as is formulated by scheme (4.236). Thus, scheme (4.236) is forbidden because it would lead to a series of fluctuations: $\mathbf{p}, E$-fluctuations of $\mathbf{p}, E$-fluctuations of

---

[180] See scheme (4.236).
[181] Clearly, all structures are defined by $\mathbf{p}, E$-fluctuations.

$\mathbf{p}, E$-fluctuations etc. according to

$$\begin{aligned}
&\Delta \mathbf{p}_1, \Delta E_1, \\
&\Delta \mathbf{p}_2, \Delta E_2 \leftarrow \Delta \Delta \mathbf{p}_1, \Delta \Delta E_1, \\
&\Delta \mathbf{p}_3, \Delta E_3 \leftarrow \Delta \Delta \mathbf{p}_2, \Delta \Delta E_2 = \Delta \Delta \Delta \mathbf{p}_1, \Delta \Delta \Delta E_1, \\
&\vdots \\
&\Delta \mathbf{p}_n, \Delta E_n \leftarrow \Delta \Delta \mathbf{p}_{n-1}, \Delta \Delta E_{n-1} \\
&\qquad = \Delta \cdots \Delta \Delta \mathbf{p}_1, \Delta \cdots \Delta \Delta E_1.
\end{aligned} \qquad (4.237)$$

The fluctuation series defined by the various terms of scheme (4.237) are not defined (realizable) because such a scheme would lead to contradictions. Thus, also scheme (4.236) cannot be defined and, therefore, there should be no basis for a "hierarchy of the parts in a part" within projection theory. On the other hand, we have to clarify how the organization of matter can be described under these restrictions (conditions) dictated by projection theory. In the next section we will discuss the situation.

4.13.2.4. Independent $\mathbf{p}, E$-fluctuations

The feature that there cannot be $\mathbf{p}, E$-fluctuations of $\mathbf{p}, E$-fluctuations etc., has to be considered as a basic fact and must be taken into account in the construction of complex systems. In fact, we know that matter is able to form complex aggregates and, therefore, we have to assume that the subsystems, which belong to such a complex aggregate, have the ability to interact with various subsystems simultaneously and not only with one. That is, instead of $\mathbf{p}, E$-fluctuations of $\mathbf{p}, E$-fluctuations etc., we have to assume that the subsystems are able to interact simultaneously with various other subsystems where these interactions, i.e., $\mathbf{p}, E$-fluctuations, have to be considered as independent of each other. Only in this way we are able to describe complex aggregates, which we actually observe in nature. In conclusion, each subsystem of the ensemble interact simultaneously with other subsystems and the interactions, i.e., $\mathbf{p}, E$-fluctuations, must be independent from each other since $\mathbf{p}, E$-fluctuations of $\mathbf{p}, E$-fluctuations etc., are not allowed within projection theory. In Sections 4.1–4.7 we treated many-body systems just in this way.

This in particular leads to the following scenario. Let us consider a certain subsystem, say 1, which is characterized by specific internal $\mathbf{p}, E$-fluctuations symbolized by $\Delta\mathbf{p}_1, \Delta E_1$.[182] A second subsystem, say 2, interacts with subsystem 1 defining $\mathbf{p}, E$-fluctuations (symbolized by $\Delta\mathbf{p}_2, \Delta E_2$) that are independent of the fluctuations $\Delta\mathbf{p}_1, \Delta E_1$, that is, we have the co-existence of independent $\mathbf{p}, E$-fluctuations $\Delta\mathbf{p}_1, \Delta E_1$ and $\Delta\mathbf{p}_2, \Delta E_2$, and we would like to express this property simply by $\{\Delta\mathbf{p}_1, \Delta E_1, \Delta\mathbf{p}_2, \Delta E_2\}$. In this way processes like $\Delta\Delta\mathbf{p}_1, \Delta\Delta E_1$ are avoided.[183] This principle of independent $\mathbf{p}, E$-fluctuations must of course also be valid for the whole complex aggregate, i.e., in the case of n subsystems we come to the following scheme[184]:

$$\{\Delta\mathbf{p}_1, \Delta E_1\},$$
$$\{\Delta\mathbf{p}_1, \Delta E_1; \Delta\mathbf{p}_2, \Delta E_2\},$$
$$\{\Delta\mathbf{p}_1, \Delta E_1; \Delta\mathbf{p}_2, \Delta E_2; \Delta\mathbf{p}_3, \Delta E_3\},$$
$$\vdots$$
$$\{\Delta\mathbf{p}_1, \Delta E_1; \Delta\mathbf{p}_2, \Delta E_2; \Delta\mathbf{p}_3, \Delta E_3; \cdots ; \Delta\mathbf{p}_n, \Delta E_n\}. \quad (4.238)$$

Then, the corresponding wave functions $\Psi_1(\mathbf{p}, E), \Psi_2(\mathbf{p}, E), \ldots, \Psi_n(\mathbf{p}, E)$, are also independent from each other and scheme (4.236) must be replaced by

$$\Psi_1(\mathbf{p}, E),$$
$$\Psi_1(\mathbf{p}, E)\Psi_2(\mathbf{p}, E),$$
$$\Psi_1(\mathbf{p}, E)\Psi_2(\mathbf{p}, E)\Psi_3(\mathbf{p}, E), \quad (4.239)$$
$$\vdots$$
$$\Psi_1(\mathbf{p}, E)\Psi_2(\mathbf{p}, E)\Psi_3(\mathbf{p}, E)\cdots\Psi_n(\mathbf{p}, E).$$

Then, the whole aggregate of matter forms the state $\Psi_{agg}(\mathbf{p}, E)$:

$$\Psi_{agg}(\mathbf{p}, E) = \Psi_1(\mathbf{p}, E)\Psi_2(\mathbf{p}, E)\Psi_3(\mathbf{p}, E)\cdots\Psi_n(\mathbf{p}, E). \quad (4.240)$$

---

[182] Clearly, subsystem 1 must consist of more than one part in order to be able to define these $\mathbf{p}, E$-fluctuations at all.
[183] See scheme (4.237).
[184] Instead of scheme (4.237).

In connection with the analysis given here, the following point is relevant. Because no $\mathbf{p}, E$-fluctuation is more fundamental than the others, all kinds of $\mathbf{p}, E$-fluctuations are equivalent, we always remain at the same level and, therefore, the basis for a "hierarchy of the parts in a part" is not given. No doubt, this feature is relevant for the description of matter organization.

4.13.2.5. *Conclusion*

Again, the principle "hierarchy of the parts in a part" is based on the assumption that there are certain units in the world (elementary particles, strings) and levels of description, respectively, that are more fundamental than others. This concept is mainly based on the belief that a human is able to recognize that what is often called ultimate (absolute) truth. However, as we have outlined in Section 2.2, the ultimate truth can principally not be perceived by a human being when the laws of evolution are true. Within this conception a basic reality or an ultimate truth cannot be denied but means that it remains hidden to a human observer. This is a fundamental point.

Human beings like to argue in terms of an ultimate truth without knowing what it really is. We find this peculiarity in almost all fields. Robert B. Laughlin wrote:

> *Thus it is in our natures to orient ourselves using absolute truth but to be confused and conflicted over exactly what it is.*[185]

Projection theory is more modest here because it does not deny the term absolute truth but asserts that it remains hidden for man.

In fact, projection theory is based on the insight that we are principally not able to perceive basic reality and the processes within it. This was the reason why we had to construct a fictitious reality that is embedded in $(\mathbf{p}, E)$-space and we came to the conclusion that within the projection principle a "hierarchy of the parts in a part" is not definable. The function $\Psi_{agg}(\mathbf{p}, E)$, defined by equation (4.240), reflects a certain level of description and it is important in this

---

[185] Refer to Laughlin, R., (2005) A different universe: Basic Books.

connection that the function $\Psi_{agg}(\mathbf{p}, E)$ cannot be explained by laws that are more basic than those dictated by $\mathbf{p}, E$-fluctuations leading to $\Psi_{agg}\mathbf{p}, E$. Within projection theory the level of $\mathbf{p}, E$-fluctuations is existent and nothing else. This is (almost) in accord with that what Laughlin said:

> We also know that while a simple and absolute law, such as hydrodynamics, can evolve from the deeper laws underneath, it is at the same time independent of them, in that it would be the same even if the deeper laws are changed.[186]

This statement obviously means that the laws of hydrodynamics have to be considered as basic even when there exist simultaneously deeper, underlying laws. In projection theory, there are however no laws that are more fundamental than those defined at the level of $\mathbf{p}, E$-fluctuations.

4.14. Granular Space-Time Structures

We have decomposed the "world" into N interacting subsystems where the interaction was assumed to be distance-dependent, and we have denoted the corresponding wave function by $\Psi(\mathbf{r}_{12}, \mathbf{r}_{13}, \ldots, \mathbf{r}_{(N-1)N}, t_{12}, t_{13}, \ldots, t_{(N-1)N})$. The details in connection with distance-dependent interactions are given in the Sections 4.1–4.7. Each subsystem may have a certain form (shape)[187] and comes into play by a distance-independent interaction, i.e., each space-time position $\mathbf{r}_k, t_k$ with $k = 1, \ldots, N$, that appears in $\Psi(\mathbf{r}_{12}, \mathbf{r}_{13}, \ldots, \mathbf{r}_{(N-1)N}, t_{12}, t_{13}, \ldots, t_{(N-1)N})$, is connected to a certain wave function $\Psi_F(\mathbf{r}_k, t_k)$ which characterizes the form of subsystem k, that is, we have

$$\mathbf{r}_k, t_k \rightarrow \Psi_F(\mathbf{r}_k, t_k), \quad k = 1, \ldots, N. \quad (4.241)$$

The probability density $\Psi^{\bullet}_{Fk}(\mathbf{r}_k, t_k)\Psi_{Fk}(\mathbf{r}_k, t_k)$ defines the form of subsystem k.

---

[186] Refer to Laughlin, R., (2005) A different universe: Basic Books.
[187] See Section 4.8.

4.14.1. Combined Interactions

In Section 4.4 we have treated both interaction types (distance-dependent and distance-independent interaction) independent from each other. In fact, we may assume that the $\mathbf{p}, E$-fluctuations in connection with both interactions do not influence each other; the character of the distance-dependent interaction is basically different from the distance-independent interaction so that the independence of both interactions from each other is justified. Then, we may simply extend the function $\Psi(\mathbf{r}_{12}, \mathbf{r}_{13}, \ldots, \mathbf{r}_{(N-1)N}, t_{12}, t_{13}, \ldots, t_{(N-1)N})$ by the form wave functions $\Psi_{F1}(\mathbf{r}_1, t_1) \Psi_{F2}(\mathbf{r}_2, t_2), \ldots, \Psi_{FN}(\mathbf{r}_N, t_N)$ as follows:

$$\Phi(\mathbf{r}_{12}, \mathbf{r}_{13}, \ldots, \mathbf{r}_{(N-1)N}, t_{12}, t_{13}, \ldots, t_{(N-1)N}, \mathbf{r}_1,$$
$$t_1, \mathbf{r}_2, t_2, \ldots, \mathbf{r}_N, t_N)$$
$$= \Psi(\mathbf{r}_{12}, \mathbf{r}_{13}, \ldots, \mathbf{r}_{(N-1)N}, t_{12}, t_{13}, \ldots, t_{(N-1)N})$$
$$\times \Psi_{F1}(\mathbf{r}_1, t_1) \Psi_{F2}(\mathbf{r}_2, t_2) \cdots \Psi_{FN}(\mathbf{r}_N, t_N). \quad (4.242)$$

Such an ansatz leads in general to a granular $\mathbf{r}, t$-structure. Let us give an example and let us suppose that the "world" (system) consists of three subsystems, i.e., we have $N = 3$. Then, instead of equation (4.242), we get

$$\Phi(\mathbf{r}_{12}, \mathbf{r}_{13}, \mathbf{r}_{23}, t_{12}, t_{13}, t_{23}, \mathbf{r}_1, t_1, \mathbf{r}_2, t_2, \mathbf{r}_3, t_3)$$
$$= \Psi(\mathbf{r}_{12}, \mathbf{r}_{13}, \mathbf{r}_{23}, t_{12}, t_{13}, t_{23}) \Psi_{F1}(\mathbf{r}_1, t_1)$$
$$\times \Psi_{F2}(\mathbf{r}_2, t_2) \Psi_{F3}(\mathbf{r}_3, t_3), \quad (4.243)$$

and the probability density is expressed by

$$\Phi^\bullet(\mathbf{r}_{12}, \mathbf{r}_{13}, \mathbf{r}_{23}, t_{12}, t_{13}, t_{23}, \mathbf{r}_1, t_1, \mathbf{r}_2, t_2, \mathbf{r}_3, t_3)$$
$$\times \Phi(\mathbf{r}_{12}, \mathbf{r}_{13}, \mathbf{r}_{23}, t_{12}, t_{13}, t_{23}, \mathbf{r}_1, t_1, \mathbf{r}_2, t_2, \mathbf{r}_3, t_3)$$
$$= \Psi^\bullet(\mathbf{r}_{12}, \mathbf{r}_{13}, \mathbf{r}_{23}, t_{12}, t_{13}, t_{23}) \Psi(\mathbf{r}_{12}, \mathbf{r}_{13}, \mathbf{r}_{23}, t_{12}, t_{13}, t_{23})$$
$$\times \Psi_{F1}^\bullet(\mathbf{r}_1, t_1) \Psi_{F1}(\mathbf{r}_1, t_1) \Psi_{F2}^\bullet(\mathbf{r}_2, t_2)$$
$$\times \Psi_{F2}(\mathbf{r}_2, t_2) \Psi_{F2}^\bullet(\mathbf{r}_3, t_3) \Psi_{F2}(\mathbf{r}_3, t_3). \quad (4.244)$$

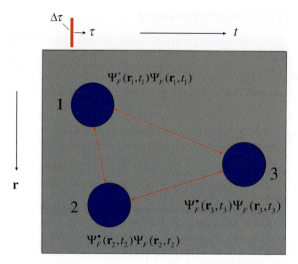

Fig. 4.5. Granular space-time structure in connection with a "world" consisting of only three subsystems. The three circles correspond to $\Psi^{\bullet}_{Fk}(\mathbf{r}_k, t_k)\Psi_{Fk}(\mathbf{r}_k, t_k)$, $k = 1, 2, 3$. The arrows indicate that distance-dependent interactions are effective. There are gaps with respect to the space coordinates x, y, z as well as with respect to time t. This granular would also be detected by an external observer with a "time-feeling" in terms of τ. The role of the observer in connection with reference time τ is analyzed in Chapter 2.

The situation is sketched in Figure 4.5: The 3 circles correspond to

$$\Psi^{\bullet}_{Fk}(\mathbf{r}_k, t_k)\Psi_{Fk}(\mathbf{r}_k, t_k), \quad k = 1, 2, 3, \qquad (4.245)$$

within $(\mathbf{r}, t)$-space, and the distance-dependent interaction is indicated by arrows. In Figure 4.5, a granular $\mathbf{r}, t$-structure is recognizable, that is, we have specific gaps in $(\mathbf{r}, t)$-space where the function $\Phi^{\bullet}\Phi$ defined by (4.244) becomes zero.

There are not only certain intervals with respect to x, y, z with $\mathbf{r} = (x, y, z)$, where the functions $\Psi^{\bullet}_{Fk}(\mathbf{r}_k, t_k)\Psi_{Fk}(\mathbf{r}_k, t_k)$, $k = 1, 2$, become zero, but we also have gaps in connection with time t, that is, also here we have

$$\Psi^{\bullet}_{Fk}(\mathbf{r}_k, t_k)\Psi_{Fk}(\mathbf{r}_k, t_k) = 0, \quad k = 1, 2, 3. \qquad (4.246)$$

In nature we observe certain gaps with respect to the space-coordinates x, y, z[188] but we do not observe gaps in connection with time t. However, this might be an effect at the macroscopic level (level of everyday life) and must not appear at the microscopic level, which is not accessible with our sense organs.

4.14.2. *Selections*

The system ("world") consisting of three subsystems, characterized in Figure 4.5 by three circles, fluctuates statistically between all possible $\mathbf{r}, t$-configurations t, defined by the ranges of $\Psi^{\bullet}_{Fk}(\mathbf{r}_k, t_k)\Psi_{Fk}(\mathbf{r}_k, t_k)$, $k = 1, 2, 3$. The $\mathbf{r}, t$-structure defined by all three circles is created by single events, and it is a feature of real numbers that the complete $(\mathbf{r}, t)$-structure, characterized by $\Psi^{\bullet}_{Fk}(\mathbf{r}_k, t_k)\Psi_{Fk}(\mathbf{r}_k, t_k)$, $k = 1, 2, 3$, can already be created within a time interval ε that can be infinitesimal but must be different from zero, and this is because an infinite number of events occur within ε.[189]

On the other hand, this system with three subsystems is observed by means of the reference time τ, described in Figure 4.5 by the vertical beam τ, which has a width of $\Delta\tau$, that is, a rectangular form for the reference time distribution is used in Figure 4.5.[190] At reference time t_B, with $\tau - \Delta_\tau/2 \leq t_B \leq \tau + \Delta_\tau/2$, all the configurations t, that is, the complete information with respect to the functions $\Psi^{\bullet}_{Fk}(\mathbf{r}_k, t_k)\Psi_{Fk}(\mathbf{r}_k, t_k)$, $k = 1, 2, 3$ are realized simultaneously, so to speak, but only those at time t_B can be observed $\Psi^{\bullet}_{Fk}(\mathbf{r}_k, t_k)\Psi_{Fk}(\mathbf{r}_k, t_k)$, $k = 1, 2, 3$. As we have outlined in Section 2.3.9, the reference time τ behaves almost non-statistically, i.e., the width of the beam in Figure 4.5 is $\Delta\tau$, and $\Delta\tau$ is very small; τ is defined by our clocks that we use in everyday life going monotonically from the past to the future. In this way the time structures given by $\Psi^{\bullet}_{Fk}(\mathbf{r}_k, t_k)\Psi_{Fk}(\mathbf{r}_k, t_k)$, $k = 1, 2, 3$ are systematically scanned by the reference time beam. In this way, we continuously observe $\mathbf{r}$-structures (sun, moon, cars, trees,

---

[188] There is, for example, space between sun and moon.
[189] This feature has been called ε property in Section 2.3.8.4.
[190] For details, see Section 2.3.9.

etc.) in the course of time τ, and there are no gaps in connection with such observation where no **r**-structure appears.

4.14.3. *The Unified Whole*

The world seems to be built up of subsystems, which are normally assumed to be independent from the interaction between them, indicated by the arrows in Figure 4.5. In other words, we suppose anyhow that there can really exist certain subsystems without mutual interactions.[191] It is assumed that each of the subsystems can exist in an isolated state and that this existence is not dependent on certain external units and parameter, respectively. Our observations in everyday life suggest such a model. However, we have to be careful.

This picture of independent units (subsystems) assumes that, at a certain time τ_{begin}, the interaction can be switched on and, in principle, that it can be switched off at time τ_{end}. However, we do not observe such a switching-on/switching-off scenario in nature. In particular, it assumes a mechanism that is hostile against projection theory because it can obviously not be based on **p**, E-fluctuations.

We know from the findings in Section 4.9 that the three subsystems represented in Figure 4.5 cannot exist in an isolated state, i.e., each of the three systems cannot exist without other systems. The geometrical forms $\Psi^{\bullet}_{Fk}(\mathbf{r}_k, t_k)\Psi_{Fk}(\mathbf{r}_k, t_k)$, $k = 1, 2, 3$, can only be different from zero if there are interaction processes with other systems. These interactions processes must not but can take place exclusively between the three subsystems on the basis of distance-independent interactions but may take place without the presence of distance-dependent interactions. Then, the situation in Figure 4.5 is characterized as follows. There are no arrows between the units, and the form of the units must be identical, that is, we have

$$\Psi^{\bullet}_{F1}(\mathbf{r}_1, t_1)\Psi_{F1}(\mathbf{r}_1, t_1) = \Psi^{\bullet}_{F2}(\mathbf{r}_2, t_2)\Psi_{F2}(\mathbf{r}_2, t_2)$$
$$= \Psi^{\bullet}_{F3}(\mathbf{r}_3, t_3)\Psi_{F2}(\mathbf{r}_3, t_3). \quad (4.247)$$

---

[191] In Figure 4.5 it is the situation without the arrows.

In conclusion, none of the three subsystems may exist in an isolated state.

Therefore, instead of a world with a lot of superimposed interactions, which is characterized by equation (4.242), we should have only one process forming a unified whole, and we would like to describe it by the function $\Psi(\mathbf{r}, t)$ where the two variables $\mathbf{r}, t$ describe all the details of the whole cosmos. In other words, we have

$$\Phi(\mathbf{r}_{12}, \mathbf{r}_{13}, \mathbf{r}_{23}, t_{12}, t_{13}, t_{23}, \mathbf{r}_1, t_1, \mathbf{r}_2, t_2, \mathbf{r}_3, t_3) \Rightarrow \Psi(\mathbf{r}, t). \quad (4.248)$$

There is however another big difference between the representation

$$\Phi(\mathbf{r}_{12}, \mathbf{r}_{13}, \mathbf{r}_{23}, t_{12}, t_{13}, t_{23}, \mathbf{r}_1, t_1, \mathbf{r}_2, t_2, \mathbf{r}_3, t_3),$$

and

$$\Psi(\mathbf{r}, t).$$

While the system $\Phi(\mathbf{r}_{12}, \mathbf{r}_{13}, \mathbf{r}_{23}, t_{12}, t_{13}, t_{23}, \mathbf{r}_1, t_1, \mathbf{r}_2, t_2, \mathbf{r}_3, t_3)$ can exist without other external units, the system $\Psi(\mathbf{r}, t)$ cannot. The global wave function $\Psi(\mathbf{r}, t)$ describes *one* system and, therefore, $\Psi(\mathbf{r}, t)$ describes the form (shape) of this system. Thus, instead of $\Psi(\mathbf{r}, t)$ we would like to use the notation $\Psi_F(\mathbf{r}, t)$. We have treated such systems in Section 4.8 in connection with distance-independent interactions, and it is a general feature that such systems cannot exist in an isolated state, i.e., we need for the existence of our system with $\Psi_F(\mathbf{r}, t)$ a counterpart, and the mutual interaction, in this case distance-independent, between the system and its counterpart leads to the existence of both systems. In the case of a $\Psi_F(\mathbf{r}, t)$-description we have not recourse to the artificial switching-on/switching-off scenario as we discussed in connection with $\Phi(\mathbf{r}_{12}, \ldots, t_{12}, \ldots, \mathbf{r}_1, t_1, \ldots)$.[192]

We have already discussed in Section 2.3.9 that the complete form of the system, directly described by $\Psi_F^\bullet(\mathbf{r}, t)\Psi_F(\mathbf{r}, t)$, is permanently new created between each infinitesimal time interval $\Delta\tau$ around τ. In other words, there is a permanent new creation of the whole system, described by $\Psi_F^\bullet(\mathbf{r}, t)\Psi_F(\mathbf{r}, t)$, within $\Delta\tau$ around each time τ, and

---

[192] See equation (4.242).

the reference time τ goes strictly from the past to the future. Again, $\Delta\tau$ may be infinitesimal but must be different from zero. Thus, the entire information for $\Psi_F^\bullet(\mathbf{r},t)\Psi_F(\mathbf{r},t) \neq 0$ is given at time τ, or more precisely within the time interval $\tau \pm \Delta\tau/2$, if the $\mathbf{r},t$-range of $\Psi_F^\bullet(\mathbf{r},t)\Psi_F(\mathbf{r},t)$ remains finite:

$$\lim_{\mathbf{r},t\to\infty} \Psi_F^\bullet(\mathbf{r},t)\Psi_F(\mathbf{r},t) = 0. \tag{4.249}$$

Nevertheless, the human observer can only make statements about $\Psi_F^\bullet(\mathbf{r},t)\Psi_F(\mathbf{r},t)$ for times t, which are within the time interval $\Delta\tau$. In connection with Figure 4.6, the observer can only detect the structure $\Psi_F^\bullet(\mathbf{r},t_1)\Psi_F(\mathbf{r},t_1)$ with $t_1 = \tau$. The structures $\Psi_F^\bullet(\mathbf{r},t_2)\Psi_F(\mathbf{r},t_2)$ and $\Psi_F^\bullet(\mathbf{r},t_3)\Psi_F(\mathbf{r},t_3)$ exist at time τ but are not accessible to the observer because we have $t_2 < \tau \pm \Delta\tau/2$ and $t_3 > \tau \pm \Delta\tau/2$.

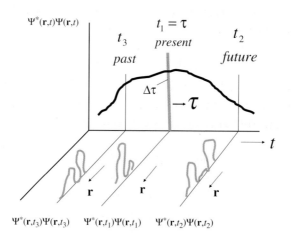

Fig. 4.6. The space-time structure described by the wave function $\Phi(\mathbf{r}_{12},\ldots,t_{12},\ldots,\mathbf{r}_1,t_1,\ldots)$[193] can in principle also be described by a function $\Psi(\mathbf{r},t)$ having only two variables, that is, in this case the function $\Psi(\mathbf{r},t)$ must contain all the details of $\Phi(\mathbf{r}_{12},\ldots,t_{12},\ldots,\mathbf{r}_1,t_1,\ldots)$. Then, we have a system that forms a unified whole, and we have not to recourse to the artificial switching-on/switching-off scenario.

---

[193] See equation (4.244).

4.14.4. Simple Cosmological Considerations

If we transfer this picture to the whole cosmos, we have the following situation. Let us denote the space-time points of the cosmos in $(\mathbf{r}, t)$-space by $\mathbf{r}_c, t_c$ and its wave function by $\Phi_{FC}(\mathbf{r}_c, t_c)$. In other words, instead of $\Psi_F(\mathbf{r}, t)$ we study the function $\Phi_{FC}(\mathbf{r}_c, t_c)$:

$$\Psi_F(\mathbf{r}, t) \rightarrow \Phi_{FC}(\mathbf{r}_c, t_c). \qquad (4.250)$$

$\Phi_{FC}(\mathbf{r}_c, t_c)$ describes the form of the cosmos. Let us denote the Fourier transform of $\Phi_{FC}(\mathbf{r}_c, t_c)$ by $\Phi_{FC}(\mathbf{p}_c, E_c)$ where $\mathbf{p}_c, E_c$ are the variables of the cosmos in $(\mathbf{p}, E)$-space. The existence of the cosmos assumes a counterpart, and for the corresponding wave functions and variables we have chosen the following markings: $\Phi_{FCC}(\mathbf{r}_{cc}, t_{cc})$ is the wave function in $(\mathbf{r}, t)$-space and the variables $\mathbf{r}_{cc}, t_{cc}$ are the space-time points. $\Phi_{FCC}(\mathbf{p}_{cc}, E_{cc})$ is the Fourier transform of $\Phi_{FCC}(\mathbf{r}_{cc}, t_{cc})$ and $\mathbf{p}_{cc}, E_{cc}$ are the variables of the counterpart of the cosmos in $(\mathbf{p}, E)$-space.

We have outlined in Section 4.14.3 that each of both systems (cosmos and its counterpart) can only exist if there is second system, which interact with each other. None of the systems can exist without the other, that is, the cosmos cannot exist without the counterpart and vice versa. In other words, there are $\mathbf{p}, E$-fluctuations $\Delta \mathbf{p}, \Delta E$ between the cosmos and the counterpart. Through these fluctuations the cosmos and the counterpart come into existence and we get

$$\Phi_{FC}(\mathbf{p}_c, E_c) \neq 0, \qquad (4.251)$$

and

$$\Phi_{FCC}(\mathbf{p}_{cc}, E_{cc}) \neq 0. \qquad (4.252)$$

The projections (Fourier transforms) of $\Phi_{FC}(\mathbf{p}_c, E_c)$ and $\Phi_{FCC}(\mathbf{p}_{cc}, E_{cc})$ onto $(\mathbf{r}, t)$-space leads to $\Phi_{FC}(\mathbf{r}_c, t_c)$ and $\Phi_{FCC}(\mathbf{r}_{cc}, t_{cc})$. The situation is summarized in scheme (4.253):

$$\begin{array}{ccc} \Phi_{FC}(\mathbf{p}_c, E_c) & \underset{\Delta \mathbf{p}, \Delta E}{\leftrightarrow} & \Phi_{FCC}(\mathbf{p}_{cc}, E_{cc}) \\ \downarrow & & \downarrow \\ \Phi_{FC}(\mathbf{r}_c, t_c) & & \Phi_{FCC}(\mathbf{r}_{cc}, t_{cc}) \end{array} \qquad (4.253)$$

The basic equation in the case of a distance-independent interaction is given by equation (4.218). In analogy to this equation we get for the cosmos and its counterpart the following formulations:

$$i\hbar\frac{\partial}{\partial t}\Phi_{FC}(\mathbf{r}_c, t_c) = -\frac{\hbar^2}{2m_0}\Delta\Phi_{FC}(\mathbf{r}_c, t_c)$$
$$+ V_{FC}(x_c, y_c, z_c, t_c)\Phi_{Fc}(\mathbf{r}_c, t_c), \qquad (4.254)$$

$$i\hbar\frac{\partial}{\partial t}\Phi_{FCC}(\mathbf{r}_{cc}, t_{cc}) = -\frac{\hbar^2}{2m_0}\Delta\Phi_{FCC}(\mathbf{r}_{cc}, t_{cc})$$
$$+ V_{FCC}(x_{cc}, y_{cc}, z_{cc}, t_{cc})\Phi_{FCC}(\mathbf{r}_{cc}, t_{cc}). \qquad (4.255)$$

Since the cosmos and its counterpart must be identical[194] both potential functions must fulfil the condition

$$V_{FC}(x_c, y_c, z_c, t_c) = V_{FCC}(x_{cc}, y_{cc}, z_{cc}, t_{cc}). \qquad (4.256)$$

Equivalent equations can be formulated for $(\mathbf{p}, E)$-space.

4.14.5. *Arbitrary Motions through Space and Time*

We may shift the functions $\Phi_{FC}(\mathbf{r}_c, t_c)$ and $\Phi_{FCC}(\mathbf{r}_{cc}, t_{cc})$ relative to $(\mathbf{r}, t)$-space without to change the interaction,[195] i.e., the law for the $\mathbf{p}, E$-fluctuations, which are described by the probability densities $\Phi_{FC}^{\bullet}(\mathbf{p}_c, E_c)\Phi_{FC}(\mathbf{p}_c, E_c)$ and $\Phi_{FCC}^{\bullet}(\mathbf{p}_{cc}, E_{cc})\Phi_{FCC}(\mathbf{p}_{cc}, E_{cc})$, is not changed when $\mathbf{r}_c$ and t_c are shifted by the quantities $\Delta_{c,\mathbf{r}}$ and $\Delta_{c,t}$:

$$\begin{aligned}\mathbf{r}_c &\to \mathbf{r}_c - \Delta_{c,\mathbf{r}},\\ t_c &\to t_c - \Delta_{c,t},\end{aligned} \qquad (4.257)$$

and when $\mathbf{r}_{cc}$ and t_{cc} are shifted by the quantities $\Delta_{cc,\mathbf{r}}$ and $\Delta_{cc,t}$:

$$\begin{aligned}\mathbf{r}_{cc} &\to \mathbf{r}_{cc} - \Delta_{cc,\mathbf{r}},\\ t_{cc} &\to t_{cc} - \Delta_{cc,t}.\end{aligned} \qquad (4.258)$$

---

[194] This is as in the case of the systems i and j; see Section 4.8.
[195] See in particular the discussion in Section 4.8.3.

Interactions

In analogy to that which is outlined in Section 4.8.3 we obtain for the variables $\Delta_{c,\mathbf{r}}$ and $\Delta_{c,t}$ the following equation of motion for an observer who is resting relative to $(\mathbf{r}, t)$-space:

$$\Psi(\Delta_{c\mathbf{r}}, \Delta_{ct}) = C_c \Psi(\mathbf{p}_{0c}, E_{0c}) \exp\left\{\frac{i}{\hbar}(\mathbf{p}_{0c} \cdot \Delta_{c\mathbf{r}} - E_{0c}\Delta_{ct})\right\}. \quad (4.259)$$

We obtain for the variables $\Delta_{c,\mathbf{r}}$ and $\Delta_{c,t}$ an analogous equation with exactly the same structure:

$$\Psi(\Delta_{cc\mathbf{r}}, \Delta_{cct}) = C_{cc} \Psi(\mathbf{p}_{0cc}, E_{0cc}) \exp\left\{\frac{i}{\hbar}(\mathbf{p}_{0cc} \cdot \Delta_{cc\mathbf{r}} - E_{0cc}\Delta_{cct})\right\}. \quad (4.260)$$

The quantities $\mathbf{p}_{0c}, E_{0c}$ and $\mathbf{p}_{0cc}, E_{0cc}$ in equations (4.259) and (4.260) are the total momenta and the total energies of the cosmos and the counterpart of the cosmos. We have outlined in Section 4.4.2 that the $\mathbf{p}, E$-fluctuations[196] are independent of the actual values $\mathbf{p}_{0c}, E_{0c}$ and $\mathbf{p}_{0cc}, E_{0cc}$, i.e., the $\mathbf{p}, E$-values fluctuate around these values but there are no $\mathbf{p}, E$-fluctuations of $\mathbf{p}_{0c}, E_{0c}$, and $\mathbf{p}_{0cc}, E_{0cc}$. In other words, there is no coupling between $\mathbf{p}_{0c}, E_{0c}$, and $\mathbf{p}_{0cc}, E_{0cc}$ and the $\mathbf{p}, E$-fluctuations. Because there are no interactions between external systems the values $\mathbf{p}_{0c}, E_{0c}$, and $\mathbf{p}_{0cc}, E_{0cc}$ remain constant in the course of time τ.

We know from Chapter 3 that both equations, that is, equations (4.259) and (4.260), are the equations for free (non-interacting) systems. Thus, we must have[197]:

$$\Psi(\Delta_{c\mathbf{r}}, \Delta_{ct}) = 0, \quad (4.261)$$

$$\Psi(\mathbf{p}_{0c}, E_{0c}) = 0, \quad (4.262)$$

with

$$\mathbf{p}_{0c} = 0, E_{0c} = 0, \quad (4.263)$$

---

[196] In this case we have internal fluctuations.
[197] See Chapter 3.

and

$$\Psi(\Delta_{cc\,\mathbf{r}}, \Delta_{cc\,t}) = 0, \qquad (4.264)$$

$$\Psi(\mathbf{p}_{0cc}, E_{0cc}) = 0, \qquad (4.265)$$

with

$$\mathbf{p}_{0cc} = 0, E_{0cc} = 0. \qquad (4.266)$$

That is, both systems, i.e., cosmos and its counterpart, move arbitrarily through $(\mathbf{r}, t)$-space, denoted here by frame of reference S. This is the case for observers who are resting relative to S. However, both systems the cosmos and its counterpart, are not observable by the observers, resting relative to S, and this is because we have

$$\Psi^{\bullet}(\Delta_{c\,\mathbf{r}}, \Delta_{c\,t})\Psi(\Delta_{c\,\mathbf{r}}, \Delta_{c\,t}) = 0, \qquad (4.267)$$

$$\Psi^{\bullet}(\Delta_{cc\,\mathbf{r}}, \Delta_{cc\,t})\Psi(\Delta_{cc\,\mathbf{r}}, \Delta_{cc\,t}) = 0. \qquad (4.268)$$

All these features have been discussed in Section 4.8.9.

In conclusion, the cosmos and its counterpart are not detectable for observers who are resting relative to S, i.e., relative to $(\mathbf{r}, t)$-space. However, the cosmos with $\Phi_{FC}(\mathbf{p}_c, E_c)$ is observable for an observer who is resting within this cosmos and, according to what we have outlined in Section 4.8.9, for this resting observer we must have

$$\Psi^{\bullet}(\Delta_{c\,\mathbf{r}}, \Delta_{c\,t})\Psi(\Delta_{c\,\mathbf{r}}, \Delta_{c\,t}) = 1. \qquad (4.269)$$

It is easy to recognize that this observer, resting relative to the cosmos, is not able to detect the counterpart of the cosmos, that is, also here we have

$$\Psi^{\bullet}(\Delta_{cc\,\mathbf{r}}, \Delta_{cc\,t})\Psi(\Delta_{cc\,\mathbf{r}}, \Delta_{cc\,t}) = 0, \qquad (4.270)$$

as in the case of the observer who is resting relative to S.[198]

On the other hand, the counterpart of the cosmos with $\Phi_{FCC}(\mathbf{p}_{cc}, E_{cc})$ is observable for an observer who is resting within this

---

[198] See equation (4.268).

counterpart, and in this case we have[199]

$$\Psi^\bullet(\Delta_{cc\,\mathbf{r}}, \Delta_{cc\,t})\Psi(\Delta_{cc\,\mathbf{r}}, \Delta_{cc\,t}) = 1. \qquad (4.271)$$

However, this observer, resting relative to the counterpart, is not able to detect the cosmos, and we have also here

$$\Psi^\bullet(\Delta_{c\,\mathbf{r}}, \Delta_{c\,t})\Psi(\Delta_{c\,\mathbf{r}}, \Delta_{c\,t}) = 0, \qquad (4.272)$$

as in the case of the observer who is resting relative to S.[200]

In connection with scheme (4.253), it should be mentioned that the conservation laws for the momentum and the energy require that at each time τ the following relationships are fulfilled

$$\Phi^\bullet_{FC}(\mathbf{p}_c, E_c)\Phi_{FC}(\mathbf{p}_c, E_c)$$
$$= \Phi^\bullet_{FCC}(-\mathbf{p}_{cc}, -E_{cc})\Phi_{FCC}(-\mathbf{p}_{cc}, -E_{cc}), \qquad (4.273)$$

with

$$\Phi^\bullet_{FC}(\mathbf{p}_c, E_c)\Phi_{FC}(\mathbf{p}_c, E_c)$$
$$= \Phi^\bullet_{FC}(-\mathbf{p}_c, -E_c)\Phi_{FC}(-\mathbf{p}_c, -E_c), \qquad (4.274)$$

and

$$\Phi^\bullet_{FCC}(\mathbf{p}_{cc}, E_{cc})\Phi_{FCC}(\mathbf{p}_{cc}, E_{cc})$$
$$= \Phi^\bullet_{FCC}(-\mathbf{p}_{cc}, -E_{cc})\Phi_{FCC}(-\mathbf{p}_{cc}, -E_{cc}). \qquad (4.275)$$

The details for the requirement of this symmetry are outlined above in Section 4.4.5.

Equation (4.273) means the following. If, for example, the cosmos takes at time τ the values $\mathbf{p}_a, E_a$, the counterpart of the cosmos must simultaneously be in the state of $-\mathbf{p}_a, -E_a$. Note that the total momenta and total energise are expressed by $\mathbf{p}_{0c} = 0, E_{0c} = 0,$[201] and $\mathbf{p}_{0cc} = 0, E_{0cc} = 0.$[202]

---

[199] This is in analogy to equation (4.269).
[200] See equation (4.267).
[201] See equation (4.263).
[202] See equation (4.266).

4.14.6. Decomposition of the Cosmos

When we accept that the cosmos forms a unified whole described by a wave function of the form $\Phi_{FC}(\mathbf{r}_c, t_c)$ we may decompose this structure, and we may, for the purpose of description, introduce subsystems that interact with each other. Therefore, instead of $\Phi_{FC}(\mathbf{r}_c, t_c)$ we would obtain for example a function like[203]

$$\Phi(\mathbf{r}_{12}, \mathbf{r}_{13}, \ldots, \mathbf{r}_{(N-1)N}, t_{12}, t_{13}, \ldots, t_{(N-1)N},$$
$$\mathbf{r}_1, t_1, \mathbf{r}_2, t_2, \ldots, \mathbf{r}_N, t_N).$$

That is, for the purpose of description we would have

$$\Phi_{FC}(\mathbf{r}_c, t_c) \rightarrow \Phi(\mathbf{r}_{12}, \mathbf{r}_{13}, \ldots, \mathbf{r}_{(N-1)N}, t_{12}, t_{13}, \ldots, t_{(N-1)N},$$
$$\mathbf{r}_1, t_1, \mathbf{r}_2, t_2, \ldots, \mathbf{r}_N, t_N). \quad (4.276)$$

However, the transition (4.276) does not change reality, which in principle remains unified, characterized by $\Phi_{FC}(\mathbf{r}_c, t_c)$. Clearly, the transition (4.276) also does not eliminate the counterpart of the cosmos that we have characterized by the function $\Phi_{FCC}(\mathbf{r}_{cc}, t_{cc})$.

4.15. Summary and Final Remarks

The most basic information about the world outside is represented in the form of pictures in front of us; all essential things are represented within the frame of pictures. These pictures contain geometrical structures, which are given as a function of the variables $\mathbf{r}$ and t. Within projection theory, these pictures correspond to the probability density $\Psi^{\bullet}(\mathbf{r}, t)\Psi(\mathbf{r}, t)$, that is, the wave function $\Psi(\mathbf{r}, t)$ is the most fundamental element. $\Psi(\mathbf{r}, t)$ can be transferred from $(\mathbf{r}, t)$-space to $(\mathbf{p}, E)$-space without any loss on information.[204] Thus, the wave function $\Psi(\mathbf{p}, E)$ of $(\mathbf{p}, E)$-space is as fundamental as $\Psi(\mathbf{r}, t)$ of $(\mathbf{r}, t)$-space, and the density distribution $\Psi^{\bullet}(\mathbf{p}, E)\Psi(\mathbf{p}, E)$ is relevant for the $\mathbf{p}, E$-distribution in $(\mathbf{p}, E)$-space.

---

[203] See equation (4.242).
[204] See equation (2.5).

There is however a big difference between the variables $\mathbf{r}, t$ and $\mathbf{p}, E$. The variables $\mathbf{p}, E$ behave quite differently from the variables $\mathbf{r}, t$. For $\mathbf{p}$ and E conservation laws are valid. This inevitably means that a change in $\mathbf{p}$ and E of the system under investigation must be accompanied with a change of $\mathbf{p}$ and E of another system outside. In other words, the $\mathbf{p}, E$-fluctuations in connection with the system under investigation mean that there is an interaction process between the system and another system outside (environment).

Such conservation laws do not exist for the variables $\mathbf{r}, t$ leading to the feature that $\mathbf{r}, t$ play a completely passive role. This feature supports our assumption that $\mathbf{r}$ and t are exclusively the elements of pictures; the $(\mathbf{r}, t)$-space is the frame on which reality is projected.

In conclusion, the most basic information in connection with the human observations is expressed by the wave function $\Psi(\mathbf{r}, t)$ and the probability density $\Psi^{\bullet}(\mathbf{r}, t)\Psi(\mathbf{r}, t)$, respectively. The functions $\Psi(\mathbf{p}, E)$ and $\Psi^{\bullet}(\mathbf{p}, E)\Psi(\mathbf{p}, E)$ represent the equivalent information in $(\mathbf{p}, E)$-space. The interactions, i.e., $\mathbf{p}, E$-fluctuations take place in $(\mathbf{p}, E)$-space; the result is projected onto $(\mathbf{r}, t)$-space and we obtain a picture of reality.

The function $\Psi(\mathbf{p}, E)$ can be described by interaction potentials on the basis of equation (2.40):

$$E\Psi(\mathbf{p}, E) = \frac{\mathbf{p}^2}{2m_0}\Psi(\mathbf{p}, E)$$
$$+ V\left(i\hbar\frac{\partial}{\partial p_x}, i\hbar\frac{\partial}{\partial p_y}, i\hbar\frac{\partial}{\partial p_z}, -i\hbar\frac{\partial}{\partial E}\right)\Psi(\mathbf{p}, E).$$

This equation can be completely expressed in terms of the variables $\mathbf{r}$ and t for the determination of $\Psi(\mathbf{r}, t)$ and $\Psi^{\bullet}(\mathbf{r}, t)\Psi(\mathbf{r}, t)$, respectively, and we get equation (2.35):

$$i\hbar\frac{\partial}{\partial t}\Psi(\mathbf{r}, t) = -\frac{\hbar^2}{2m_0}\Delta\Psi(\mathbf{r}, t) + V(x, y, z, t)\Psi(\mathbf{r}, t).$$

In this way, we transfer the $\mathbf{p}, E$-fluctuations (interactions) from $(\mathbf{p}, E)$-space to $(\mathbf{r}, t)$-space, and these interactions are reflected in

$(\mathbf{r}, t)$-space as correlations between the various $\mathbf{r}, t$-points described by $\Psi^{\bullet}(\mathbf{r}, t)\Psi(\mathbf{r}, t)$.

Both equations[205] are general in character and are not depictable in $(\mathbf{p}, E)$-space and $(\mathbf{r}, t)$-space, respectively, but they are positioned at a "level of reality" which is above the "level of observation" where exclusively only specific solutions of (2.35) and (2.40) appear. A detailed discussion is given in connection with "levels of reality".[206]

Due to the equivalence of $(\mathbf{p}, E)$-space and $(\mathbf{r}, t)$-space, the potential functions can also be expressed in terms of the variables $\mathbf{r}$ and t. This does however not mean that there exist a real potential energy between two positions in $(\mathbf{r}, t)$-space. There is nothing else than geometrical structures in $(\mathbf{r}, t)$-space without any material and/or energetic aspect. Not the potential energy described by $V(\mathbf{r}, t)$ is embedded in $(\mathbf{r}, t)$-space, but only the geometrical structures that are based on the wave function $\Psi(\mathbf{r}, t)$ and the probability density $\Psi^{\bullet}(\mathbf{r}, t)\Psi(\mathbf{r}, t)$, respectively.

Within projection theory the interaction potential can be interpreted at best as auxiliary element without any imaginable background.[207] This point has been discussed extensively in Section 4.9.

In other words, not the potential is the primary quantity but the $\mathbf{p}, E$-fluctuations and $\mathbf{r}, t$-correlations, respectively. In $(\mathbf{r}, t)$-space we have exclusively geometrical structures, and no real objects are embedded here. Therefore, it makes no sense to assume that there is an exchange of energy between the geometrical positions.

As we have outlined in Section 2.3.8.3 an isolated consideration of $\Psi^{\bullet}(\mathbf{r}, t)\Psi(\mathbf{r}, t)$ makes not much sense but only in connection with real processes which however take place in $(\mathbf{p}, E)$-space. In particular, we stated in Section 2.3.8.3 the following. One of the possible values for $\mathbf{p}$ and for E is present in the intervals $\mathbf{r}, \mathbf{r} + \Delta \mathbf{r}$ and $t, t + \Delta t$ with the probability density $\Psi^{*}(\mathbf{r}, t)\Psi(\mathbf{r}, t)$. This statement is consistent with the general fact that we can only say something about the

---

[205] See equations (2.35) and (2.40).
[206] Interested readers might refer to Schommers, W., Cosmic secrets: World Scientific, in preparation.
[207] There is no information or energy transfer through $(\mathbf{r}, t)$-space.

space-time elements x, y, z, t in connection with masses and physically real processes.[208]

In conclusion, the projection principle leads to fundamental new aspects in connection with the notion "interaction". It is important to underline that we have here to distinguish between distance-dependent and distance-independent correlations, calculated by distance-dependent and distance-independent interactions. While the distance-independent interactions define the form (shape) of a system, that is, its geometrical structure, distance-dependent interactions describe the correlations between the various systems in $(\mathbf{r}, t)$-space.

On the basis of this basic conception the following points have been treated in this chapter:

How basic is the notion "interaction"?

The notion "interaction" should be considered as basic effect because it was not able in the history of science to explain this notion satisfactorily by the elements of other theoretical pictures.

Classical force laws

Two interpretations are of particular relevance: "action-at-a-distance" and the so-called "proximity effect".

Delocalized systems

The whole interaction scenario in physics, i.e., in classical physics, usual quantum theory, quantum field theory, string theory, is based on processes between localized systems in space, i.e., in $(\mathbf{r}, \tau)$-space. Since the interactions in projection theory take place in $(\mathbf{p}, E)$-space[209] none of these specific processes can be applied here. It makes not much sense to explain these $\mathbf{p}, E$-fluctuations by means of the "mechanisms" used in conventional physics but we have to accept that they simply "emerge".

---

[208] See Section 2.2.3.
[209] Here we have $\mathbf{p}, E$-fluctuations.

Space-time limiting interactions

There can exist space-time limiting global interactions, which are completely determined by the parameters $(\pm \mathbf{r}_a, \pm t_a)$, which determine the $\mathbf{r}, t$-limitations.

Mutual (distance-dependent) interactions

Certain $\mathbf{p}, E$-fluctuations can lead to distance-dependent interactions having the effect that there are distance-dependent correlations in connection with the $\mathbf{r}, t$-points.

The $\mathbf{p}, E$-concert

If the $\mathbf{p}, E$-fluctuations between all pairs of the ensemble, consisting of N subsystems, are exactly the same at time τ, we can speak of a perfect "concert".

Individual processes

In connection with the entire system, consisting of N subsystems, we have $N(N-1)$ $\mathbf{p}, E$-processes, which are all independent of each other, and each process is described by the function $\Psi(\mathbf{p}, E)$ and $\Psi^*(\mathbf{p}, E)\Psi(\mathbf{p}, E)$, respectively. In the course of time τ, the N subsystems equally run through all possible $\mathbf{p}, E$-states with a certain probability which is described by the probability density $\Psi^*(\mathbf{p}, E)\Psi(\mathbf{p}, E)$. Then, we have $N(N-1)$ individual wave functions $\Psi_{12}(\mathbf{p}, E), \Psi_{13}(\mathbf{p}, E), \ldots, \Psi_{(N-1)N}(\mathbf{p}, E)$.

Pair distributions

All pairs $k, l = 1, \ldots, N, k \neq l$ of an ensemble, described by $\Psi(\mathbf{r}_{kl}, t_{kl})$, contain exactly the same information about the interaction processes, i.e., $\mathbf{p}, E$-fluctuations.

Collective effects

All pairs k, l of the ensemble are equally described by the function $\Psi(\mathbf{p}, E)$. However, the $(\mathbf{p}, E)$-fluctuations with respect to all pairs are strictly correlated and, therefore, the ensemble forms a collective $(\mathbf{p}, E)$-system.

Interactions with past and future events

Not only the space-structure at time τ is described within projection theory but also effects with respect to the system-specific time t, that is, the interactions of the subsystems with past and future events. The formalism of usual physics only allows us to investigate the space-structure at time τ because the system-specific time t is not defined here. Within projection theory the formalism is extended and the space-time structure at time τ is concerned.

Distance-independent interactions (form interaction)

We have treated a system consisting of N interacting subsystems, and we assumed that a distance-dependent interaction between the subsystems is effective, that is, the mutual interactions between the subsystems are dependent on the space-time distances $\mathbf{r}_{ki} - \mathbf{r}_l, t_k - t_l$ with $k, l = 1, \ldots, N, k \neq l$ where $\mathbf{r}_1, \mathbf{r}_2, \ldots, \mathbf{r}_N, t_1, t_2, \ldots, t_N$ are the positions and system-specific times in $(\mathbf{r}, t)$-space at time τ. However, projection theory opens the possibility for another kind of interaction, which is not dependent on the space-time distances and, therefore, we have called it "distance-independent interactions". In other words, there can be interactions between two systems, say i and j, whose strength is not dependent on the space-time distances $\mathbf{r}_i - \mathbf{r}_j, t_i - t_j$. These distance-independent interactions furnish a system with a specific form (shape). Therefore, we have called this kind of interaction also "form interaction". In this case there are no space-time correlations between the interacting systems.

Arbitrary jumps

If we have for example two systems i and j, which interact with each other only through a distance-independent interaction, then both systems, i.e., their geometrical positions, jump arbitrarily from one space-time position to another and, in principle, their effective velocities relative to $(\mathbf{r}, t)$-space can be close to infinity.

The meaning of the potential functions

In the case of form (distance-independent) interactions the "classical" potential picture is not applicable. Instead of one potential for both systems (distance-dependent interaction) here we have two potentials $V_F(x,y,z,t)_k$, $k = i,j$,[210] one for system i and another for system j. Thus, the usual interpretation in terms of more or less classical notions can no longer be applied in projection theory. In fact, the potential takes in $(\mathbf{p}, E)$-space the form of an operator

$$V_F\left(i\hbar\frac{\partial}{\partial p_x}, i\hbar\frac{\partial}{\partial p_y}, i\hbar\frac{\partial}{\partial p_z}, -i\hbar\frac{\partial}{\partial E}\right),$$

and this expression is not accessible to visualizable pictures. Therefore, interaction potentials should be considered as abstract quantities for the determination of the $\mathbf{p}, E$-fluctuation in $(\mathbf{p}, E)$-space which appear in $(\mathbf{r}, t)$-space as correlations between the various space-time positions $(\mathbf{r}, t)$, i.e., we have $\mathbf{r}, t$-correlations. In other words, not the potential is the primary quantity but the $\mathbf{p}, E$-fluctuations and $\mathbf{r}, t$-correlations, respectively. In $(\mathbf{r}, t)$-space, we have exclusively geometrical structures, and no real objects are embedded here. Therefore, it is wrong to assume that there is an exchange of energy $V_F(x,y,z,t)_k$, $k = i,j$ between the (geometrical) structures i and j.

Self-creating interaction processes

We have considered two systems i and j, which interact through a distance-independent interaction. The cause for the existence of both systems in $(\mathbf{r}, t)$-space are the $\mathbf{p}, E$-fluctuations in $(\mathbf{p}, E)$-space; existence in $(\mathbf{r}, t)$-space does not mean "material existence" in $(\mathbf{r}, t)$-space but "geometrical existence". The interaction between system i and system j are the reason for their existence in both spaces, i.e., in $(\mathbf{r}, t)$-space as well as in $(\mathbf{p}, E)$-space. These interactions have to be considered as a self-creating processes. System i cannot exist without system j and vice versa.

Relativistic effects

We have also to consider that the conceptions of projection theory and Theory of Relativity are quite different from each other. Within Theory of Relativity, the real bodies are embedded in space-time and

---

[210] See equation (4.218).

the real masses actually move through space; within projection theory we have projections onto space-time, i.e., $(\mathbf{r}, t)$-space and the resulting geometrical structures do not really move through $(\mathbf{r}, t)$-space but the projections appear as jumps between two arbitrary space-time positions, where the $\mathbf{r}, t$-points between these two arbitrary positions are not involved. It is therefore not surprising that the principles of relativity are not applicable in their basic form to projection theory. However, relativistic effects must also be of relevance in connection with projection theory.

Hierarchy of the parts in a part

Within usual physics in most cases the following concept of matter is used. There exist elementary particles (strings, branes), i.e., particles without substructure. Then, it is assumed that all matter in the cosmos is made from these basic building blocks (elementary particles). In particular, it is assumed within this concept that each of these elementary units can exist independently from other things in the cosmos. Matter is formed by putting the basic building blocks systematically together, and we have that what we have called above "hierarchy of the parts in a part". Such a conception, i.e., "hierarchy of the parts in a part", is obviously not possible within projection theory. The reason is given as follows. Within projection theory interactions are described by $\mathbf{p}, E$-fluctuations. However, the existence of various specific $\mathbf{p}, E$-curves does not mean that there are interactions which are qualitatively different from each other. There are no $\mathbf{p}, E$-fluctuations which are more fundamental than others. All $\mathbf{p}, E$-values are equivalent and there is no $\mathbf{p}, E$-value which would be more fundamental than another $\mathbf{p}, E$-value.

The unified whole

The world seems to be built up of subsystems that are normally assumed to be independent from the interaction between them. In other words, we suppose anyhow that there can really exist certain subsystems without mutual interactions. It is assumed that each of the subsystems can exist in an isolated state and that this existence is not dependent on certain external units and parameter, respectively. Our observations in everyday life suggest such a model. However, we have to be careful. This picture of independent units (subsystems) assumes that, at a certain time τ_{begin}, the interaction can be switched on and, in principle, that it can be switched off at time τ_{end}. However, we do not observe such a switching-on/switching-off scenario in nature. In particular, it assumes a mechanism that is hostile against projection theory because it can obviously not be based on $\mathbf{p}, E$-fluctuations. Therefore, instead of a world with a lot of superimposed interactions we should have only one process forming a unified whole, and we have to describe it by the function $\Psi(\mathbf{r}, t)$ where the two variables $\mathbf{r}, t$ describe all the details of the whole cosmos.

In the next chapter,[211] we will discuss the particle-wave dualism in connection with conventional quantum theory and projection theory. The second part of Chapter 5 deals with the role of the observer.

---

[211] See Chapter 5.

5

Some Basic Questions

Preliminary Remarks

In this chapter, we would like to give some brief remarks on basic questions which have been extensively discussed in connection with conventional quantum theory. Here the "particle-wave question" and the "role of the observer" are of particular importance. What is the viewpoint of projection theory and what are the differences compared to conventional quantum theory? Both points, the "particle-wave question" and the "role of the observer", are today still the subject of many controversial debates. However, we will recognize that projection theory is able to give positive contributions to these questions. In particular, we are able to find out why conventional quantum theory can only give unsatisfactory answers in connection with the basic understanding of certain quantum phenomena.

5.1. The Particle-Wave Question

We have discussed in Section 1.6.1 that in usual quantum theory, a non-interacting system can be a particle as well as a wave, depending on the chosen experimental arrangement by the observer. According to the principle of complementarity[1] only *one* of the two incompatible possibilities (wave, particle) can be realized, and this implies that the experimental arrangements that determine those properties must be similarly mutually exclusive, otherwise we have a problem since we are confronted with a logical contradiction. Just this problem

---
[1] See Section 1.6.1.

appears in connection with the new kind of experiments,[2] which have been performed by two groups.[3] Both experimental groups came to the same conclusion. Light showed both particle and wave aspects *simultaneously*, and this is in contrast to the principle of complementarity that is one of the very basics of modern (usual) quantum theory (Copenhagen Interpretation).

5.1.1. *No Need for an Experimental Arrangement*

In usual quantum theory, a non-interacting system can be a point-like system with constant momentum $\mathbf{p}_0 = p_x$ and constant energy E_0 *or* a wave, having the wave-length λ ($\lambda = \hbar/p_x$) and the frequency ω ($\omega = \hbar/E_0$). Not the impact of the experimental apparatus on a *well-defined* system decides about the nature of the system (wave or particle) during the contact with the experimental apparatus, but we have the following situation in usual quantum theory. Before the experimental contact, the system is *not defined* and the decision about its nature (wave or particle) is dependent on *how* the experimental apparatus has been arranged.

Within projection theory, we do not need an experimental arrangement that decides about the nature of a system (wave and/or particle). The system is completely described by the *interaction* of the system with its environment. No interaction means within projection theory that the system does not exist.[4] When can we expect that a system is described by a wave? Let us briefly discuss this point.

As already outlined several times, within the investigation given here there can be no point-like particle with constant momentum $\mathbf{p}_0$ and constant energy E_0. Within projection theory, the form (shape) of a system is always created by distance-independent interactions

---

[2] Interested readers might refer to Mizobuchi, Y., and Othake, Y., (1992) Phys. Lett. A **168**, 1 and to Brida, G., Genovese, M., Gramegna, M., and Predazzi, E., (1996) Phys. Lett. A **328**, 313.
[3] Interested readers might refer to the Mizobuchi-Ohtake experiment and the experiment by Brida, Genovese, Gramegna and Predazzi in the references in the above footnote. For details, see Section 1.8.
[4] See Chapter 3.

Some Basic Questions 363

and is defined by the probability distribution $\Psi_F^*(\mathbf{r},t)\Psi_F(\mathbf{r},t)$.[5] This form-interaction, discussed in the last chapter, is not known in classical physics and also not in usual quantum theory.

Also certain limitations with respect to space and time, say ($\pm\mathbf{r}_a$, $\pm t_a$), create such kind of interaction. In this case we have[6]

$$\Psi(\mathbf{r},t) = \Psi_F(\mathbf{r},t)$$
$$= C \exp\left\{\frac{i}{\hbar}(\mathbf{p}_0 \cdot \mathbf{r} - E_0 t)\right\}, \quad -\mathbf{r}_a, -t_a \leq \mathbf{r}, t \leq \mathbf{r}_a, t_a.$$
(5.1)

The system described by (5.1) can also be considered as a particle which is however as large as the universe if ($\pm\mathbf{r}_a, \pm t_a$) are the limitations of the universe.[7]

The case $\Psi(\mathbf{r},t) = 0$ (free system) is exactly fulfilled if there are no limitations concerning space and time: $-\infty \leq \mathbf{r} \leq \infty, -\infty \leq t \leq \infty$. However, $\Psi(\mathbf{r},t)$ is not exactly zero if space and time are cosmologically limited by ($\pm\mathbf{r}_a, \pm t_a$). The values for $\mathbf{r}_a$ and t_a may take any possible values but must be different from infinity, so that $\Psi(\mathbf{r},t) = 0$ is fulfilled to any degree of accuracy. In this case the function $\Psi(\mathbf{r},t)$ for a free system is close to

$$\Psi(\mathbf{r},t) = \Psi_F(\mathbf{r},t)$$
$$= C \exp\left\{\frac{i}{\hbar}(\mathbf{p}_0 \cdot \mathbf{r} - E_0 t)\right\} \cong 0, \quad -\mathbf{r}_a, -t_a \leq \mathbf{r}, t \leq \mathbf{r}_a, t_a,$$
(5.2)

with any degree of accuracy, where $\mathbf{p}_0$ is the mean momentum of the system and E_0 its mean energy. However, because ($\pm\mathbf{r}_a, \pm t_a$) are considered to be very large, the $\mathbf{p}, E$-fluctuations[7] are almost zero and we have $\mathbf{p} \cong \mathbf{p}_0, E \cong E_0$, that is, both values ($\mathbf{p}_0$ and E_0) are constant with any degree of accuracy if $\mathbf{r}_a$ and t_a are sufficiently large.

---

[5] See Section 4.8.
[6] See Section 4.4.1.
[7] See Section 4.4.1.

In conclusion, if there is (almost) no interaction, the system is always a *wave*.[8] In this case the momentum and the energy are given in good approximation by $\mathbf{p}_0$ and E_0. That is, such a system is definitely a wave in $(\mathbf{r}, t)$-space and has nothing to do with a localized particle; localized particle means that the system takes a point-like form in $(\mathbf{r}, t)$-space. Within projection theory the form is given by

$$\Psi_F^*(\mathbf{r}, t)\Psi_F(\mathbf{r}, t) = C^2, \quad -\mathbf{r}_a, -t_a \leq \mathbf{r}, t \leq \mathbf{r}_a, t, \quad (5.3)$$

in the case of (5.1) or (5.2), and this form is extended over the whole space-time region if $\mathbf{r}_a$ and t_a are the space-time limitations of the universe. In connection with (5.2) and (5.3), respectively, point-like material objects are not defined.

In contrast to this picture, within usual quantum theory, a non-interacting system can be a point-like system with constant momentum $\mathbf{p}_0$ and constant energy E_0 *or* a wave, having the wave-length λ ($\lambda = \hbar/p_x$) and the frequency ω ($\omega = \hbar/E_0$), and everything is embedded in space (space-time). In accordance with Bohr (Copenhagen interpretation) logical contradictions can only be avoided in connection with suitable experimental arrangements. Clearly, the principles of projection theory lead to a completely other conception. Here logical contradictions do not appear without experimental arrangement. With increasing $\mathbf{p}, E$-fluctuations in $(\mathbf{p}, E)$-space, the quantity $\Psi_F^*(\mathbf{r}, t)\Psi_F(\mathbf{r}, t)$ becomes more and more localized in $(\mathbf{r}, t)$-space and the systems tends more and more towards a point-like particle. In other words, the situation in this theory is quite different from that in usual quantum theory. There is definitely no particle-wave duality in the sense of Bohr. Whether or not a system behaves like a wave or a particle is dependent on the interaction of the system with its environment. An (almost) free particle can never be point-like because the function $\Psi_F^*(\mathbf{r}, t)\Psi_F(\mathbf{r}, t)$ is extended over the whole space-time.

In conclusion, in accordance with the experiments done by Mizobuchi and Ohtake[9] and, on the other hand, by Brida, Genovese,

---

[8] This is described by equation (5.2).
[9] Interested readers might refer to Mizobuchi, Y., and Ohtake, Y., (1992) Phys. Lett. A **168**, 1.

Gramegna and Predazzi,[10] within projection theory there is no particle-wave duality. More details will be given,[11] also in connection with the famous double-slit experiment, which can be described without any problem within projection theory.

5.1.2. What do We Measure?

5.1.2.1. Situation in conventional quantum theory

Within usual quantum theory, any detector-click is identified with the detection of a point-like particle as, for example, an electron. Within projection theory, the "click" of the detector has nothing to do with a point-like particle.

Born's Interpretation

The assumption that we detect a point-like material object with our detectors goes back on Max Born.[12] Born rejected Schrödinger's interpretation of the wave function $\psi(\mathbf{r}, \tau)$ in terms of real matter waves in analogy to water waves, where $\psi(\mathbf{r}, \tau)$ is the wave function within conventional quantum theory.[13] He argued as follows: The corpuscular nature of quantum objects was not just a convention, because individual objects can be counted by experimental devices (e.g., Geiger counters); that is, they behaved like true corpuscles. Therefore, in Born's opinion the wave described by Schrödinger's equation were not material, and he interpreted the probability density $\psi^*(\mathbf{r}, \tau)\psi(\mathbf{r}, \tau)$ in connection with a point-like material object which moves statistically through space, which is part of $(\mathbf{r}, t)$-space.[14] As we will discuss below, Born's picture is not free of problems, but let us first give some critical remarks on the quantum weirdness in connection with the famous double-slit experiment.

---

[10] Interested readers might refer to Brida, G., Genovese, M., Gramegna, M., and Predazzi, E., (1996) Phys. Lett. A **328**, 313.
[11] Interested readers might refer to Schommers, W., in preparation.
[12] See, in particular, Section 1.3 and refer to Schommers, W. (Ed.), (1989) Quantum theory and pictures of reality: Springer Verlag.
[13] See Section 1.4.
[14] See also the discussion in Section 1.4.

Double-Slit Experiment

Besides the particle-wave dualism, the quantum weirdness is particularly reflected in many situations as, for example, in connection with the famous double-slit experiment, symbolically represented in Figure 5.1(a). Let us consider a point-like object which is created at time τ_1 at position B_A, "moves" through space and is registered at time $\tau_2 > \tau_1$ at a screen S_A. Between B_A and S_A, we position a double-slit. We know that the probability density at the screen S_A is like an interference pattern, which is identical with that obtained in connection with material water waves. There are however no matter waves but wave probabilities, but we observe at position S_A the same point-like material object that has been created at position B_A. If we admit that the point-like material object is existent between B_A and S_A, that is, for times τ with $\tau_1 < \tau < \tau_2$, we would not obtain the observed pattern for the probability distribution. This is a serious problem, and it is well-known that this problem is solved within the Copenhagen interpretation of quantum theory by a specific construction.

Within this construction the object exists at time τ_1 at position B_A and, furthermore, it exists too at position S_A at time τ_2 but it is not defined for times τ with $\tau_1 < \tau < \tau_2$. Within the Copenhagen interpretation, we can only make statements about the real world, which is the point-like material object in connection with observations. For times τ with $\tau_1 < \tau < \tau_2$, only the wave function $\psi(\mathbf{r}, \tau)$ is existent between the positions B_A and S_A but not the material object. This has been well-formulated by Pagels:

> *The quantum weirdness lies in the realization that as long as you are not actually detecting an electron, its behaviour is that of a wave of probability. The moment you look at the electron it is a particle. But as soon as you are not looking it behaves like a wave again. That is rather weird, and the ordinary idea of objectivity can accommodate it.*[15]

---

[15] Refer to Pagels, H.P., (1983) The cosmic code: Michael Joseph.

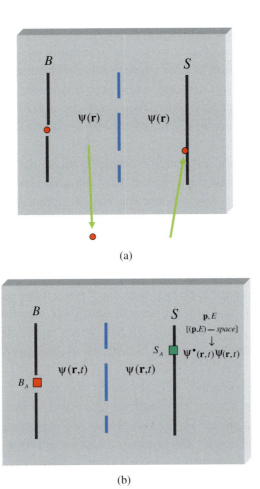

Fig. 5.1. (a) Double-slit experiment. Within Bohr's interpretation of usual quantum theory a point-like material object is only at position B and position S existent but not between B and S. Between B and S only the wave function $\psi(\mathbf{r}, \tau)$ is definable.[15] Then, one can take the view that the material object has left the space but it enters the space again at position S. (b) Within projection theory, such strange constructions are not needed for the description of the double-slit experiment.[16] Here material objects are never embedded in $(\mathbf{r}, t)$-space (not between B and S and also not at S). B_A is the picture of the device, which creates the object and S_A is the picture of the detector which registers the object, i.e., its momentum $\mathbf{p}$ and its energy E at time τ, and $\mathbf{p}$ and E do not belong to $(\mathbf{r}, t)$-space but exclusively to $(\mathbf{p}, E)$-space.

---

[16] Interested readers might refer to Schommers, W., in preparation.

A lot of people cannot accept such a construction; for example, Albert Einstein never accepted usual quantum theory together with Bohr's interpretation as a complete theory. Without any doubt, Bohr's idea is ingenious but not only a few people consider the Copenhagen interpretation of quantum theory as a stop gap. Clearly, Bohr's point of view is definitely wrong when we base our considerations on projection theory.

Within Born's probability interpretation the wave function $\psi(\mathbf{r}, \tau)$ is tightly linked to a (point-like) material object. However, for times τ with $\tau_1 < \tau < \tau_2$ only the wave function $\psi(\mathbf{r}, \tau)$ is existent between the positions B_A and S_A but not the material object, that is, the wave function $\psi(\mathbf{r}, \tau)$ can exist without material object and this is in a certain sense against Born's basic interpretation of the wave function.

One can take the view that the object has left the space for times τ with $\tau_1 < \tau < \tau_2$ and only the wave function $\psi(\mathbf{r}, \tau)$ remains. However, at position S_A (time τ_2) the material object enters the space again. This (ill and non-convincing) construction is equivalent to Bohr's conception.[17]

All these problems[18] are obviously due to the fact that within usual quantum theory the material quantum object is embedded in space (space-time). Projection theory delivers a solution for the double-slit experiment and similar situations without contradictions and strange interpretations, respectively.[19] There are never material objects embedded in $(\mathbf{r}, t)$-space in projection theory not at position B_A, not between B_A and S_A and also not at S. The situation is schematically illustrated in Fig. 5(b).

Material Objects in Space?

In fact, within usual quantum theory, the object with mass m_0 really exists in space and only the variables $\mathbf{r}, \mathbf{p}, E$, which characterize the

---

[17] See, in particular, Figure 5.1(a).
[18] Refer to the particle-wave duality and the interpretation of the double-slit experiment.
[19] Interested readers might refer to Schommers, W., in preparation.

quantum object, are considered to be uncertain.[20] However, within this conception nothing is said about the mass m_0; m_0 has obviously to be considered also here as the inertial mass, as is the case in Newton's mechanics. The mass m_0 appears in Schrödinger's equation, and this equation could not be derived in conventional quantum theory, i.e., the source for the mathematical structure of Schrödinger's equation remains hidden within conventional quantum theory. Therefore, also the quantum aspect of the mass m_0 cannot be formulated and only what can be done is to assume that m_0 behaves like a classical mass which defines inertia in the sense of Newton. However, we will recognize from the following discussion that this can hardly be accepted when we work within the framework of conventional quantum theory. In connection with typical quantum observations here we have the following situation:

1. *Motion without Inertia*

Let us consider the particle-problem in terms of the stationary probability density $\psi^*(\mathbf{r},\tau)\psi(\mathbf{r},\tau) = \psi^*(\mathbf{r})\psi(\mathbf{r})$, and let $\psi^*(\mathbf{r})\psi(\mathbf{r}) \neq 0$ in the range Δx and let A and B be two positions within Δx; at time τ_A we register the particle with a detector at position A, at time τ_B we register the same particle at position B. Because there is no physical law that tells us *when* the real particle with mass m_0 arrives at position B after it has left position A, the time interval $\Delta \tau = \tau_B - \tau_A$ can take any value, for example, it may be infinitesimal small but different from zero. Then, the seeming velocity $v = \Delta x / \Delta \tau$ can take any value:

$$v = \frac{\Delta x}{\Delta \tau} \to \infty. \tag{5.4}$$

In particular, v may be larger than the velocity of light c. Equation (5.4) reflects a motion without inertia. Within classical mechanics motion is a

> *continuous blend of changing positions. The object moves in a flow from one point to another. Quantum mechanics failed to reinforce that picture. In fact, it indicated that motion could not take place in that way. They jumped from one place to another,*

---

[20] See, for example, Pagels, H.P., (1983) The cosmic code: Michael Joseph.

seemingly without effort and without bothering to go between two places.[21]

This statement by Fred Alan Wolf suggests that within usual quantum theory the particle moves without inertia.

In conclusion, within usual quantum theory a motion without inertia can take place within the volume where $\psi^*(\mathbf{r})\psi(\mathbf{r})$ is different from zero. However, the mass m_0 is nevertheless defined, and this is because $\psi(\mathbf{r})$ and $\psi^*(\mathbf{r})\psi(\mathbf{r})$, respectively, is based on Schrödinger's equation of usual quantum theory where the mass $m_0 \neq 0$ appears as parameter. The "motion without inertia in connection with a non-vanishing mass m_0"[22] has to be considered as a contradiction within usual quantum theory because within this approach the mass m_0 still reflects the effect of inertia and nothing else.

2. Paradoxical Situation

There is a further problem when we assume that the quantum object is embedded as real material object in space (space-time). We have already outlined in Section 1.4 the following facts:

> *We know that the particle was at time τ_A at position A and we also know that the same particle was at time τ_B at position B. But also in connection with these detections no velocity and no trajectory may be defined. This has the following consequence: A trajectory and velocity can obviously only be avoided if the particle (for example, detected at position B) is simultaneously existent and non-existent at position B. The problem obviously only appears when we connect the detection with an individual (point-like particle). An individual demands a minimum description since this notion has been chosen with respect to our intuitive demands for visualization. The individual arrives position B at time τ_B and it leaves it at exactly the same time τ_B; otherwise a trajectory would be defined — a horizontal line in the (x, τ)-diagram; the corresponding velocity would be zero but is defined. But this picture*

---

[21] Refer to Wolf, F.A., (1981) Taking the quantum leap: Harper and Row.
[22] Reflected by equation (5.4).

reflects a paradoxical situation as we have outlined above. The particle is simultaneously existent and non-existent![23]

As in the case of mass m_0, also here usual quantum theory leads to an unsatisfactory situation.

5.1.2.2. *Situation in projection theory*

All these problems within conventional quantum theory do not appear within projection theory and the reason is simply given by the fact that there are no (point-like) material objects embedded in space-time[24] in projection theory. This seems to be the key for the solution of all the problems that appear in connection with usual quantum theory.

In projection theory, the "detector-clicks" are needed in order to be able to verify certain geometrical structures $\Psi_F^*(\mathbf{r},t)\Psi_F(\mathbf{r},t)$ in $(\mathbf{r},t)$-space. Such a geometrical structure can be a wave (a particle of large extension) or a strict localized particle. Each detector-click merely means that certain values for $\mathbf{p}$ and E have been detected at a certain instant τ, and these $\mathbf{p}, E$-values cannot be connected to a point-like mass. That is, the "click" of the detector is not activated by a point-like particle, which is embedded in $(\mathbf{r},t)$-space, but exclusively by one of the possible $\mathbf{p}, E$-pairs where $\mathbf{p}$ and E belong to $(\mathbf{p},E)$-space.

In other words, a detector-click has nothing to do with the geometrical structure in $(\mathbf{r},t)$-space; the geometrical structure, expressed by $\Psi_F^*(\mathbf{r},t)\Psi_F(\mathbf{r},t)$, merely decides at which space-time point the detector can be active.[25] Such "clicks" always take place independent of the interaction that determines the form of the system. The geometrical structure can be, for example, described by a wave (a system of large extension) or by a strict localized system, which we can call "particle". That is, within the theoretical picture given here any geometrical structure can only be verified by means of detector-clicks.

Furthermore, there can be no $\mathbf{p}, E$-transport through $(\mathbf{r},t)$-space but only projections on it. Two systems, say 1 and 2, having at time τ

---

[23] See Section 1.4.
[24] See Section 2.8.10.
[25] See Section 2.3.8.3.

the positions $\mathbf{r}_1$ and $\mathbf{r}_2$ (and times t_1 and t_2) and the values $\mathbf{p}_1, E_1$ and $\mathbf{p}_2, E_2$ for the momenta and energies, have at the next possible time τ' the values $\mathbf{r}'_1, t'_1, \mathbf{r}'_2, t'_2$ and $\mathbf{p}'_1, E'_1, \mathbf{p}'_2, E'_2$. That is all we can say about the elementary steps. For example, the transitions from $\mathbf{p}_1$ to $\mathbf{p}'_1$, $\mathbf{p}_2$ to $\mathbf{p}'_2$ and those from E_1 to E'_1, E_2 to E'_2 cannot be analyzed further. The same holds for the elements of $(\mathbf{r}, t)$-space. Interaction processes are explained in terms of these elementary steps. Within such a principle it gives no sense to explain these elementary steps by further processes. There is no "process" of the "process". A more detailed analysis will be given.[26]

Clearly, both, the detector and the system under investigation, are equally positioned in basic reality where the interactions take place. Both objects are equally projected onto $(\mathbf{r}, t)$-space and appear as geometrical structures (picture of reality) that we have in front of us in connection with our observations in everyday life. We observe the interaction of the detector with the quantum system as changes of the detectors geometrical structure.[27] This conception is quite different from Born's conception.

5.2. The Role of the Observer

5.2.1. *Compatible with the Principles of Evolution*

How do we experience the world in everyday life? We see certain objects in front of us and when we approach and touch them we experience a material effect and not a shadow or something like that. Thus, we conclude that all the objects are embedded in space and that they exist objectively there. But this is obviously a fallacy when we take the results of projection theory into account.

Within conventional quantum theory, the material world is still embedded in space but it has, within Copenhagen Interpretation, no longer an objective existence. How can objects be embedded in space without existing objectively? This and similar questions has led to

---

[26] Interested readers might refer to Schommers, W., in preparation.
[27] For example, the pointer of the detector moves from one position to another.

a never-ending scientific debate reflecting the fact that conventional quantum theory is very often to be felt as fundamentally inadequate,[28] although its empirical success is without any doubt impressive.

The situation within the theoretical picture given within projection theory can be summarized as follows. We experience objects such as quantum objects and also those we experience in everyday life as, for example, a stone, at certain positions in space, although these things are actually not positioned there. The objects and also the observers are not embedded in space (space-time). The situation is analyzed in Section 2.2.

Clearly, at space-time points $\mathbf{r}, t$, where $\Psi^*(\mathbf{r}, t)\Psi(\mathbf{r}, t) \neq 0$, we register detector-clicks or we observe the deflection of its pointer,[29] but all these effects take place in $(\mathbf{r}, t)$-space and belong to the "picture of the detector". The actual interaction, however, takes place in $(\mathbf{p}, E)$-space and is characterized exclusively by $\mathbf{p}, E$-fluctuations. But also photons will interact with the system, defined by $\Psi^*(\mathbf{r}, t)\Psi(\mathbf{r}, t)$, at those space-time points $\mathbf{r}, t$ where $\Psi^*(\mathbf{r}, t)\Psi(\mathbf{r}, t) \neq 0$ and will be scattered and registered by the eyes of the observer. However, also the scattering and registration reflect processes take place in $(\mathbf{p}, E)$-space and the observer gets the impression that there is something real located in space. If $\Psi^*(\mathbf{r}, t)\Psi(\mathbf{r}, t)$ describes, for example, a stone we are firmly convinced that there is actually a real material object positioned at a certain position in $(\mathbf{r}, t)$-space. This is supported by the fact that we really feel the stone as material object when we touch it, and we immediately know that it is not merely a shadow or something like that. This is however a fallacy from the point of view of projection theory. When we touch the stone we feel it and this is entirely due to an interaction process in $(\mathbf{p}, E)$-space and has nothing to do with a material object (stone) resting passively at a certain position in space. Both, the stone as well as the observer, equally represent pictures in space-time. Only the pictures of the observer and the stone in $(\mathbf{r}, t)$-space are recognizable. Their structures in basic reality remain hidden.[30] The

---

[28] See, in particular, Section 5.1.
[29] See Section 2.3.8.3.
[30] See Section 2.2.

interaction process, responsible for our object feeling, take place in basic reality and are described by $\mathbf{p}, E$-fluctuations in $(\mathbf{p}, E)$-space but the whole recognizable scenario is projected onto $(\mathbf{r}, t)$-space.

This scientific position is the most realistic viewpoint and is compatible with the principles of evolution.[29] It is completely superfluous that there are non-interacting objects resting passively in space which only interact now and then, as is assumed in classical mechanics as well as in usual quantum theory. Such objects would be useless and not allowed from the point of view of evolution. Evolution requires as little outside world as possible, only as much as is absolutely necessary. A resting passive object would be useless and, therefore, its existence is not allowed from the viewpoint of evolution.[31]

The scientific conception of projection theory is compatible with the principles of evolution. Here is that what we experience as "stone", let us define it by $\Psi^*_{stone}(\mathbf{r}, t)\Psi_{stone}(\mathbf{r}, t)$, is *not* a "resting, passive object", and this has nothing to do with the observers interaction during the observation process. The property $\Psi^*_{stone}(\mathbf{r}, t)\Psi_{stone}(\mathbf{r}, t) \neq 0$ means that there is an interaction between the stone and its environment and it is just this process that defines the stone leading to $\Psi^*_{stone}(\mathbf{r}, t)\Psi_{stone}(\mathbf{r}, t) \neq 0$, that is, the stone can never be an isolated object like a passively resting object in $(\mathbf{r}, t)$-space, but it is involved in processes in the outside world, just leading to $\Psi^*_{stone}(\mathbf{r}, t)\Psi_{stone}(\mathbf{r}, t) \neq 0$.

The passive case is given within projection theory by a free system which has the property $\Psi^*(\mathbf{r}, t)\Psi(\mathbf{r}, t) = 0$.[32] Such a system does not exist within projection theory, and this is in accordance with the principles of evolution. The existence of the curve $\Psi^*(\mathbf{r}, t)\Psi(\mathbf{r}, t)$ gives only sense in connection with an observer because $(\mathbf{r}, t)$-space comes into play through the observers brain functions.[33] All the objects in front of us, or in principle, the whole universe within $(\mathbf{r}, t)$-space, only exist in connection with an observer, although there must be

---

[31] Interested readers might refer to Schommers, W., (1994) Space and time, matter and mind: World Scientific.
[32] See Chapter 3.
[33] See Section 2.2.

an objective reality outside, represented in $(\mathbf{p}, E)$-space. This result is close to Wheeler's viewpoint:

> *No theory of physics that deals only with physics will ever explain physics. I believe that as we go on trying to understand the universe, we are at the same time trying to understand man [...] The physical world is in some deep sense tied to the human being.*[34]

In conclusion, any theory must precisely reflect the observers' role in nature, and the observer developed in accordance with the principles of evolution. Any other kind of observer as, for example, an animal, might come to another conclusion about the world. We have discussed this point in Section 2.2.9 in connection with the so-called chick-experiment. Such kind of experiments can help to find our own position in the world.

Within the theoretical investigation given here, everything is a matter of processes and the existence of static, non-interacting objects in space have been eliminated. Like the philosophers Whitehead and Bergson, within projection theory it is argued for the primacy of process.

5.2.2. *Configurations in Space-Time*

Any system is characterized by $\Psi^*(\mathbf{r}, t)\Psi(\mathbf{r}, t)$ and a *human observer* registers this curve by means of $\Psi^*_{ref}(\tau - t)\Psi_{ref}(\tau - t)$.[35] The curve $\Psi^*(\mathbf{r}, t)\Psi(\mathbf{r}, t)$ is systematically scanned by $\Psi^*_{ref}(\tau - t)\Psi_{ref}(\tau - t)$ and only those values of t can be observed which correspond with the reference time τ.[36] Such selection processes come into play by human observers or other biological systems. However, we have to assume that such mechanisms, described by $\Psi^*_{ref}(\tau - t)\Psi_{ref}(\tau - t)$, are in general different for different biological systems.

---

[34] Refer to von Baeyer, H.C., (2004) Information, the language of science: Harvard University Press.
[35] See Section 2.3.9.1.
[36] See also Figure 2.10.

In the case of detectors we have no such selection mechanisms and this means that functions like $\Psi_{ref}^*(\tau-t)\Psi_{ref}(\tau-t)$ are not defined. In other words, it gives no sense to assume any kind of reference time τ in connection with detectors or similar apparatuses. Whereas the human observer selects, leading to a restricted view of the world and also to different configurations at different times τ, detectors do not select. In the case of a human observer, the configurations

$$\Psi^*(\mathbf{r},t_1)\Psi(\mathbf{r},t_1), \Psi^*(\mathbf{r},t_2)\Psi(\mathbf{r},t_2),\ldots, \Psi^*(\mathbf{r},t_k)\Psi(\mathbf{r},t_k),\ldots \quad (5.5)$$

are selected at times $\tau_1,\tau_2,\ldots,\tau_k,\ldots$, and we obtain those picture sequences which we observe in everyday life. On the other hand, detectors always register the complete reality $\Psi^*(\mathbf{r},t)\Psi(\mathbf{r},t)$ at times τ_k,[37] and not only restricted versions[38] of it, that is, instead of (5.5) we have

$$\begin{aligned}\tau_1 &: \Psi^*(\mathbf{r},t)\Psi(\mathbf{r},t)\\ \tau_2 &: \Psi^*(\mathbf{r},t)\Psi(\mathbf{r},t)\\ &\vdots\\ \tau_k &: \Psi^*(\mathbf{r},t)\Psi(\mathbf{r},t)\\ &\vdots\end{aligned} \quad (5.6)$$

with

$$-\infty \leq t \leq \infty$$

for each τ. Here we obtain a static picture. However, a detector has no space-time feeling and, therefore, the information (5.6) cannot be experienced in $(\mathbf{r},t)$-space but in $(\mathbf{p},E)$-space in the form of $\Psi^*(\mathbf{p},E)\Psi(\mathbf{p},E)$, that is, we have to replace $\Psi^*(\mathbf{r},t)\Psi(\mathbf{r},t)$ in equation (5.6) by $\Psi^*(\mathbf{p},E)\Psi(\mathbf{p},E)$. Again, both distributions, $\Psi^*(\mathbf{r},t)\Psi(\mathbf{r},t)$ and $\Psi^*(\mathbf{p},E)\Psi(\mathbf{p},E)$ are completely equivalent concerning their information content; $\Psi(\mathbf{r},t)$ and $\Psi(\mathbf{p},E)$ are connected by a Fourier transformation.[39] Therefore, the observer is able to

---

[37] See Section 2.3.9.7.
[38] See equation (5.5).
[39] See Section 2.3.4.3.

measure the whole information, at least in principle, at each time τ_k although the world in front of him is a restricted one at the same time τ_k.[40] This is consistent with the facts we have outlined in Section 2.3.9.7.

The observer can comprehend only one instant, say t_k, of time at each moment τ, and the static display (5.6) is converted into motion and activity described by equation (5.5). John Wheeler's statement[41]

Time is nature's way of keeping everything from happening at once

is therefore not applicable to detectors or to any other elementary interactions in nature that are characterized by (5.6). The situation is rather comparable with that we have discussed above in connection with Feynman diagrams[42]:

To a creature capable of comprehending the whole span of time as we comprehend the span of space, the activity of annihilation and creation represented in this diagram is not activity at all. It is a stationary display, a picture painted in space and time.[43]

In conclusion, the situation described by the static picture (5.6) in connection with detectors and other elementary processes is comparable to that of Feynman's diagrams.

Within classical mechanics and usual quantum theory, any interaction is constructed on the basis of the observers time-feeling, based on the reference time τ. The interaction law for two particles, having the positions $\mathbf{r}_1$ and $\mathbf{r}_2$ at time τ, is formulated as a function of $\mathbf{r}_1$ and $\mathbf{r}_2$. For example, the well-known Coulomb potential is proportional to $1/|\mathbf{r}_2 - \mathbf{r}_1|$. This situation can be symbolically expressed by

$$\tau : \mathbf{r}_1 \leftrightarrow \mathbf{r}_2. \tag{5.7}$$

The system-specific time t is not defined here.

---

[40] See equation (5.5).
[41] Refer to Al-Khalili, J., (1999) Black holes, worm holes and time machines: Institute of Physics.
[42] See also Figure 2.11 and the text of Section 2.3.9.12.
[43] Refer to Ford, K.W., (1963) Elementary particles: Blaisdell Publishing Company.

Let us consider two systems which are characterized by $\Psi_1^*(\mathbf{r}_1, t_1)$ $\Psi_1(\mathbf{r}_1, t_1)$ and $\Psi_2^*(\mathbf{r}_2, t_2)\Psi_2(\mathbf{r}_2, t_2)$ with $-\infty \leq \mathbf{r}_1, t_1, \mathbf{r}_2, t_2 \leq \infty$. According to projection theory, both systems interact in $(\mathbf{p}, E)$-space, and these interactions, i.e., the $\mathbf{p}, E$-fluctuations, are completely transferred to $(\mathbf{r}, t)$-space and appear here as correlations. In contrast to the case (5.7), there is not only a correlation between the positions $\mathbf{r}_1$ and $\mathbf{r}_2$ ($\mathbf{r}_1 \neq \mathbf{r}_2$) at time τ but also between the times t_1 and t_2 ($t_1 \neq t_2$) and we get

$$\tau : \mathbf{r}_1, t_1 \leftrightarrow \mathbf{r}_2, t_2. \tag{5.8}$$

Therefore, such systems "interact" at time τ not only in space but also with the past, present and the future. The same is the case for detectors etc., although the observer registers, due to the selection, only one configuration $\Psi^*(\mathbf{r}, t_k)\Psi(\mathbf{r}, t_k)$ at time τ, which is expressed by equation (5.5).

5.3. Summary

Conventional Quantum Theory

In conventional quantum theory, we have the following situation. A non-interacting system can be a point-like system with constant momentum $\mathbf{p}_0 = p_x$ and constant energy E_0 *or* a wave, having the wavelength λ ($\lambda = \hbar/p_x$) and the frequency ω ($\omega = \hbar/E_0$). The nature of the system (particle or wave) is *not defined* without experimental device, and the decision about its nature (wave or particle) is dependent on *how* the experimental apparatus has been arranged. In other words, a non-interacting system can be a particle as well as a wave, depending on the chosen experimental arrangement by the observer.

According to the principle of complementarity, only *one* of the two incompatible possibilities (wave, particle) can be realized and this implies that the experimental arrangements that determine those properties must be similarly mutually exclusive, otherwise we have a problem since we are confronted with a logical contradiction. However, it has been demonstrated by new experiments that light showed both particle and wave aspects *simultaneously*, and this is in contrast to the principle of complementarity which is one of the very basics of modern quantum theory (Copenhagen Interpretation).

Within usual quantum theory, any detector-click is identified with the detection of a point-like particle as, for example, an electron.

Projection Theory

Within projection theory, the situation is different. Here we do not need an experimental arrangement that decides about the nature of a system. The system is completely described by the *interaction* of the system with its environment. No interaction means within projection theory that the system does not exist.

If there is (almost) no interaction the system is always a *wave*. In this case the momentum and the energy are given in a good approximation by $\mathbf{p}_0$ and E_0. With increasing $\mathbf{p}, E$-fluctuations in $(\mathbf{p}, E)$-space, the quantity $\Psi_F^*(\mathbf{r}, t)\Psi_F(\mathbf{r}, t)$, which describes the form (shape) of a system in $(\mathbf{r}, t)$-space, becomes more and more localized in $(\mathbf{r}, t)$-space and the system tends more and more towards a point-like particle. In other words, the situation in this theory is quite different from that in usual quantum theory. There is definitely no particle-wave duality in the sense of Bohr. This is the reason why we do not need here an experimental arrangement that decides about the nature of a system (wave or particle). Whether or not a system behaves like a wave or a particle is dependent on the interaction of the system with its environment. An (almost) free particle can never be point-like because the function $\Psi_F^*(\mathbf{r}, t)\Psi_F(\mathbf{r}, t)$ is extended over the whole space-time.

In projection theory, the "detector-clicks" are needed in order to be able to verify certain geometrical structures in $(\mathbf{r}, t)$-space as, for example, $\Psi_F^*(\mathbf{r}, t)\Psi_F(\mathbf{r}, t)$. Such a geometrical structure can be a wave, i.e., a particle of large extension, or a strict localized particle. Each detector-click merely means that certain values for $\mathbf{p}$ and E have been detected at a certain instant τ, and these $\mathbf{p}, E$-values cannot be identified with a point-like mass. That is, the "click" of the detector is not activated by a point-like system, which is embedded in $(\mathbf{r}, t)$-space, but exclusively by one of the possible $\mathbf{p}, E$-pairs where $\mathbf{p}$ and E belong to $(\mathbf{p}, E)$-space. Clearly, both, the detector and the system under investigation, are equally positioned in basic reality where the interactions take place. Both objects are equally projected onto $(\mathbf{r}, t)$-space and appear as geometrical structures, i.e., as picture of reality, that we have in front of us in connection with our observations in everyday life.

6

Summary

Within the memory of man all essential things are represented within the frame of pictures. This is the most basic statement. All of what we experience in everyday life are pictures in front of us. Such pictures contain the objects such as the sun, moon, etc., the space and the time. We feel the objects, that is, we observe them with our five senses and also with measuring instruments. Thus, we believe in the concrete existence of these things and we call them "material objects". But what about space and what about time? It is normally assumed that the space is a container filled with matter and time is just that which we measure with our clocks. However, there are some reasons to take another viewpoint and to consider this container conception as unrealistic, as prejudice so to say. Already the philosopher Immanuel Kant pointed out this serious problem.[1]

We cannot put a "piece of space" on the table and we can also not put a "piece of time" on the table. Such pieces do not exist. We are not able to observe the elements of space, that is, its coordinates x, y, z and we are also not able to observe the elements of time τ. We have no senses for that and, furthermore, the development of measuring methods for the detection of x, y, z and τ are not even thinkable.

How are space and time treated within modern conventional physics? Within Newton's mechanics, space and time are independent of each other and also independent of matter. Einstein developed this conception further; within Theory of Relativity, space and time are no longer independent of each other but form a space-time. Then, the following question arises. Can we put a "piece of space-time" on the table? No, we definitely cannot. As in the cases of space and time,

---

[1] Interested readers might refer to Schommers, W., (2008) Advanced Science Letters **1**, 59 and to Schommers, W., Cosmic Secrets: World Scientific, in preparation.

also a piece of space-time does not exist in the sense of an observable quantity. In other words, also within the framework of modern physical theories the space (space-time) is considered as a container and this is problematic.

Nevertheless, we are all familiar with the space-time phenomenon although the elements of it (x, y, z, τ) are not observable with our five senses and measuring instruments. Thus, we have to conclude that the elements x, y, z, τ must be inside the observer; it makes no sense to assume that these elements (x, y, z, τ) are also outside the observer. The famous philosopher Immanuel Kant took this viewpoint, but his fundamental ideas have never been seriously used in the formulation of physical theories.

Because we definitely cannot see, hear, smell, taste or measure space and time, the elements x, y, z, τ may not be the source of a physically real effect. This requirement is the content of Mach's principle that is however not fulfilled within Newton's theory but also not within Special and General Theory of Relativity. It is obviously not possible to fulfil Mach's principle when we assume that the material bodies are embedded in space (space-time). We have discussed Mach's principle in more detail in Chapter 4.

Our experiences in connection with space and time are exclusively based on the following fact. We can only say something about distances in connection with masses, and time intervals in connection with physical processes. There is no exception!

Due to the fact that the elements x, y, z, τ of space and time are not accessible, the space-time has to be considered as a "nothing" in the physical sense but, on the other hand, we experience space and time as a real phenomenon. Clearly, physically real objects can hardly occupy this "nothing". It is problematic to embed the real world into such a metaphysical space-time. Within all modern physical theories such as Newton's mechanics, Theory of Relativity, conventional quantum theory etc., real material bodies are the contents of an "object" (space-time) that has to be considered as nothing or as a metaphysical substratum different from matter.

This seems to be a contradiction. Two kinds of objects are mixed which are obviously mutually incompatible. We can overcome this bad situation if we take the position of the philosopher Immanuel Kant: Space and time are exclusively features of our brain and the world outside is projected on it. Then, the material objects that occupy space and time can only be geometrical pictures. Projection theory[2] is based on this conception. If that is the case, the situation would be identical with that which we do in connection with our blackboards and notebooks. We draw pictures on them.

From the principles of evolution follow that the structures in the picture must be different from those in outside reality. The common or naive point of view assumes the following. The inside world which we feel to be outside us, actually exists in the outside world in exactly the same form as we perceive. According to this view, there is only one difference between the inside world and the outside world: inside there are only geometrical positions, whereas outside there are the real material bodies instead of the geometrical positions. In other words, it is normally assumed that the geometrical positions are merely replaced by material objects. However, due to the principles of evolution we have to assume that the information in the picture is only a selected part of the outside world (basic reality). Since space and time cannot be considered as elements of basic reality, its structure must be different from that in the picture. In other words, from the principles of evolution we have to conclude that basic reality contains much more information than an observer can depict in space and time and that its structure is principally different from that in the picture; this space-time information is mainly that part of the outside world which a human observer needs for survival.

From this it directly follows that the picture of reality must be species-dependent. In other words, we have to conclude that the actions of other biological systems are in general based on a picture of reality that is different from that of the human observer.

In summary, the material objects which we observe in everyday life are not embedded in space (space-time). Space and time have to be considered as auxiliary elements for the representation of the selected information about the outside world (basic reality).

---

[2] See Chapter 2.

The chick experiment demonstrated that these pictures must be species-dependent, and this is confirmed by the principles of evolution. In other words, the physical reality is projected onto space and time, and the space-time elements x, y, z, τ cannot be seized with our five senses or with specific measuring instruments. These features of space and time must be considered in the formulation of physical theories. We have introduced a certain projection principle for the description of quantum phenomena.

Situation in Conventional Quantum Theory

The essential features of the mathematical formalism of (non-relativistic) quantum theory were constructed by Heisenberg and Schrödinger in 1925–1926. On the basis of this quantum-mechanical formalism, an enormous number of effects in atomic physics, chemistry, solid-state physics, etc., could be predicted and explained. However, more than 80 years after its formulation the interpretation of this formalism is

> by far the most controversial of problem of current research in the foundations of physics and divides the community of physicists and philosophers of science into numerous opposing schools of thought.[3]

That the differences between the various interpretations are not only marginal can be recognized by comparison of the so-called *Copenhagen Interpretation* and the *Many-Worlds Theory*. Both interpretations lead to pictures of reality which are completely different from each other.

Both, the Copenhagen Interpretation and the Many-World Theory give predictions in accord with experience and, therefore, they are equivalent from the experimental point of view. However, both interpretations lead to pictures of reality which are completely different from each other, and it is legitimate at the present stage of conventional quantum theory to choose the picture of reality which one finds

---

[3] Interested readers might refer to Jammer, M., (1974) The philosophy of quantum mechanics: Wiley.

satisfactory and pleasing. In contrast to usual quantum theory, within classical mechanics we have one real world, and each part (system) of it can be thought of as having an independent existence regardless of whether it is observed or not. This classical picture of reality is not in conflict with our intuitive concepts.

The question is then whether a formalism, i.e., the present form of usual quantum theory, with a diversity of completely different interpretations is convincing? It is certainly not convincing. A formalism, for example that of usual quantum theory, which describes the real world well but is, on the other hand, not able to produce a consistent world view, has to be considered as fundamentally inadequate even when this formalism describes the experimental material impressive, as in the case of usual quantum theory. This point has often been discussed in literature.

Such a situation means the following. Since we do not know the definite source of quantum phenomena, we are hardly able to develop this formalism further in a systematic manner. But no progress in basic science also means that there can be no or only restricted progress with respect to applications, just in connection with quantum nanotechnology.

The standard interpretation of usual quantum theory is the Copenhagen interpretation, and we discussed this world view in connection with the "collapse of the wave function", the "wave-particle duality", the "principle of complementarity" etc. However, new experiments showed[4] that particle and wave aspects can occur simultaneously and this is in contrast to Bohr's principle of complementarity that is one of the very basics of usual quantum theory.

In conclusion, the situation within usual quantum theory is unsatisfactory. A consistent world view could not be achieved; there is an immense diversity of opinion and a huge diversity of interpretations. Furthermore, there are certain facts which cannot be explained by conventional quantum theory, for example, the collapse of the wave function. Moreover, the new experiments in connection with wave-particle dualism can force us to give up essential parts of it. We have

---

[4]See Section 2.

possibly to find new physical theories and conceptions, respectively. In this monograph we propose a new ansatz: projection theory.

There are obviously some specific serious problems in connection with usual quantum theory. These problems are often ignored, and this is mainly due to the fact that the empirical success of usual quantum theory is impressive. But due to the new experiments,[5] which have been performed by two independent groups, we are confronted with a new situation, and we have to analyze critically the basics of conventional quantum theory, i.e., its mathematical apparatus as well as its interpretation. In Chapter 2, we have quoted some critical points which appear in connection with usual quantum theory. Let us repeat the main facts as follows:

1. Schrödinger's equation could not be deduced in usual quantum theory but was assumed on the basis of "reasonable arguments". In other words, we do not really know the physical reality on which Schrödinger's equation is based.
2. The *collapse of the wave function* cannot be explained in usual quantum theory; one would expect that Schrödinger's equation, which is the fundamental tool within non-relativistic quantum theory, is able to describe such a process but this is unfortunately not the case. The collapse of the wave function and the assignment of statistical weights are not explained; within usual quantum theory

   > *they are consequences of an external a priori metaphysics, which is allowed to intervene at this point and suspend the Schrödinger equation.*[6]

   In other words, the situation in connection with the collapse of the wave function is unsatisfactory and has to be solved.
3. Time τ is still a classical parameter in usual quantum theory. While the coordinates are *statistical* quantities, time does not behave statistically. Time τ remains unchanged when we go from classical

---

[5] Interested readers might refer to Mizobuchi, Y., and Othake, Y., (1992) Phys. Lett. A **168**, 1 and to Brida, G., Genovese, M., Gramegna, M., and Predazzi, E., (1996) Phys. Lett. A **328**, 313.

[6] Interested readers might refer to de Witt, B., (1970) Physics Today, September issue.

mechanics to quantum theory; this is clearly reflected in the fact that the coordinates can be *operators*; time is always a simple *parameter*. This is the reason why there is no uncertainty relation for the time τ and the energy E that would agree in its physical content with the position–momentum uncertainty relation.

4. The use of a particle defined as a local existent, which is assumed within the frame of usual quantum theory, seems also to be problematic and this is probably due to the fact that we transfer a picture used in everyday life to the microscopic realm. Can a particle, i.e., point-like individual, exist without trajectory, as in the usual quantum theory? We argued that such a concept is also problematic because an individual demands a minimum description since this notion has been chosen with respect to our intuitive demands for *visualization*. This is due to our experiences in everyday life. The concept of a point-like individual without trajectory obviously leads to a paradoxical situation.

In Chapter 2, we discussed these points in connection with projection theory; it turned out that the point of view of conventional quantum theory cannot simply be extended, but we have to include further fundamental facts, so far not directly considered in the basic description of physical systems. We in particular recognized that we can learn something new about the relationship between the "observer" and that which we call "reality" when we consider certain facts from biological evolution and behavior research. The projection principle is based on the insight that the material objects such as the sun, moon, etc., are not embedded in space (space-time), but the physical reality is projected onto space (space-time) within the frame of the projection principle.

Situation in Projection Theory

Reality is projected on space-time, i.e., on $(\mathbf{r}, t)$-space, where the variable t is a system-specific time, which is not known in conventional quantum theory. In conventional quantum theory only the classical, external parameter τ is used for the definition of time, where τ is measured with clocks. This time τ is applied in projection theory as reference time. The observed pictures are the most fundamental information. How basic reality is structured can principally not be said. The description of these fundamental pictures is done on the basis of fictitious realities which are constructed on the space-time elements **r** and t. The variables of fictitious realities are the momentum **p** and the energy E. Thus, we have two spaces in projection theory: $(\mathbf{r}, t)$-space and $(\mathbf{p}, E)$-space.

There is however a big difference between the variables $\mathbf{r}, t$ and $\mathbf{p}, E$. The variables $\mathbf{p}, E$ behave quite differently from the variables $\mathbf{r}, t$. For **p** and E, conservation laws are valid. This inevitably means that a change in **p** and E of the system under investigation must be accompanied with a change of **p** and E of another system outside. In other words, the **p**, E-fluctuations in connection with the system under investigation mean that there is an interaction process between the system and another system outside (environment).

Such conservation laws do not exist for the variables $\mathbf{r}, t$ leading to the feature that $\mathbf{r}, t$ play a completely passive role. This feature supports our assumption that **r** and t are exclusively the elements of pictures; the $(\mathbf{r}, t)$-space is the frame on which reality is projected.

In conclusion, the physical processes take place in $(\mathbf{p}, E)$-space and this information is projected onto $(\mathbf{r}, t)$-space; $(\mathbf{r}, t)$-space does not contain real material objects but only geometrical structures. In particular, $(\mathbf{r}, t)$-space cannot be an element of the outside world but is exclusively positioned in the brain of the observer and appears simultaneously with the objects, or more precisely with the geometrical structures of them. Again, this viewpoint is close to that which the philosopher Immanuel Kant (1724–1804) proposed.

Space and time have to be considered as auxiliary elements for the representation of the selected information about the outside world (basic reality). We came to the conclusion that the material objects that we observe in everyday life are not embedded in space (space-time). This is quite a general statement. The projection principle, introduced in connection with $(\mathbf{r}, t)$-space and $(\mathbf{p}, E)$-space, is compatible with this general finding.

Both spaces, i.e., $(\mathbf{r}, t)$-space and $(\mathbf{p}, E)$-space, are connected by a Fourier transform, and we were able to deduce the basic equations for the determination of the wave functions: $\Psi(\mathbf{p}, E)$ for $(\mathbf{p}, E)$-space and $\Psi(\mathbf{r}, t)$ for $(\mathbf{r}, t)$-space. These basic equations are more general than those discussed in conventional quantum theory. The functions $\Psi^*(\mathbf{r}, t)\Psi(\mathbf{r}, t)$ and $\Psi^*(\mathbf{p}, E)\Psi(\mathbf{p}, E)$ have to be interpreted as probability densities with respect to both sets of variables, that is, for $\mathbf{r}, t$ and $\mathbf{p}, E$; this situation is analogous to that in usual quantum theory. However, in usual quantum theory, only the variable $\mathbf{r}$ behaves statistically but not the time, which is given by τ in conventional quantum theory, and has to be considered as a classical external parameter; the system-specific time t is not defined within conventional quantum theory. Furthermore, it is not possible to deduce the basic equation (Schrödinger's equation) for the determination of the wave function in conventional quantum theory. As in classical physics, within usual quantum theory, everything is embedded in space (space-time).

The functions $\Psi(\mathbf{r}, t)$ and $\Psi(\mathbf{p}, E)$ can be described by interaction potentials on the basis of equations (2.35) and (2.40). However, in contrast to conventional physics within projection theory, the interaction potential can be interpreted at best as auxiliary element without any imaginable background. There is no information or energy transfer through $(\mathbf{r}, t)$-space.

In the case of free systems, the momentum $\mathbf{p} = \mathbf{p}_0$ and the energy $E = E_0$ remain constant in the course of time τ, and there are no $\mathbf{p}, E$-fluctuations in the case of free systems ($\Delta \mathbf{p} = 0$, $\Delta E = 0$). As we have discussed in Chapter 3, such non-interacting systems are completely useless and their existence would be against the principles of evolution.

Our observations in everyday life suggest that the world seems to be built up of subsystems which are assumed to be independent from the interaction between them. In other words, we suppose anyhow that there are really certain subsystems possible without mutual interactions; it is assumed that each of the subsystems can exist in an isolated state and that its existence is not dependent on certain external units and parameters, respectively. However, we have to be careful.

Such free (non-interacting) systems could be considered as useful, in the sense of evolution, if, at a certain time τ_{begin}, an interaction could be switched on and, in principle, that this interaction could be switched off again at time τ_{end}. However, we do not observe such a switching-on/switching-off scenario in nature. In particular, it assumes a mechanism that is hostile against projection theory because it can obviously not be based on $\mathbf{p}, E$-fluctuations. In conclusion, free (non-interacting) systems should not exist in nature, and projection theory should be able to proof that if the projection principle contains the "principle of usefulness".

The results of Chapter 3 actually show that this is the case. We proved mathematically that free (non-interacting) systems cannot exist within projection theory. It turned out that the relevant quantities, the probability distributions $\Psi^*(\mathbf{r}, t)\Psi(\mathbf{r}, t)$ and $\Psi^*(\mathbf{p}_0, E_0)\Psi(\mathbf{p}_0, E_0)$, are exactly zero. Thus, only interacting systems are of relevance, that is, quantum processes are of fundamental importance for the material existence of the world.

The projection principle leads to fundamental new aspects in connection with the notion "interaction". It is important to underline that we have here to distinguish between distance-dependent and distance-independent correlations, calculated by distance-dependent and distance-independent interactions. While the distance-independent interactions define the form (shape) of a system, that is, its geometrical structure, distance-dependent interactions describe the correlations between the various systems in $(\mathbf{r}, t)$-space.

On the basis of this basic conception, the following points have been treated in Chapter 4:

1. *How basic is the notion "interaction"?*
2. *Classical force laws*
3. *Delocalised systems*
4. *Space-time limiting interactions*
5. *Mutual (distance-dependent) interactions*
6. *The $\mathbf{p}, E$-concert*
7. *Individual processes*
8. *Pair distributions*
9. *Collective effects*
10. *Interactions with past and future events*
11. *Distance-independent interactions (form interaction)*
12. *Arbitrary jumps*

13. *The meaning of the potential functions*
14. *Self-creating Interaction Processes*
15. *Relativistic effects*
16. *Hierarchy of the parts in a part*
17. *The world as a unified whole*

Concerning the particle-wave dualism, we have the following situation in conventional quantum theory. A non-interacting system can be a point-like system with constant momentum $\mathbf{p}_0 = p_x$ and constant energy E_0 *or* a wave, having the wave-length λ ($\lambda = \hbar/p_x$) and the frequency ω ($\omega = \hbar/E_0$). The nature of the system (particle or wave) is *not defined* without experimental device and how the experimental apparatus has been arranged. In other words, a non-interacting system can be a particle as well as a wave, depending on the chosen experimental arrangement by the observer.

According to the principle of complementarity, only *one* of the two incompatible possibilities (wave, particle) can be realized and this implies that the experimental arrangements that determine those properties must be similarly mutually exclusive, otherwise we have a problem since we are confronted with a logical contradiction. However, it has been demonstrated by new experiments that light showed both particle and wave aspects *simultaneously*, and this is in contrast to the principle of complementarity, which is one of the very basics of modern quantum theory (Copenhagen Interpretation).

Within usual quantum theory, any "detector-click" is identified with the detection of a point-like particle as, for example, an electron.

Within projection theory the situation is different. Here we do not need an experimental arrangement that decides about the nature of a system. The system is completely described by the *interaction* of the system with its environment. No interaction means within projection theory that the system does not exist. Let us repeat the main facts.

If there is (almost) no interaction, the system is always a *wave*. In this case the momentum and the energy are given in a good approximation by $\mathbf{p}_0$ and E_0. With increasing $\mathbf{p}, E$-fluctuations in $(\mathbf{p}, E)$-space, the quantity $\Psi_F^*(\mathbf{r}, t)\Psi_F(\mathbf{r}, t)$, which describes the form (shape) of a system in $(\mathbf{r}, t)$-space, becomes more and more localized in

($\mathbf{r}, t$)-space and the system tends more and more towards a point-like particle. In other words, the situation in this theory is quite different from that in usual quantum theory. There is definitely no particle-wave duality in the sense of Bohr. This is the reason why we do not need here an experimental arrangement that decides about the nature of a system (wave or particle). Whether or not a system behaves like wave or a particle is dependent on the interaction of the system with its environment. An (almost) free particle can never be point-like because the function $\Psi_F^*(\mathbf{r}, t)\Psi_F(\mathbf{r}, t)$ is extended over the whole space-time.

In projection theory, the "detector-clicks" are needed in order to be able to verify certain geometrical structures in $(\mathbf{r}, t)$-space as, for example, $\Psi_F^*(\mathbf{r}, t)\Psi_F(\mathbf{r}, t)$. Such a geometrical structure can be a wave, i.e., a particle of large extension, or a strict localized particle. Each "detector-click" merely means that certain values for $\mathbf{p}$ and E have been detected at a certain instant τ, and these $\mathbf{p}, E$-values cannot be identified with a point-like mass. That is, the "click" of the detector is not activated by a point-like particle that is embedded in $(\mathbf{r}, t)$-space but exclusively by one of the possible $\mathbf{p}, E$-pairs where $\mathbf{p}$ and E belong to $(\mathbf{p}, E)$-space. Clearly, both the detector and the system under investigation are equally positioned in basic reality where the interactions take place. Both objects are equally projected onto $(\mathbf{r}, t)$-space and appear as geometrical structures (picture of reality) which we have in front of us in connection with our observations in everyday life. Since we have principally no access to basic reality, we had to introduce "fictitious realities", which are embedded in $(\mathbf{p}, E)$-space.

Bibliography

Al-Khalili, Jim., (1999) Black Holes, Worm Holes and Time Machines: Institute of Physics, Bristol.
Born, M., (1964) Die Relativitätstheorie Einsteins: Springer-Verlag, Berlin, Heidelberg.
Brückner, R., (1977) Das schielende Kind: Schwabe Verlag, Basel, Stuttgart.
Brida, G., Genovese, M., Gramegna M. and Predazzi, E., (1996) Phys. Lett. A 328, 313.
de Broglie, L., (1943) Die Elementarteilchen: Goverts, Hamburg.
de Witt, B., (1970) Physics Today, September.
Dehnen, H., (1988) in Philosophie und Physik der Raum-Zeit, J. Audretsch und K. Mainzer (Hrgs.), Wissenschaftsverlag, Mannheim, Wien, Zürich.
Einstein, A., (1973) Grundzüge der Relativitätstheorie: Verlag Viehweg und Sohn, Braunscheig.
Finkelnburg, W., (1962) Einführung in die Atomphysik: Springer-Verlag, Berlin, New York, Heidelberg.
Gödel, K., (1947) Rev. Mod. Phys. 21, 447.
Gribbin, J., (1984) In Search of Schrödinger's Cat: Bantam books, Toronto, New York, London, Sydney, Auckland.
Ghose, P., Home, D., Agarwal, G.S., (1991) Phys. Lett. A 153, 403.
Gribbin, J., (1995) Schrödinger's Kittens and the Search for Reality: Little, Brown and Company, Boston, New York, London.
Ghose, P., (1999) Testing Quantum Mechanics on a New Ground: Cambridge University Press.
Graneau, Peter., Graneau, Neal., (2006) In the Grip of the Distance Universe: World Scientific, New Jersey, London, Singapore.
Hoffman, Banesh., (1972) Albert Einstein — Creator and Rebel: The Viking Press, New York.
Heckmann, O., (1980) Sterne, Kosmos, Weltmodelle, Deutscher Taschenbuch Verlag GmbH and Co. KG, München.
Jammer, M., (1974) The Philosophy of Quantum mechanics: Wiley, New York.
Jung, C.G., (1990) Synchronizität, Akausalität und Okkultismus: Deutscher Taschenbuch Verlag GmbH and Co, KG, München.
Kenneth, W., (1963) Ford. Elementary Particles, Blaisdell Publishing Company: New York, Toronto, London.

Kuhn, Thomas S., (1970) The Structure of Scientific Revolution: University of Chicago Press, Chicago.
Kanitscheider, B., (1984) Kosmologie, Philipp Reclam jun., Stuttgart.
Landau, L.D., Lifschitz, E.M., (1965) Quantum Mechanics: Pergamon, Oxford.
Lorenz, Konrad., (1973) *Die Rückseite des Spiegels*, Piper, München.
Laughlin, Robert., (2005) A Different Universe, Basic Books: New York.
Mizobuchi, Y., Othake, Y., (1992) Phys. Lett. A 168, 1.
Pagels, Heinz, P., (1983) The Cosmic Code, Michael Joseph: London.
Peat David, F., (1988) Superstrings and the search for the Theory of Everything, Contemporary Books: Chicago, New York.
Rubinowics, A., (1968) Quantum Mechanics: Elsevier, Amsterdam.
Schäfer, C., (1951) Einführung in die Theoretische Physik: Walter de Gruyter, Berlin.
Schommers, W., (1977) Phys. Rev. A 16, 327.
Schommers, W., (Ed.), (1989) Quantum Theory and Pictures of Reality: Springer-Verlag, Berlin, Heidelberg, New York.
Schommers, Wolfram., (1994) Space and Time, Matter and Mind: World Scientific, New Jersey, London, Singapore.
Schommers, W., (1995) Symbols, Pictures and Quantum reality: World Scientific, New Jersey, London, Singapore.
Schommers, Wolfram., (1998) The Visible and the Invisible: World Scientific, New Jersey, London, Singapore.
Schommers, Wolfram., (2008) Advanced Science Letters 1, 59.
Schommers, Wolfram., Cosmic Secrets: World Scientific, in preparation.
Schommers, Wolfram., in preparation.
Unnikrishnan, C.S., Murphy, S.A., (1996) Phys. Lett. A 221, 1.
van Hove, L., (1954) Phys. Rev. 95, 249.
von Baeyer, Hans Christian., (2004) Information. The Language of Science, Harvard University Press: Cambridge, Massachusetts.
Vilenkin, Alex., (2006) Many Worlds In One, Hill and Wang, New York.
Wolf, F.A., (1981) Taking the Quantum Leap, Harper and Row: San Francisco.

Index

absolute space-time, 30, 276, 287, 290, 314, 317–319
absolute space-time positions, 276
accelerated motion, 44, 312
act of observation, 18
action-at-a-distance, 195, 196, 354
Al-Khalili, J., 108, 109, 377
alternative realities, 18
appearances, 74, 77
arbitrary jumps, 278–280, 284, 286–288, 290, 295, 311, 320, 325, 326
arbitrary motion, 200, 347
auxiliary elements, 42, 43, 45, 113, 276, 285, 307, 383, 388

basic equations, 113, 144, 145, 168, 173, 225, 239, 246, 299, 388
basic properties, 51
basic reality, 34, 35, 38–40, 42, 49, 50, 73, 82, 94, 113, 190, 191, 330, 338, 372, 373, 379, 383, 392
basic transformation effects, 54
biological evolution, 26, 387
biological systems, 37, 40–43, 113, 375, 383
block universe, 108–110
Bohr's viewpoint, 21
Bohr, N., 2, 11, 15, 18, 21–23, 306, 364, 367, 368, 379, 385, 391
Born's interpretation, 190
Born, Max, 9, 10, 79, 313, 365
brain, 31–35, 85, 113, 290, 374, 382, 388
branes, 198, 272, 329, 358
building blocks, 200, 329, 331, 358

cerebral cortex, 31
chick, 41, 42, 375, 383
chick experiment, 42
classical description, 1, 243
classical force laws, 194, 354, 390
classical formulation, 243, 246
classical particle concept, 10
classical system, 45, 61, 122, 123, 183
classical trajectory, 12
classical wave equation, 4
collapse of the wave function, 17, 19, 21, 25, 26, 69, 72, 160, 166, 385, 386
collective effects, 224, 355, 390
concept of matter, 329, 358
concept of particle, 13
conditional wave function, 258, 259, 265, 267, 268
configurations, 83, 86, 89, 93, 96, 98, 99, 101, 227, 228, 230, 231, 233–235, 238, 254, 342, 376
consciousness, 35
conservation laws, 193, 204, 211, 218, 221, 222, 235, 245, 259–262, 273, 308, 309, 332, 350, 352, 387, 388
conservation of momentum, 244, 258
constancy phenomena, 95
convolution integral, 87, 88, 91, 92
Copenhagen interpretation, 2, 16, 17, 160, 364, 366, 368
cosmological considerations, 346
cosmos, 35, 37, 140, 142, 288, 329, 331, 332, 343, 346–351, 358, 359
Coulomb potential, 3, 377
counterpart of the cosmos, 348, 349, 351

de Sitter, W., 315
dead systems, 73, 147
decoherence, 19
decomposition of the cosmos, 351
Dehnen, H., 318
delocalized systems, 199
dependence of mass on velocity, 183, 186
detector, 11, 17, 80, 162, 191, 298, 365, 367, 369, 371–373, 376–379, 391
Dirac's equation, 7
distance-dependent interaction, 200, 280, 281, 302, 303, 321, 322, 331, 334, 339–341, 354–356, 390
distance-independent correlations, 354
distance-independent interaction, 79, 271–273, 300, 320, 327, 339, 344, 354, 356, 357, 362, 390
distribution function, 218, 242
double-slit experiment, 23, 365, 367, 368

effect of motion, 92
effective velocities, 281, 282, 284, 287, 288, 297, 311, 320, 325, 326, 356
eigenfunctions, 8, 16, 17, 153
Einstein's field equations, 314, 316, 317
Einstein, A., 312
electrical forces, 194
elementary processes, 74, 377
elementary system, 142–145, 147, 200, 272
empty space, 35, 315
energy levels, 254
environment, 19, 36, 73, 74, 95, 110, 152, 203, 237, 308, 323, 352, 362, 364, 374, 379, 388, 391
equations of motion, 1, 194, 196, 244, 245
events, 9, 17, 19, 37, 81, 94, 96, 101, 109, 225, 257, 261, 272, 286, 297, 313, 342, 355, 390
Everett III, H., 2

evolution, 26, 35–38, 40–43, 73, 83, 113, 148, 152, 300, 338, 373, 374, 383, 387, 389
exchange of momentum and energy, 204, 208, 209, 262
experimental arrangement, 15, 16, 21–23, 361, 362, 364, 378, 379, 390, 391
eye, 31, 32, 41, 373

favourable towards life, 36, 37
Feynman diagrams, 111, 377
fictitious reality, 50, 190, 191, 193, 338
fluctuations of fluctuations, 334
force, 183, 184, 194–197, 288, 289, 304, 354, 385, 390
Ford, K., 111, 112, 377
Fourier transformation, 158, 175, 194, 199
frame of references, 187, 290, 291, 321, 323–327

Gödel's solution, 316
gauge bosons, 329
geometrical optics, 95
geometrical positions, 32, 33, 37, 191, 353, 356, 383
global interaction, 201
granular space-time structure, 339, 341
gravitational law, 194–197, 304
gravity, 194, 305, 307

Hamilton's equations, 1
Hamiltonian, 16, 68, 70, 72
hard objects, 33
head of the observer, 94, 95
Heisenberg, W., 2, 9, 14, 306, 384
hierarchy of the parts in a part, 327, 329, 330, 334, 338, 358
Hoffmann, B., 315
hostile towards life, 36
human intention, 16, 23
human observer, 38–41, 43, 93, 111, 113, 191, 276, 338, 345, 375, 376, 383

individual processes, 217, 219
individual systems, 189, 190, 208, 307
inertia, 30, 43, 44, 288, 298, 299, 312–318, 334, 368–370
infinitesimal interval, 80, 81, 98, 100, 101
influence of evolution, 36
inside world, 35, 37, 383
interaction effects, 252
interaction mechanisms, 190, 260, 261, 264, 265
interaction types, 200, 329, 339
interference pattern, 366
intuitive concepts, 13, 328, 329, 384
inverse transformation, 46, 48, 76, 127, 157, 158, 167, 169, 174, 204, 208, 213, 214, 248

Jammer, M., 2, 17, 384
Jung, C. G., 35, 37

Kanitscheider, B., 319
Kant, I., 27, 28, 31, 113, 381, 382, 388
kinetic energy, 149, 184

Lagrange's equations, 1
Laughlin, Robert, B., 197, 338, 339
leptons, 329
life-time, 82, 110, 178
light, 11, 15, 21, 31, 95, 191, 273, 284, 298, 299, 320, 325, 326, 369, 378, 391
linear operators, 16–19, 69
local existent, 9, 12, 20, 26, 79, 386
logical laws, 39
Lorentz transformations, 149

Mach's principle, 29, 45, 48, 200, 279, 284–286, 311, 312, 314–316, 318–320, 382
macroscopic objects, 19
many-particle system, 46
many-worlds theory, 2
material processes, 190, 191
mean value for the energy, 177
mean values, 137, 138, 151, 163, 287

meaning of the potential functions, 299
meaning of the wave function, 77
measurement problem, 19
measurement process, 160
mechanistic world view, 2
microscopic level, 13, 328, 329, 341
mind, 13, 39–41, 50, 51, 109, 194, 281, 318, 374
momentum representation, 5, 6
motion without inertia, 369
moving observer, 290, 291

nerve-excitations, 31
Newton's equations of motion, 194, 196, 244
Newton's mechanics, 7, 28, 29, 43, 195, 288, 302, 306, 317, 318, 368, 381
non-existence, 12, 119, 134, 137, 140, 146, 147, 151, 173, 181
non-interacting systems, 73, 74, 117, 150–152, 389
non-local effects, 54, 57, 58, 253, 329
non-relativistic equation, 149
normalization condition, 134, 179

objective reality, 41, 148, 319, 374
operator, 7, 16, 17, 52–54, 59, 64, 68–70, 102, 142, 143, 150, 153, 159, 182, 183, 239, 357
operator equations, 59
operators, 4–6, 8, 16–20, 26, 51, 53, 54, 67, 69, 128, 247, 386
outside reality, 37, 383
outside world, 33, 35–38, 42, 43, 83, 95, 113, 255, 373, 374, 383, 388

Pagels, H., 366, 368
pair distributions, 221, 355, 390
pair potential, 236–239, 241, 242, 247, 248, 250, 251
paradoxical situation, 12, 13, 26, 370, 387
particle-wave question, 361
p, E-concert, 214
p, E-pool, 140, 143

p, *E*-fluctuations, 73, 252, 373, 379
p, *E*-transport, 371
perception, 31, 36, 41
perception apparatus, 41
perception processes, 31
physically real, 28–31, 33, 35, 44, 45, 48, 79, 253, 285, 312, 314, 316, 353, 382
picture of reality, 19, 20, 32, 35, 38, 40, 41, 43, 51, 55, 66, 190, 352, 372, 383
piece of space, 28, 381
Planck's constant, 46
Planck, M., 46
point-like particle, 10–12, 54, 55, 58, 81, 190, 362, 365, 371, 378, 391
potential energy, 3, 220, 237, 238, 306, 353
precise reproduction, 36
principle of complementary, 2
principle of least action, 1, 196
principle of usefulness, 35–37, 73, 147, 148, 389
principles of projection theory, 96, 191, 364
probability considerations, 133
probability distribution, 78–80, 84–86, 90, 152, 192, 241, 259, 267, 269, 362, 366, 389
probability interpretation, 10, 133, 298, 368
proximity effect, 195, 196, 304

quantum field theory, 198
quantum particles, 198
quantum reality, 10, 43, 44, 47, 48, 50, 51
quantum-mechanical level, 13
quantum-theoretical elements, 128
quantum-theoretical system, 123, 126, 324
quarks, 329

random phases, 279, 286, 334
reality outside, 27, 32–36, 39, 43, 45, 50, 74, 95, 374

reduction of the wave function, 17
reference system, 84, 85, 87, 96
registration processes, 191
relativistic case, 149
relativistic effects, 320–322, 358
rest energy, 149, 186
rest mass, 149, 184, 284, 320
resting observer, 291, 349
retina, 31, 96
role of the observer, 359
$(\mathbf{r}, t)$-space, 48, 51, 58, 79, 94, 99, 351, 353, 357, 373
$\mathbf{r}, t$-correlations, 48, 51, 58, 79, 94, 99, 351, 353, 357, 373

Schleidt, W., 41, 42
Schrödinger equation, 2
Schrödinger's theory, 3, 5
selection processes, 83, 87, 97, 375
selections, 87, 342
self-creating interaction processes, 309, 357
self-creating processes, 200, 357
space-effects, 284, 288
space-extension, 281, 299, 300
space-filling fields, 195
space-specific formulation, 60
space-time information, 38, 98, 383
space-time theories, 7, 319
specific properties, 54
static building blocks, 331
stationary case, 63, 64, 67, 72, 104, 106, 167, 168, 172, 176, 178–183
statistical fluctuations, 99, 111, 290
strings, 198, 272, 310, 328, 329, 338, 358
superposition principle, 2, 16, 17, 20, 69, 70, 72, 153, 154, 156–159
system with two subsystems, 242, 249

Taylor expansion, 185
theory of relativity, 7–9, 43, 44, 108, 187, 198, 314, 318
time-dependent Schrödinger equation, 4
time-extension, 282, 300

time-independent Schrödinger
   equation, 2, 4
total energy, 140, 142, 185, 221,
   265–267, 280
total momentum, 140, 142, 221
trajectory, 9–13, 26, 196, 298, 370,
   386, 387
transformation formulas, 324, 325
true reality, 36
turkey, 41, 42

uncertainty, 7–9, 11, 14, 15, 26, 87,
   97, 107, 108, 172, 386

uncertainty relation for time, 107
uncertainty relations, 7–9, 11, 14, 15,
   26, 107, 108, 386
unified whole, 342, 343, 345, 351,
   358, 359, 390

wave equation, 3, 4, 7
weasel, 41
Wheeler, J.A., 108, 377
Wolf, F.A., 10, 298, 369

Zenon's arrow, 12